Lecture Notes in Mathematics

Edited by A. Dold and B. Eckmann

959

Géométrie Algébrique Réelle et Formes Quadratiques

Journées S.M.F., Université de Rennes 1, Mai 1981

Edité par J.-L. Colliot-Thélène, M. Coste, L. Mahé, et M.-F. Roy

Springer-Verlag
Berlin Heidelberg New York 1982

Edité par

Jean-Louis Colliot-Thélène
Mathématiques, bâtiment 425
Université de Paris-Sud
91405 - Orsay, France

Michel Coste
Département de Mathématiques et IRMAR
Université de Niamey
B.P. 91, Niamey, Niger
Université de Rennes I
35042 - Rennes-Cedex, France

Louis Mahé
IRMAR, Université de Rennes I
35042 - Rennes-Cedex, France

Marie-Françoise Roy
Département de Mathématiques et Département de Mathématiques
Université de Niamey
B.P. 91, Niamey, Niger
Université de Paris-Nord
93439 - Villetaneuse, France

AMS Subject Classifications (1980): 10 C 04, 10 J 05, 10 J 06, 10 M 15, 14 G 30, 26 E 05, 32 C 05, 55 N 30

ISBN 3-540-11959-0 Springer-Verlag Berlin Heidelberg New York
ISBN 0-387-11959-0 Springer-Verlag New York Heidelberg Berlin

CIP-Kurztitelaufnahme der Deutschen Bibliothek
Géométrie algébrique réelle et formes quadratiques: journées SMF, Univ. de Rennes 1, Mai 1981 / ed. par J.-L. Colliot-Thélène ... - Berlin; Heidelberg; New York: Springer, 1982.
(Lecture notes in mathematics; Vol. 959)
ISBN 3-540-11959-0 (Berlin, Heidelberg, New York)
ISBN 0-387-11959-0 (New York, Heidelberg, Berlin)
NE: Colliot-Thélène, Jean-Louis [Hrsg.]; Société Mathématique de France; GT

Printing and binding: Beltz Offsetdruck, Hemsbach/Bergstr.
2146/3140-543210

On trouvera dans ce recueil deux sortes d'articles :

- Des exposés généraux de synthèse sur certains sujets centraux dans le contexte "géométrie algébrique réelle et formes quadratiques".

- Des articles originaux des participants à la rencontre.

Tous les articles publiés ont fait l'objet d'un rapport, et nous remercions vivement ceux qui nous ont aidés dans cette tâche. Nous remercions aussi I. GIORGIUTTI pour son rôle important dans la préparation du colloque et Y. BRUNEI pour son efficacité, essentielle pour le bon déroulement de ces Journées et la publication de cet ouvrage.

J.-L. COLLIOT-THÉLÈNE, M. COSTE, L. MAHÉ, M.-F. ROY.

TABLE DES MATIÈRES

Articles de synthèse.

Contributions des Participants.

Conférences données à RENNES en Mai 1981.

E. BECKER, *Valuations and real places in the theory of formally real fields.*

J. BOCHNAK, *Nash functions.*

L. BRÖCKER, *Real spectra, real divisors and distributions of signatures*

G.-W. BRUMFIEL, *Real valuation ideals.*

J.-L. COLLIOT-THÉLÈNE, *Espaces quadratiques et composantes connexes réelles.*

M. COSTE, *Ensembles semi-algébriques.*

M. COSTE, *Spectre réel, ouverts semi-algébriques et ordres sur le corps des fractions.*

H. DELFS, *Cohomology of affine semi-algebraic sets over a real closed field.*

D. W. DUBOIS, *Subordinate structure sheaves.*

G. EFROYMSON, *Some recent results in Nash ring theory.*

F. ISCHEBECK, *Binary forms and prime ideals.*

T.-Y. LAM, *On the Pythagoras number of some affine algebras.*

L. MAHÉ, *Séparation des composantes connexes réelles par des formes quadratiques.*

A. PFISTER, *On quadratic forms and abelian varieties over function fields.*

A. PRESTEL, *Pseudo real-closed fields.*

PARTICIPANTS AUX JOURNÉES

GÉOMETRIE ALGÉBRIQUE RÉELLE & FORMES QUADRATIQUES

ALONSO GARCIA Mª Emilia	Madrid : Univ. Complutense (Espagne)
ANDRADAS Carlos	Madrid : Univ. Complutense (Espagne)
BECKER Eberhard	Univ. Dortmund (B.R.D.)
BENEDETTI Riccardo	Univ. Pisa (Italie)
BOCHNAK Jacek	Vrije Univ. Amsterdam (Pays-Bas)
BRÖCKER Ludwig	Univ. Münster (Allemagne)
BRUMFIEL Gregory	Stanford University (U.S.A.)
CARRAL Michel	Univ. Toulouse - Paul Sabatier (France)
COLLIOT-THÉLÈNE Jean-Louis	Univ. Orsay (Paris-Sud)
CONDUCHE Daniel	Univ. Rennes I (France)
CONTESSA Maria	Queen's University, Kingston, (Canada)
COSTE Michel	Univ. Paris-Nord (France)
COSTE-ROY Marie-Françoise	Univ. Paris-Nord (France)
COUCHOURON Marcel	Univ. Rennes I (France)
DALALIAN Samuel	Paris (France)
DELFS Hans	Univ. Regensburg (R.F.A.)
DELZELL Charles N.	L.S.U. L.A. 70803 (U.S.A.)
DICKMANN Max	Univ. Paris VII (France)
DUBOIS Donald	Univ. of New Mexico (U.S.A.)
EFROYMSON Gustave	Univ. of New Mexico (U.S.A.)
ESCOFIER Jean-Pierre	Univ. Rennes I (France)
GAMBOA José Manuel	Madrid : Univ. Complutense (Espagne)
GAREL Emmanuelle	Univ. Rennes I (France)
CIACINTI Claudine	Univ. Rennes I (France)
GIORGIUTTI Italo	Univ. Rennes I (France)
GONDARD Danielle	Univ. Paris VI (France)
GUÉRINDON Jean	Univ. Rennes I (France)
HELLEGOUACH Yves	Univ. Caen (France)
HOUDEBINE Jean	Univ. Rennes I (France)
ISCHEBECK Friedrich	Univ. Münster (Allemagne)
JACQUEMARD Alain	Univ. Dijon (France)
LAM T. Y.	Univ. California, Berkeley (U.S.A)
LANNEAU Hervé	Univ. Rennes I (France)
MAHÉ Louis	Univ. Rennes I (France)
MARGUIN Olivier	Univ. Lyon I (France)
MARSHALL Murray	Univ. of Saskatchewan (Canada)
MEIßNER Wilfried	Univ. Dortmund (B.R.D.)
MERRIEN Jean	Univ. Rennes I (France)
PAQUES Antonio	Univ. Montpellier II (France)
PAUGAM Annette	Univ. Rennes I (France)
PFISTER Albrecht	Univ. Mainz (B.R.D.)
PRESTEL Alexander	Univ. Konstanz (B.R.D.)
RECIO Tomas	Univ. Malaga (Espagne)
RIBENBOIM Paulo	Queen's University, Kingston, Ontario (Canada)
RISLER Jean Jacques	Univ. Paris VII (France)
RIVET Roger	I.N.S.A. - Rennes (France)
ROCHE Claude	Univ. Grenoble (France)
ROLLAND Raymond	Univ. Rennes I (France)
ROBINSON Edmund	Univ. Cambridge (England)

ROSOLINI Giuseppe	Oxford Univ. (England)
SANSUC Jean-Jacques	E.N.S. Paris (France)
SCHÜLTING Heinz-Werner	Univ. Dortmund (B.R.D.)
SCHWARTZ Niels	Univ. München (B.R.D.)
SILHOL Robert	Univ. Regensburg (R.F.A.)
TOGNOLI Alberto	Univ. Tours (France)
TROTMAN David	Univ. Paris-Sud (France)
TOUGERON Jean-Claude	Univ. Rennes I (France)
VALLÉE Brigitte	Univ. Caen (France)

VALUATIONS AND REAL PLACES IN THE THEORY OF FORMALLY REAL FIELDS

Eberhard Becker

It is common use to start the theory of formally real fields by recalling their plain definition: a field K is formally real if -1 is not a sum of squares in K. In view of this definition alone, one may reasonably doubt whether this property can provide a basis for a substantial theory. But, actually, formally real fields have shown distinguished properties. In particular there is an intimate connection with real algebraic geometry as, at least, these Proceedings are demonstrating. Besides this relation one realizes that studying formally real fields means studying their valuation theory. It is only fair to state that many of the recent results would not have been obtained without the study of certain valuation rings of formally real fields.

This present note is mainly concerned with the occurrence of certain valuation rings in formally real fields and with their importance for these fields. In Section 1 the geometric background is dealt with. Then, in Section 2, arbitrary formally real fields are treated. The next two sections are devoted to special sources for valuation rings, namely the powerful representation theorem of Kadison-Dubois for Archimedean partially ordered rings and, in the last section, strongly anisotropic higher degree forms. All the results in this paper are more or less known; the main objective was to consider the occurrence of real places and valuations from a basic point of view.

1. The existence of simple real points

Given a field K, a number $n \in \mathbb{N}$ we make use of the following notations:

$$\sum_1^{\ell} K^n = \{ \sum_1^{\ell} x_i^n \mid x_1,\ldots,x_\ell \in K\} ,$$

$$\sum K^n = \bigcup_{\ell=1}^{\infty} \sum_1^{\ell} K^n .$$

Thus, K is formally real iff $-1 \notin \Sigma K^2$. A place $\lambda : K \to F \cup \infty$, where K and F are fields, is a ring homomorphism $\lambda : V \to F$ of a valuation ring V of K into F whereby we set $\lambda(x) = \infty$ if $x \notin V$.

Places $\lambda : K \to \mathbb{R} \cup \infty$ are called real places. By an order of K we understand any subset $P \subset K$ satisfying

(1.1) $\quad P+P \subset P , P \cdot P \subset P , P \cup -P = K , -1 \notin P .$

Every order P additionally satisfies $P \cap -P = \{0\}$ since the first three conditions imply that $P \cap -P$ is an ideal of K. It was Artin's basic discovery that a field is formally real iff it admits an order. We give the proof only in order to recall one main principle for the construction of orders.

If P is an order of K, then $K = P \cup -P$ implies $\Sigma K^2 \subset P$ and, in view of $1 \in P$, $P \cap -P = \{0\}$, we see $-1 \notin \Sigma K^2$. For the converse, we have to construct orders starting only with the statement $-1 \notin \Sigma K^2$. This is done by enlarging the so-called preorders to orders. By an preorder we understand any subset $T \subset K$ subject to

(1.2) $\quad T+T \subset T , T \cdot T \subset T , K^2 \subset T , -1 \notin T .$

In our situation, the set ΣK^2 is a preorder, obviously the smallest one. Now given any preorder T, choose by Zorn's lemma a maximal one, say P, over T. If $P \cup -P \neq K$, then pick $a \in K \setminus (P \cup -P)$. An easy computation shows that $P + Pa$ is a preorder properly larger than P, thus contradicting the maximality of P. Hence, $P \cup -P = K$, i.e. P is an order.

Following these arguments further on, it turns out that $T = \cap P$, where P ranges over all orders over T. To see this, consider $a \in \cap P$, P as before. If $a \notin T$, then $T - Ta$ is a preorder thus contained in an order $P_0 \supset T - Ta$. Because of $-a \in P_0$, $T \subset P_0$, we have a contradiction to the choice of a.

Now let K be an algebraic function field over $\mathbb{R}$. If V is any model of K, we denote the set of real points of V by $V(\mathbb{R})$. We are interested in the problem whether there is a model with a simple real point on it. Pursuing this question, one is led to what I would like to call the fundamental cycle of notions and ideas. Graphically it is illustrated in the following way:

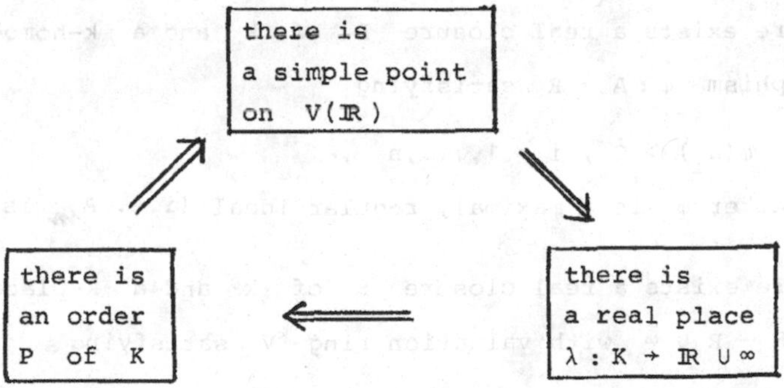

This cycle already underlay Artin's solution of the 17th problem of Hilbert as it became particularly clear by S. Lang's version of this solution.

We are now going to see that each statement in the above diagram implies the successive one as indicated by the arrows. The proof is essentially the proof of the homomorphism theorem of Artin-Lang [Lg, p. 279, Th. 5]. This theorem plays the same fundamental role for real algebraic geometry (more precisely: formally real algebraic geometry) as Hilbert's Nullstellensatz does for ordinary algebraic geometry over algebraically closed fields. Considering an algebraic variety over a formally real field k, one is interested in formally real points on it, in particular in algebraic formally real points. The theorem of Artin-Lang is concerned with their existence. We present a variant of it where special attention is given to the existence of simple formally real points.

(1.3) Theorem (Artin-Lang). Let k be a formally real field, A an integral domain, finitely generated over k and with quotient field K. Given any elements $a_1,\dots,a_n \in A\setminus\{0\}$, the following statements are equivalent:

i) there exists a real closure R of k and a k-homomorphism $\varphi : A \to R$ satisfying

1) $\varphi(a_i) > 0$, $i = 1,\dots,n$,

2) $\mathfrak{m} = \ker\varphi$ is a maximal, regular ideal (i.e. $A_{\mathfrak{m}}$ is regular)

ii) there exists a real closure R of k and a k-place $\lambda : K \to R \cup \infty$ with valuation ring V satisfying

1) $A \subset V$,

2) $\lambda(a_i) > 0$, $i = 1,\dots,n$,

iii) there exists an order P of K satisfying

$a_1,\dots,a_n \in P$.

Remark. A place $\varphi : K \to F \cup \infty$ is called a k-place if k is a common subfield of K and F and if φ is the identity on k .

Proof. We shall use the following result [Kn, proof of 2.3.] :

(1.4) Lemma. Let $\mathcal{O}$ be a noetherian regular local ring of dimension d with maximal ideal $\mathfrak{m}$ and quotient field K . Then any homomorphism $\varphi: \mathcal{O} \to F$, F a field, with $\ker \varphi = \mathfrak{m}$ extends to a place $\lambda : K \to F \cup \infty$ such that the valuation ring V of λ is discrete of rank d .

Proof of (1.4). We proceed by induction on d . If $d = 1$, then set $V := \mathcal{O}$, $\lambda = \varphi$. If $d > 1$, choose $u \in \mathfrak{m} \setminus \mathfrak{m}^2$. Then $\mathcal{O}_{(u)}$ is a discrete rank 1 valuation ring with the quotient field of $\mathcal{O}/(u)$ as its residue field. Let π denote the associated place. The homomorphism $\varphi : \mathcal{O} \to F$ factors over $\pi : \mathcal{O} \to \mathcal{O}/(u)$ and we have the following situation:

$$\begin{array}{ccccl} \mathcal{O}_{(u)} & \overset{\pi}{\dashrightarrow} & \overline{K} & & , \ \overline{K} = \mathrm{Quot}\,(\mathcal{O}/(u)) \\ \uparrow & & \uparrow & & \\ \mathcal{O} & \overset{\pi}{\dashrightarrow} & \mathcal{O}/(u) & \overset{\overline{\varphi}}{\dashrightarrow} F & , \ \overline{\varphi} \circ \pi = \varphi \ . \end{array}$$

Now, $\mathcal{O}/(u)$ is a noetherian regular local ring of dimension $d-1$. Thus, $\overline{\varphi}$ extends to a place $\overline{\lambda} : \overline{K} \to F \cup \infty$ with a valuation ring $\overline{V}$ which is discrete of rank $d-1$. Hence, the ring $\pi^{-1}(\overline{V})$ is a discrete valuation ring of rank d which belongs to the place $\overline{\lambda} \circ \pi : K \to F \cup \infty$.

We now return to the proof of (1.3). i) => ii): set $\mathfrak{p} = \ker \varphi$, then $\mathcal{O} := A_{\mathfrak{p}}$ and the natural extension $\hat{\varphi} : \mathcal{O} \to R$ of $\varphi : A \to R$

satisfy the hypothesis of (1.4). Thus there is a place $\lambda : K \to R \cup \infty$ with an associated valuation ring V satisfying $A \subset V$, $\lambda(a_i) > 0$, $i = 1,\dots,n$.

ii) ⟹ iii): We have to construct an order P. The construction makes use of the notion of a preorder and works with an arbitrary place $\lambda : K \to R \cup \infty$, where R is any ordered field. Given such a place λ, set $T := K^2\{\varepsilon \in K \mid \lambda(\varepsilon) > 0\}$. The three properties $T \cdot T \subset T$, $K^2 \subset T$, $-1 \notin T$ are easily checked. To prove $x^2\varepsilon + y^2\eta \in T$ for $x,y \in K$, $\lambda(\varepsilon)$, $\lambda(\eta) > 0$ we may assume $\lambda(x^{-1}y) \neq \infty$. But then $x^2\varepsilon + y^2\eta = x^2\varepsilon(1+(x^{-1}y)^2\varepsilon^{-1}\eta) \in T$ because of $\lambda(1+(x^{-1}y)^2\varepsilon^{-1}\eta) > 0$.

iii) ⟹ i): We start with the formulation of the homomorphism theorem of Artin-Lang as given in [Lg, p. 279, Th. 5]. Applying it to the ordered field $K(\sqrt{a_1},\dots,\sqrt{a_n})$ and the affine algebra $A(\sqrt{a_1},\dots,\sqrt{a_n})$ instead of A, we obtain by returning back to A, that A admits at least one homomorphism $\varphi : A \to R$, R a real closure of k, satisfying $\varphi(a_i) > 0$ for $i = 1,\dots,n$. Now, as A is finitely generated over a field, its singular locus is a proper closed subset of Spec (A) - note that (0) is a regular ideal [Gr, §22, (22.6.8)]. Hence we find $f \in A \setminus \{0\}$ such that $A_f \subset K$ is a regular ring. By the preceding arguments applied to A_f instead of A, we find a homomorphism $\varphi : A_f \to R$, R a real closure of k with $\varphi(a_i) > 0$, $i = 1,\dots,n$. The restriction of φ to A is the homomorphism we have been looking for.

As said before, by virtue of the theorem of Artin-Lang, the validity of the "fundamental cycle" is established. But, in fact, much more has been shown. The third statement in (1.3) only refers to a property of K, not of A. Thus, A can be chosen as any integral affine algebra with quotient field K . Interpreted in geometric terms, this means that the statement

"a given model V has a simple real point"

is actually equivalent to either of the much stronger properties

"for a given model V the simple real points form a Zariski-dense subset"

or

"any model of K has a simple real point" .

Each of the three statements we have been studying must, of course, imply the preceding one as they are all equivalent. Therefore, it is interesting to find direct proofs.

The first task is to construct an order of K provided a simple real point is given. More generally, one asks for geometric constructions of orders of a function field by means of a given model. Such constructions have been found, in particular by Bröcker (see his contribution in these Proceedings). Therefore, we are only giving some indications.

Let V be an affine variety over $\mathbb{R}$ with integral coordinate ring $A = \mathbb{R}[T_1,\ldots,T_n]/\mathfrak{a}$ and function field K . Then $V(\mathbb{R}) = \mathrm{Hom}_{\mathbb{R}}(A,\mathbb{R})$ is a closed subset of $\mathbb{R}^n$. We impose on $V(\mathbb{R})$ the subspace topology, from now on being referred to as the "strong topology". Note that V carries its natural Zariski topology which induces on $V(\mathbb{R})$ a coarser topology than the strong one. We shall add the adjective "strong" to any topological notion if its refers to the strong topology.

The construction of an order is based on the following result, compare with [D-E, p. 134, Th. 4.9].

<u>(1.5) Lemma.</u> Let U be a strong open subset of $V(\mathbb{R})$. If U contains a simple point, then U is Zariski dense in V.

<u>Proof.</u> Let $x \in U$ be a simple point, $x = (x_1, \ldots, x_n)$. Denote the coordinate function on V by $t_1, \ldots, t_n$. Then $(t_i - x_i)(x) = 0$ for $i = 1, \ldots, n$. Now choose $\varepsilon > 0$ such that $B(x,\varepsilon) \cap V(\mathbb{R}) \subset U$ where $B(x,\varepsilon)$ is the open ball with center x and radius ε. Set $f_i = t_i - (x_i - \frac{\varepsilon}{n})$, $g_i = (x_i + \frac{\varepsilon}{n}) - t_i$ for $i = 1, \ldots, n$. Then $f_i(x) > 0$, $g_i(x) > 0$ for $i = 1, \ldots, n$. Applying the theorem of Artin-Lang, we find an order P of K with $f_1, \ldots, f_n, g_1, \ldots, g_n \in P$. Now assume $f \in A$ vanishes on U. We have to show $f = 0$. If $f \neq 0$, we may assume $f \in P$. By again applying Artin-Lang, we find a simple point $x' \in V(\mathbb{R})$ subject to $f_i(x')$, $g_i(x') > 0$ and $f(x') > 0$. But from $f_i(x')$, $g_i(x') > 0$ for $i = 1, \ldots, n$ we conclude $x' \in B(x,\varepsilon) \cap V(\mathbb{R}) \subset U$, contradicting $f(x') > 0$.

Now we are ready to find an order of K provided a simple real point is given. In fact, one can even weaken the hypothesis and, at the same time, get more.

<u>(1.6) Theorem.</u> $x \in V(\mathbb{R})$ lies in the strong closure of the set of simple real points iff there is an order P of K which satisfies

if $f \in A$, $f(x) > 0$, then $f \in P$.

<u>Proof.</u> Assume first that x lies in the closure. Set

$T := \{f \in A \mid f \geq 0$ on some strong open neighbourhood of x in $V(\mathbb{R})\}$. One easily checks the properties $T+T \subset T$, $T \cdot T \subset T$, $A^2 \subset T$, $-1 \notin T$. Next we show $T \cap -T = 0$. But this follows from (1.5) since $f \in T \cap -T$ has to vanish on a strong open set containing a simple real point. We try to extend T to a preorder T' of K by setting $T' := \{ab^{-1} \mid a,b \in T, b \neq 0\}$. In fact, T' turns out to be a preorder. Hence, there is an order P of K with $P \supset T$. Now, if $f \in A$ is given with $f(x) > 0$, then obviously $f \geq 0$ on a strong open neighbourhood, hence $f \in P$. For the converse we adopt the notation of the proof of (1.5). Then, for any $\varepsilon > 0$, we have $f_i(x), g_i(x) > 0$, $i = 1,\dots,n$. By hypothesis, $f_1,\dots,f_n,g_1,\dots,g_n \in P \setminus \{0\}$. Then the theorem of Artin-Lang shows the existence of a simple real point x' with $f_i(x'), g_i(x') > 0$, i.e. $x' \in B(x,\varepsilon) \cap V(\mathbb{R})$.

The last result describes to some extent how orders can be constructed geometrically. As said before, there is a comprehensive theory of geometric constructions, namely by means of ultrafilter of open semialgebraic sets (see L. Bröcker's contribution).

We now turn to the construction of a real place from a given order. Already Artin-Schreier [A-S, p. 94] and, more detailed, Baer [Ba] have described this construction. As it works for arbitrary ordered fields, we consider an arbitrary field K with an order P on it. Then set

$$A(P) := \{a \in K \mid r \pm a \in P \text{ for some } r \in \mathbb{Q},\ r > 0\}.$$

$A(P)$ is called the ring of finite elements; in $A(P)$ we have the ideal

$$I(P) := \{a \in K \mid r \pm a \in P \text{ for all } r \in \mathbb{Q},\ r > 0\}$$

of the infinitely small elements. That $A(P)$ is a ring with ideal $I(P)$ follows at once from the identity

$$(u+v)(u'+v') + (u-v)(u'-v') = 2[uu'+vv'] .$$

Moreover we have:

<u>(1.7) Theorem.</u>

i) $A(P)$ is a valuation ring with maximal ideal $I(P)$,

ii) $\overline{P} := \{a + I(P) \mid a \in P \cap A(P)\}$ is an Archimedean order of the residue field $A(P)/I(P)$.

<u>Proof.</u> Given $x \in K \setminus A(P)$ we may assume $x \in P$. Then $r - x \notin P$ for all $r \in \mathbb{Q}$, $r > 0$. Hence, $x - r \in P$ and $\frac{1}{r} - x^{-1}$, $\frac{1}{r} + x^{-1} \in P$ for all such r's . This means $x^{-1} \in I(P)$, and i) is proved. Obviously, $\overline{P} + \overline{P} \subset \overline{P}$, $\overline{P} \cdot \overline{P} \subset \overline{P}$, $\overline{P} \cup -\overline{P} = A(P)/I(P)$, and for every $\overline{a} \in A(P)/I(P)$ there is $r \in \mathbb{Q}$, $r > 0$ with $r \pm \overline{a} \in \overline{P}$. Thus, $-1 \notin \overline{P}$ remains to be shown. But $-1 \in \overline{P}$ implies $-1 - u \in I(P)$ for some $u \in P \setminus \{0\}$ and, by the definition of $I(P)$, $-u = 1 + (-1-u) \in P$, a contradiction.

To proceed further, we make use of the well known result that every Archimedean ordered field admits a unique order embedding into $\mathbb{R}$, considered as an ordered field. In our situation we have an uniquely determined order embedding $i_{\overline{P}} : i_{\overline{P}} : A(P)/I(P) \to \mathbb{R}$, $i_{\overline{P}}(\overline{P}) \subset \mathbb{R}^2$. Then setting $\lambda_P := i_{\overline{P}} \circ \pi$, π the natural homomorphism $A(P) \to A(P)/I(P)$, we end up having the following result.

<u>(1.8) Theorem.</u> An order P of K gives rise to a real place $\lambda_P : K \to \mathbb{R} \cup \infty$ with valuation ring $A(P)$.

<u>(1.9) Corollary.</u> A field K is formally real if and only if it admits a real place $\lambda : K \to \mathbb{R} \cup \infty$.

<u>Proof.</u> If K is formally real, it has an order; hence, a real place. Conversely, any real place comes from an order as shown before. Thus K is formally real.

In our geometric situation we finally have to deal with the construction of simple real points provided a real place is given. We keep up our previous notations. Let $\lambda : K \to \mathbb{R} \cup \infty$ be a place which is finite on the coordinate ring A; it then induces an $\mathbb{R}$-homomorphism $\varphi : A \to \mathbb{R}$ - note: λ has to be trivial on $\mathbb{R}$ - thus a point $x \in V(\mathbb{R})$. The correspondence is given by $\lambda(f) = f(x)$ for any $f \in A$. The point x is called the center of λ. The next theorem describes the set of the centers and should be seen together with theorem (1.6).

<u>(1.10) Theorem.</u> [D4, p. 62, Th. 3] $x \in V(\mathbb{R})$ is a center of a place $\lambda : K \to \mathbb{R} \cup \infty$ which is finite on A if and only if x lies in the strong closure of the set of simple real points.

<u>Proof.</u> Assume first x is contained in the closure. By (1.6) we find an order P of K with $f \in P$ for any $f \in A$ satisfying $f(x) > 0$. If $f \in A$ is arbitrary, then $f(x) \in \mathbb{R}$ and thus there is $r \in \mathbb{Q}$ with $r \pm f(x) > 0$. This means $r \pm f \in P$ and $A \subset A(P)$. If $f \in A$ is given with $f(x) = 0$, then for every $r \in \mathbb{Q}$, $r > 0$ we have $(r \pm f)(x) > 0$ and thus $f \in I(P)$. Now given any $g \in A$, we have $(g - g(x))(x) = 0$, hence $\lambda_P(g - g(x)) = 0$. But we have $\lambda_P(g - g(x)) = \lambda_P(g) - g(x)$, and x is the center of λ_P. Conversely, assume that x is the center of a place $\lambda : K \to \mathbb{R} \cup \infty$ which is finite on A. Given $\varepsilon > 0$, consider $f_i, g_i \in A$, $i = 1,\dots,n$ as in the proof of (1.5) and those following. Then

$\lambda(f_i)$, $\lambda(g_i) > 0$, $i = 1,\ldots,n$. The theorem of Artin-Lang now shows the existence of a simple real point in $B(x,\varepsilon) \cap V(\mathbb{R})$.

In general, a real place $\lambda : K \to \mathbb{R} \cup \infty$ need not be finite on A ; hence, need not have a center on $V(\mathbb{R})$. If we pass, however, to projective varieties $\tilde{V}$ over $\mathbb{R}$, then any real place has indeed a center on $\tilde{V}(\mathbb{R})$. Theorem (1.10) remains true in this general situation; we only have to drop the condition "finite on A" which now has lost its meaning.

(1.11) Remark. One final remark is in order. The topology on $V(\mathbb{R})$ seems to depend on the chosen representation of A as a factor ring $\mathbb{R}[T_1,\ldots,T_n]/\mathfrak{a}$. But, actually, the strong topology can be defined in terms of A alone. More precisely, $\mathrm{Hom}_{\mathbb{R}}(A,\mathbb{R})$ admits a topology such that the natural bijection $\mathrm{Hom}_{\mathbb{R}}(A,\mathbb{R}) \xrightarrow{\sim} \{x \in \mathbb{R}^n \mid \forall_{F \in \mathfrak{a}}\ F(x) = 0\}$ is a homeomorphism. A look at the theorem of Artin-Lang suggests a possible topology: given $a_1,\ldots,a_n$, regard the set $D(a_1,\ldots,a_n) :=$ $:= \{\varphi \in \mathrm{Hom}_{\mathbb{R}}(A,\mathbb{R}) \mid \forall_i\ \varphi(a_i) > 0\}$ as an open set. It is easily checked that the sets $D(a_1,\ldots,a_n)$ for $n \in \mathbb{N}$, $a_1,\ldots,a_n \in A$ may serve as a basis for a topology. Using this topology we get the required homeomorphism between $\mathrm{Hom}_{\mathbb{R}}(A,\mathbb{R})$ and $\{x \in \mathbb{R}^n \mid \forall_{F \in \mathfrak{a}}\ F(x) = 0\}$.

With this topology on $\mathrm{Hom}_{\mathbb{R}}(A,\mathbb{R})$ we stand directly in front of the topological space $\mathrm{Spec}_r(A)$: the real spectrum of A , (see M. Coste, M.-F. Roy : "La topologie du spectre réel", to appear Contemp. Math. 1 (1981)). It is namely readily verified that under the canonical mapping $\mathrm{Hom}_{\mathbb{R}}(A,\mathbb{R}) \longrightarrow \mathrm{Spec}_r(A)$ our just defined topology is nothing else but the subspace topology inherited from $\mathrm{Spec}_r(A)$.

2. Arbitrary formally real fields

We have seen in the last section that the property of a function field of being formally real is equivalent to the existence of geometrically well distinguished points: the simple real points. If arbitrary fields are now considered, there is, of course, no chance to interpret the property of being formally real in geometric terms. But in the geometric situation of the last section we have seen that the existence of simple real points is furthermore equivalent to the existence of either orders or real places. Dropping in this series of equivalent properties the existence of simple real points, we get the following result already proved in the first section.

(2.1) **Theorem.** Given a field K, the following statements are equivalent:

i) K is formally real,

ii) K admits an order,

iii) K admits a real place.

In consequence, we approximate the geometric situation as closely as possible if we study for a formally real field K the set $X(K)$ of all orders on K and the set $M(K)$ of all real places $K \to \mathbb{R} \cup \infty$.

As already proved, there is a natural mapping:

$$(2.2) \qquad \begin{cases} X(K) \to M(K) \\ P \mapsto \lambda_P \end{cases}$$

In case K is a function field of a projective $\mathbb{R}$-variety V we further have the center mapping:

$$(2.2)\quad \begin{cases} M(K) \to V(\mathbb{R}) \\ \lambda \mapsto \text{center of } \lambda \end{cases}$$

We are going to impose, for an arbitrary formally real K, topologies on both space $X(K)$ and $M(K)$ which render the mappings (2.1) and (2.2) continuous.

<u>The topological space $X(K)$</u>

We shall write X instead of $X(K)$ when no confusion is to be suspected. Given $a \in K^\times$, we have the function

$$(2.3)\quad \hat{a} : X \to \mathbb{Z},\ P \mapsto \operatorname{sgn}_P(a) := \begin{cases} 1 & , a \in P \\ -1 & , a \in -P \end{cases}$$

X is given the weak topology with respect to all the functions $\hat{a}$, $a \in K^\times$, whereby $\mathbb{Z}$ has the discrete topology. The resulting topology is called the Harrison-topology having as a basis the sets

$$(2.4)\quad D(a_1,\dots,a_n) := \{P \in X \mid a_i \in P \ \text{ for } i = 1,\dots,n\}, \quad \text{for } n \in \mathbb{N},\ a_1,\dots,a_n \in K^\times .$$

Thus, we have $X(K) = \operatorname{Spec}_r(K)$. (Compare this with the paper by M.-F. Coste-Roy).

As first noticed by L. Bröcker, for any ring A the set of closed points of $\operatorname{Spec}_r(A)$ form a compact Hausdorff space. As to $X(K)$, it is more common to derive this result in the following way. Set $G = K^\times/(\Sigma K^2)^\times$, letting $\hat{G}$ be its compact Pontrjagin character group. Then $X \to \hat{G}$, $P \mapsto \operatorname{sgn}_P$ is a homeomorphic embedding of X onto a closed subspace of $\hat{G}$; hence,

(2.5) Theorem. $X(K)$ is a totally disconnected compact Hausdorff space.

X is easily seen to be totally disconnected.

The functions $\hat{a}$, $a \in K^{\times}$, generate a subring $W_{red}(K)$ of $C(X,\mathbb{Z})$. The reason for this notation will become clear at once. Sylvester's law of inertia shows that there is an epimorphism - the total signature -

(2.6) $\quad \text{sgn}: W(K) \to W_{red}(K)$, $[\rho] \mapsto (P \longmapsto \text{sgn}_P(\rho))$

where $W(K)$ denotes the Witt ring of K . (See, e.g., [L] .)

The kernel of the total signature was determined by Pfister [Pf] :

(2.7) Theorem. The torsion subgroup $W_t(K)$, the nil radical of $W(K)$ and the kernel of $\text{sgn}: W(K) \to W_{red}(K)$ coincide.

To study $W_{red} \subset C(X,\mathbb{Z})$ thus means to investigate quadratic forms over fields modulo torsion, which is the subject of the so-called "reduced theory of quadratic forms". (See [B-K] , [M] , [Br-M] .) One main task is to characterize W_{red} as a subring of $C(X,\mathbb{Z})$. This has been successfully done and we are going to describe the result.

For this purpose, let T be any preorder of K . Then set $X_T(K) := \{P \in X(K) \mid T \subset P\}$. $X_T(K)$ is a closed subspace, as is easily checked. Hence, $X_T(K)$ is a compact Hausdorff space. Then set $W_T(K) := \text{res}(W_{red}(K))$ where $\text{res}: C(X,\mathbb{Z}) \to C(X_T,\mathbb{Z})$ is the restriction map. $W_T(K)$ is called the reduced Witt ring modulo T and is generated by the restrictions $\hat{a}|_{X_T}$, $a \in K^{\times}$.

There is a class of preorders for which W_T is easily computable. The characterization of W_{red} in $C(X,\mathbb{Z})$ makes use of these special preorders because of the simple fact: if $f \in W_{red}$, then $f|_{X_T} \in W_T$ for every preorder T.

To get to these preorders, first note that we have a topological embedding $X_T \to \widehat{K^\times/T^\times}$, $P \mapsto \mathrm{sgn}_P$, $\widehat{K^\times/T^\times}$ the Pontrjagin character group. In fact, the image $\widehat{X_T}$ of X_T is contained in the set of all characters χ satisfying $\chi(-1) = -1$. Then let us define: T is called a <u>fan</u> if every character χ with $\chi(-1) = -1$ lies in the image, i.e. is of the form $\chi = \mathrm{sgn}_P$. (Compare with [B-K].) We have

<u>(2.8) Lemma.</u> T is a fan if and only if we have $T + Ta = T \cup Ta$ for any $a \in K^\times \setminus -T$.

<u>Proof.</u> First note that $T + Ta$ is a preorder for $a \notin -T$. Hence, $T + Ta = \cap P$, P ranging over all orders containing T and a. First assume $T + Ta = T \cup Ta$ for $a \notin -T$. If then $\chi \in \widehat{K^\times/T^\times}$ is given with $\chi(-1) = -1$, one checks that $P := \ker \chi$ is an order and thus $\chi = \mathrm{sgn}_P$. Conversely, for $a \notin -T$ the set $U = (T+Ta) \setminus \{0\}$ is a group. Hence, the annihilator $U^\perp$ of U in $\widehat{K^\times/T^\times}$ is topologically generated by the finite products of the sgn_P, $P \in X_T$, $a \in P$. We claim: $U^\perp = \{\chi \in \widehat{K^\times/T^\times} \mid \chi(a) = 1\}$, which would imply $T + Ta = T \cup Ta$. To prove the claim, pick χ with $\chi(a) = 1$. If $\chi(-1) = -1$, we are done. Then assume $\chi(-1) = 1$. Because of $a \notin -T$, there is χ_0 with $\chi_0(a) = 1$, $\chi_0(-1) = -1$. Then $\chi = \chi_0(\chi_0\chi)$. Because of $\chi_0(-1) = (\chi_0\chi)(-1) = -1$, we are through.

Now assume T is a fan. Choose $\chi_0 \in \widehat{X_T}$ and denote the group

$K^\times/(T \cup -T)^\times$ by G. Then we have a homeomorphism

(2.9) $\varphi : X_T \to \hat{G}$, $P \mapsto \chi_O^{-1} \cdot \mathrm{sgn}_P$

and an isomorphism

(2.9)' $\varphi_* : C(\hat{G}, \mathbb{Z}) \to C(X_T, \mathbb{Z})$.

We proceed by working with $C(\hat{G}, \mathbb{Z})$ instead of $C(X_T, \mathbb{Z})$. The advantage lies in the fact that the functions $C(\hat{G}, \mathbb{C})$, $\mathbb{C}$ with discrete topology, can be expressed as $\sum_{a \in G} \langle f|\tilde{a}\rangle \tilde{a}$, by virtue of the Fourier inversion theorem. Hereby, $\tilde{a}$ means the character $\tilde{a} : \hat{G} \to \mathbb{C}^\times$, $\eta \mapsto \eta(a)$. Pulling back the representation to $C(X_T, \mathbb{Z})$ and $K^\times/T^\times$, we have, by writing μ for the normalized Haar measure on $\hat{G}$,

(2.10) $$f = \sum_{a \in G} \left(\int_{\hat{G}} f(\chi_O \eta)\, \hat{a}\, (\chi_O \eta)\, \mu\, (d\eta) \right) \hat{a}$$

for $f \in C(X_T, \mathbb{Z})$

This immediately yields

(2.11) Theorem. Given a fan T and $f \in C(X_T, \mathbb{Z})$ the following statements are equivalent:

i) $f \in W_T$,

ii) $\int_{\hat{G}} f(\chi_O \eta)\, \hat{a}\, (\chi_O \eta)\, \mu\, (d\eta) \in \mathbb{Z}$ for all $a \in K^\times/T^\times$.

This criterion takes on a particularly simple form if the index $[K^\times : T^\times]$ is finite. Then ii) can be replaced by the so-called fan criterion

(2.12) $$\sum_{P \in X_T} f(P)\, \operatorname{sgn}_P(a) \equiv 0 \bmod |X_T| .$$

Note that $|X_T| = \frac{1}{2}[K^\times : T^\times]$ holds.

For $a = 1$ (2.12) reduces to

(2.13) $$\sum_{P \in X_T} f(P) \equiv 0 \bmod |X_T| .$$

We are now ready to state the characterization theorem for W_{red} as a subring of $C(X,\mathbb{Z})$. It was first proved by Becker-Bröcker [B-B] and then extended by Marshall to a semi-local ring [M, Chapt. 7]. Marshall's proof was carried out in his theory of abstract Witt rings. The proof of Becker-Bröcker made use of a certain local-global principle which, in turn, had been proved by methods which will be dealt with in the last section of this paper. There is a third proof, by Brown-Marshall [Br-M], the essential idea of which will be indicated here.

<u>(2.14) Theorem.</u> Assume $f \in C(X(K),\mathbb{Z})$. Then f is contained in $W_{red}(K)$ if and only if the fan criterion (2.13) is satisfied for every fan of finite index in $K^\times$.

<u>Outline of the proof.</u> Assume f satisfies (2.13) but does not lie in W_{red}. Note that $W_{red} = W_{\Sigma K^2}$. Choose by Zorn's lemma a maximal preorder T such that $f|_{X_T} \notin W_T$ holds. Now a detailed study of the set of the real places $\lambda : K \to \mathbb{R} \cup \infty$ satisfying $\lambda(T) \in \mathbb{R}_+ \cup \{\infty\}$ finally shows that T is a fan. But the validity of (2.13) for all fans of finite index implies that $f|_{X_T}$ lies in W_T, a contradiction.

At first glance, criterion (2.13) seems to be of little or no importance since the knowledge of all fans is required. But, actually, fans can be regarded as known objects once the valuation theory of that field is known. In [Br 2] Bröcker proved the theorem that every fan can be obtained, via a certain valuation ring, from a "trivial fan". Hereby, a fan is called trivial if it is either an order or the intersection of two orders. Note that an intersection of at most two orders is a fan.

(2.15) Theorem. Let T be a fan of K, denote by A_T the compositum of all valuation rings $A(P)$, $P \in X_T$. Then A_T is a valuation ring with formally real residue field. Let I denote its maximal ideal and set $\overline{T} := \{a + I \mid a \in T \cap A_t\}$. Then $\overline{T}$ is a trivial fan of A_T/I.

A proof of this theorem, in a generalized version, can also be found in [B 5].

The representation theorem (2.14), together with the supplement (2.15), finds applications in real algebraic geometry. Details are to be found in the paper of either Bröcker or Schülting. The applications are based on the mappings

$$X(K) \to M(K) \to V(\mathbb{R})$$

provided a projective $\mathbb{R}$-variety with function field K is given.

Using the composed map $X(K) \to V(\mathbb{R})$, one may pull back questions on $V(\mathbb{R})$ to questions on $X(K)$ and treat them by means of results on $X(K)$. For instance, some problems concerning the components of $V(\mathbb{R})$ can be solved by the representation theorem (2.14). (See the above-mentioned contributions of Bröcker and

Schülting.)

The topological space M(K)

As in the case of $X(K)$, we introduce the topology on $M(K) = \{\lambda : K \to \mathbb{R} \cup \infty\}$ as the weak topology with respect to a certain family of functions on $M(K)$.

To understand the ideas leading to the following definitions and notations, consider for a moment the field F of meromorphic functions on a Riemann surface S. (See, e.g., [C].) Each point $x \in S$ induces a $\mathbb{C}$-place $\lambda_x : F \to \mathbb{C} \cup \infty$ and, conversely, every $\mathbb{C}$-place $\lambda : F \to \mathbb{C} \cup \infty$ is of the form $\lambda = \lambda_x$ for a unique $x \in S$. Thus, S can be identified with the set $M_{\mathbb{C}}(F)$ of all $\mathbb{C}$-places $F \to \mathbb{C} \cup \infty$. In this way, $M_{\mathbb{C}}(F)$ even becomes a topological space. It turns out that this topology is the coarsest one such that all functions $\hat{f} : M_{\mathbb{C}}(F) \to \mathbb{C} \cup \infty$, $\lambda \mapsto \lambda(f)$ for $f \in F$ are continuous, i.e., the weak topology with respect to all $\hat{f}$, $f \in F$. Hereby, $\mathbb{C} \cup \infty$ is, as usual, regarded as the 1-point-compactification of $\mathbb{C}$.

In our situation of a formally real field we proceed in an analogous way. Consider $\hat{\mathbb{R}} := \mathbb{R} \cup \infty$ as the 1-point-compactification of $\mathbb{R}$. For $a \in K$ set

$$\hat{a} : \begin{cases} M(K) \to \hat{\mathbb{R}} \\ \lambda \mapsto \lambda(a) \end{cases}$$

Then impose on $M = M(K)$ the weak topology with respect to all the functions $\hat{a}$, $a \in K$. Thus, the elements of K operate on K as "meromorphic functions". A function $\hat{a}$ is "holomorphic" if $\hat{a}(M) \subset \mathbb{R}$ is satisfied, i.e., if $\lambda(a) \in \mathbb{R}$ for all $\lambda \in M$. Denoting the valuation ring of λ by V_λ, we see that $\hat{a}$ is holomorphic if and only if $a \in \bigcap_{\lambda \in M} V_\lambda$ holds. We therefore call

$$H(K) = \bigcap_{\lambda \in M(K)} V_\lambda$$ the <u>real holomorphy ring</u> of K.

This ring has been thoroughly investigated in [D3, §6], [Sch 1], [Sch 2] where the following results can also be found.

We are now going to prove some basic results on $H = H(K)$, an immediate consequence of which will be that $M(K)$ is a compact Hausdorff space. In accordance with the definition of $A(P)$, $P \in X(K)$, we more generally set

$$A(T) = \{a \in K \mid r \pm a \in T \quad \text{for some} \quad r \in \mathbb{Q}\ ,\ r > 0\}$$

for an arbitrary preorder T of K.

<u>(2.16) Theorem.</u>

i) $H(K) = A(\Sigma K^2)$,

ii) $H(K)$ is generated as a ring by the elements $\frac{1}{1+q}$, $q \in \Sigma K^2$,

iii) $H(K)$ is a Prüfer ring with quotient field K,

iv) if V is a valuation ring of K, then $H(K) \subset V \iff V$ has a formally real residue field,

v) $H(K) = \cap V$, V ranging over all valuation rings of K with a formally real residue field.

<u>Proof.</u> Denote by H_o the ring which is generated by the

elements $\frac{1}{1+q}$, $q \in \Sigma K^2$. We first show $H_0 \subset A(\Sigma K^2) \subset H$. Because of $1 \pm \frac{1}{1+q} \in \Sigma K^2$, $H_0 \subset A(\Sigma K^2)$ is shown. Assume $n \pm a \in \Sigma K^2$, but $\lambda(a) = \infty$ for some $\lambda \in M$. Then $\lambda(a^{-1}) = 0$, $\frac{n+a}{n-a} = -\frac{1+na^{-1}}{1-na^{-1}}$ and $\lambda(\frac{n+a}{n-a}) = -1$ follows. We know $\lambda = \lambda_P$ for some $P \in X$; by the construction of λ_P we see $\frac{n+a}{n-a} \in -P$, a contradiction to $n \pm a \in \Sigma K^2$. Next we prove that H_0 is a Prüfer ring (for the definition see [G]); here we follow [Dr]. We have to show that every localization $(H_0)_m$, m a maximal ideal, is a valuation ring. Let m be any maximal ideal of H_0; given $x \in K^\times$, we derive from $1 = \frac{1}{1+x^2} + \frac{1}{1+x^{-2}}$ that either $\frac{1}{1+x^2}$ or $\frac{1}{1+x^{-2}}$ is a unit in $(H_0)_m$. This shows that $x^2 \in (H_0)_m$ or $x^{-2} \in (H_0)_m$ holds. In the case of $(x+1)^2$, $x^2 \in (H_0)_m$ we get $2x = (x+1)^2 - x^2 \in (H_0)_m$. Because of $2 \in (H_0)_m^\times$ we see $x \in (H_0)_m$. The assumption $(x+1)^2$, $x^{-2} \in (H_0)_m$ leads to $1 + 2x^{-1} + x^{-2} \in (H_0)_m$ and hence $x^{-1} \in (H_0)_m$. The remaining cases are equally dealt with. Thus, $(H_0)_m$ is a valuation ring. As a Prüfer ring H_0 is the intersection of its valuation overrings. If $H_0 \subset V$, V a valuation ring, then the residue field of V has to be formally real as $\frac{1}{1+q} \in V$ for any $q \in \Sigma K^2$. Conversely, if the residue field of V is formally real, then, for $q \in \Sigma K^2$, we have $q = x^2\varepsilon$, $x \in K$, $\varepsilon \in V^\times \cap \Sigma V^2$ which implies $\frac{1}{1+q} \in V$ and hence $H_0 \subset V$. Now, given any valuation ring with formally real residue field k, choose an order $\overline{P}$ of k and pull it back to an order P of K by means of the construction used in the proof of (1.3), ii) => iii). One then checks $V \supset V_\lambda$ for $\lambda = \lambda_P$. Hence, $H_0 = H(K)$ is proved; as a consequence, all other statements are shown likewise.

Remark. In [Bru] Brumfiel has introduced the notion of a semi-integral extension of partially ordered rings. Using his characterization 6.4.1 on p. 126, one can easily check that the semi-integral closure of $\mathbb{Q}$ with its unique order in $(K,\Sigma K^2)$ is just the real holomorphy ring $H(K)$.

Given $a \in H$, we have the continuous function $\hat{a} : M \to \mathbb{R}$. Thus, the weak topology on M with respect to the functions $\hat{a}$, $a \in H$ is coarser than the one we have introduced above. But, in fact, they coincide. To settle the case $\hat{a}(\lambda) = \infty$ first note that $\varphi : \hat{\mathbb{R}} \to \hat{\mathbb{R}}$, $x \mapsto \frac{x}{1+x^2}$, $\infty \mapsto 0$ defines a homeomorphism between $\{x \in \mathbb{R} \mid |x| > 1\} \cup \{\infty\}$ and $\{y \in \mathbb{R} \mid |y| < \frac{1}{2}\}$; further check the identity $(\varphi \circ \lambda)(a) = \lambda(\frac{a}{1+a^2})$ and finally show $\frac{a}{1+a^2} \in H$ for any $a \in K$.

This new description of our topology, together with (2.16), allows a reinterpretation which, in turn, immediately leads to the proof that $M(K)$ is a compact Hausdorff space. If $\lambda \in M(K)$ is given, it then induces a homomorphism $\lambda|_H : H \to \mathbb{R}$. Conversely, given a ring homomorphism $\varphi : H \to \mathbb{R}$, then, putting $m := \ker \varphi$, it naturally extends to a homomorphism $\hat{\varphi} : H_m \to \mathbb{R}$. But, as H is a Prüfer ring, H_m turns out to be a valuation ring; in fact, a valuation ring with a formally real residue field contained in $\mathbb{R}$. Thus, $\hat{\varphi}$ is a real place of K and we have a natural identification between M and $\mathrm{Hom}\,(H,\mathbb{R})$. This mapping is even a homeomorphism, as just seen, if we give $\mathrm{Hom}\,(H,\mathbb{R})$ its Stone-topology, i.e., the weak topology with respect to the evaluation maps $\hat{a} : \mathrm{Hom}\,(H,\mathbb{R}) \to \mathbb{R}$, $\varphi \mapsto \varphi(a)$. Now, the embedding $\mathrm{Hom}\,(H,\mathbb{R}) \to \mathbb{R}^H$, $\varphi \mapsto (\varphi(a))_{a\in H}$ is seen to be a homeomorphism onto a closed subspace which, in turn, is contained in the compact space $\prod_{a\in H} [-n_a, n_a]$ whereby n_a satisfies $n_a \pm a \in \Sigma K^2$. Summarizing, we can state:

(2.17) Theorem.

i) $M(K)$ is a compact Hausdorff space,

ii) the mapping $\begin{cases} M(K) \to \mathrm{Hom}\,(H(K),\mathbb{R}) \\ \lambda \mapsto \lambda|_{H(K)} \end{cases}$ is a homeomorphism.

Remark. Because of the fact $H(K) = A(\Sigma K^2)$, the partial order $P_O := \Sigma K^2 \cap H$ is seen to be an Archimedean partial order of H, i.e., for every $a \in H$ there is $n \in \mathbb{N}$ with $n - a \in P_O$. Actually, we even have $P_O = \Sigma H^2$. Because of $H = \cap V_\lambda$, it follows at once from the following observation: if the valuation ring V has a formally real residue field and if $\sum_1^n x_i^2 \in V$, then $x_i \in V$ for all i. This can be used to show that $M(K)$ or $\mathrm{Hom}\,(H(K),\mathbb{R})$ is naturally identified as a topological space with the set of all closed points in $\mathrm{Spec}_r H(K)$. Namely, the canonical embedding $\mathrm{Hom}\,(H(K),\mathbb{R}) \to \mathrm{Spec}_r H(K)$ can easily be shown to yield a homeomorphism between $\mathrm{Hom}\,(H(K),\mathbb{R})$ and the compact Hausdorff space of the closed points of $\mathrm{Spec}_r H(K)$. This was first observed by Bröcker and Schülting.

(2.18) Theorem. Let K be a formally real field. Then the following statements hold:

i) $\begin{cases} X(K) \to M(K) \\ P \mapsto \lambda_P \end{cases}$ is continuous and surjective,

the topology on $M(K)$ is the quotient topology of $X(K)$,

ii) if K is additionally a function field over $\mathbb{R}$ and V a projective model of K, then the mapping

$$\begin{cases} M(K) \to V(\mathbb{R}) \\ \lambda \mapsto \text{center of } \lambda \end{cases} \quad \text{is continuous.}$$

<u>Proof.</u> Let λ be a real place of K; on the residue field $\lambda(V_\lambda) = k$ we have the order $\overline{P} := k \cap \mathbb{R}^2$. By the process used in the proof of (1.3), ii) => iii), we can pull $\overline{P}$ back to an order P of K. Since $\overline{P}$ is Archimedean, one proves $\lambda = \lambda_P$. Thus, the mapping $X \to M$ is surjective. Next we show that the functions $\hat{a}$, $a \in H$ are continuous in the quotient topology of X on M; this shows that the compact Hausdorff topology is coarser than the compact quotient topology which, of course, implies that both coincide and we are done. So, pick $a \in H$. We have to show that $X \to \mathbb{R}$, $P \mapsto \lambda_P(a)$ is continuous. If $\lambda_P(a) \neq 0$, say $\lambda_P(a) > 0$; and if a neighbourhood U of $\lambda_P(a)$ is given, choose $r, s \in \mathbb{Q}$ with $0 < r < \lambda_P(a) < s$ and $[r,s] \subset U$. Then $\{P' \in X \mid a - r, s - a \in P'\}$ is a neighbourhood of P with image in U. If $\lambda_P(a) = 0$, then for $\varepsilon \in \mathbb{Q}$, $\varepsilon > 0$ the neighbourhood $\{P' \in X \mid \varepsilon \pm a \in P'\}$ has image in $[-\varepsilon, \varepsilon]$. Hence, i) is proved. To prove ii) we first check that it is sufficient to prove the statement for an affine $\mathbb{R}$-variety V and the open subspace of all $\lambda \in M$ which are finite on the coordinate ring A of V. But, in this case, all follows from the relation $\lambda(f) = f(x)$ for $f \in A$, x the center of λ, when applied to the coordinate functions.

In the case of $X(K)$, the functions $\hat{a} : X(K) \to \mathbb{Z}$, $P \mapsto \operatorname{sgn}_P(a)$ led us to define the reduced Witt ring $W_{red}(K)$, by definition the ring which is generated in $C(X(K),\mathbb{Z})$ by these functions $\hat{a}$, $a \in K^{\times}$. We proceed with $M(K)$ in a corresponding manner. This time we even have a representation

$$(2.19) \qquad \Phi : \begin{cases} H(K) \to C(M(K),\mathbb{R}) \\ a \mapsto \hat{a} \end{cases}$$

Let $C^{+}(M,\mathbb{R}) = \{f \in C(M,\mathbb{R}) \mid f(x) \geq 0 \text{ for every } x \in M\}$. The determination of $\Phi^{-1}(C^{+})$, $\ker \Phi$, $\operatorname{im} \Phi$ is the subject of the following theorem which, in turn, is only a rather special case of the Kadison-Dubois-theorem of the next section.

<u>(2.20) Theorem.</u>

i) The image of Φ is dense in $C(M(K),\mathbb{R})$,

ii) $\Phi^{-1}(C^{+}(M(K),\mathbb{R})) = \{a \in H(K) \mid \forall_{r \in \mathbb{Q}, r>0}\ r + a \in \Sigma K^2\}$,

iii) $\ker \Phi = \{a \in H(K) \mid \forall_{r \in \mathbb{Q}, r>0}\ r \pm a \in \Sigma K^2\}$,

iv) if K is a function field over $\mathbb{R}$, then Φ is injective.

<u>Proof.</u>

i) This immediately follows from the Stone-Weierstraß theorem since $M(K)$ is a compact Hausdorff space and $\mathbb{Q} \subset H(K)$.

ii) If $r + a \in \Sigma K^2$, then $r + a \in H \cap \Sigma K^2 = \Sigma H^2$ and hence for $\lambda \in M(K)$: $\lambda(r+a) = r + \lambda(a) \geq 0$. This holds for all $r \in \mathbb{Q}$, $r > 0$; thus, $\lambda(a) = \hat{a}(\lambda) \geq 0$, $\hat{a} \in C^{+}$. Conversely, if $\hat{a} \in C^{+}$, then for $r \in \mathbb{Q}$, $r > 0$ we have $\lambda(r+a) > 0$ for any $\lambda \in M$. We know for $P \in X(K)$ that $\operatorname{sgn}_P(b) = \operatorname{sgn}_{\mathbb{R}^2} \lambda_P(b)$ holds provided

$\lambda_P(b) \neq 0$. This proves $r + a \in \cap P$, $P \in X(K)$; hence $r + a \in \Sigma K^2$.

iii) follows from ii) .

To prove

iv) choose any affine model V of K with coordinate ring A . Write $a = \frac{f}{g^2}$, $f, g \in A$. Then $r \pm a \in \Sigma K^2$ implies $rg^2 \pm f \in A \cap \Sigma K^2$. In general it is not true that $rg^2(x) \pm f(x) \geq 0$ follows for $x \in V(\mathbb{R})$. But this is true at least for all simple points $x \in V(\mathbb{R})$ as we shall see in a moment. Taking this for granted, we conclude $f(x) = 0$ for every simple point $x \in V(\mathbb{R})$ since r may be an arbitrary positive rational number. By (1.5), we get $f = 0$, i.e., $a = 0$. The following lemma remains to be proved.

(2.21) Lemma. Let $\mathcal{O}$ be a noetherian regular local ring with a formally real residue field k ; let $\pi : \mathcal{O} \to k$ be the natural epimorphism. If $a \in \mathcal{O} \cap \Sigma K^2$ is given, then $\pi(a) \in \Sigma k^2$.

Proof. By (1.4), π extends to a place $\lambda : K \to k \cup \infty$. Since k is formally real, we have $\lambda(\Sigma K^2) = \Sigma k^2 \cup \infty$ as seen in the proof of (2.16). This means $\pi(a) = \lambda(a) \in \Sigma k^2$.

In the geometric situation of a function field K over $\mathbb{R}$, $H(K)$ can thus be regarded as a ring of functions on $M(K)$. Moreover, if a projective model V is given, we have the continuous mapping $M \to V(\mathbb{R})$. In view of this, it is obvious that the real holomorphy ring must have some importance for the geometry of $V(\mathbb{R})$. These geometric applications are dealt with in Schülting's contribution.

3. The representation theorem of Kadison-Dubois

This theorem is concerned with Archimedean partially ordered rings. It is to be considered as a far-reaching generalization of the well known result that every field with an Archimedean order admits a unique order embedding into $\mathbb{R}$. The theorem of Kadison-Dubois is not restricted to partial orders which contain the squares; rather, general partial orders are allowed. It is exactly this feature which makes the theorem so powerful.

For the sake of simplicity, assume that the partially ordered ring (R,P) contains $(\mathbb{Q},\mathbb{Q}_+)$. By a partial order P , we mean any subset $P \subset R$ subject to the axioms

(3.1) $$P + P \subset P \; , \; P \cdot P \subset P \; , \; 0,1 \in P \; , \; -1 \notin P \; .$$

Furthermore, P is assumed to be Archimedean, i.e.,

(3.2) $$\forall_{a \in R} \; \exists_{r \in \mathbb{Q}} \; r > 0 \; , \; r - a \in P \; .$$

Set $X = X(R,P) := \{\varphi \in \mathrm{Hom}(R,\mathbb{R}) \mid \varphi(P) \subset \mathbb{R}_+\}$, X is the set of order preserving ring homomorphisms. Impose on X the weak topology with respect to all evaluation functions $\hat{a} : X \to \mathbb{R}$, $\varphi \mapsto \varphi(a)$. X is a compact Hausdorff space as the embedding $X \to \prod_{a \in R} [-n_a, n_a]$, $\varphi \mapsto (\varphi(a))_{a \in R}$ - hereby $n_a \pm a \in P$ - induces a homeomorphism with a closed subspace.

So far this procedure only extends the manner we worked with the real holomorphy ring $H(K)$ and its Archimedean partial order ΣH^2. In that case we had $X(H(K)) \simeq M(K) \neq \emptyset$. In general, one has mainly to show that $X(R,P)$ contains sufficiently many elements. It is essentially done by methods of functional analysis. As to the proof of the following theorem, the representation theorem of Kadison-Dubois, see [K], [D1], [D2] and [B3].

(3.2) Theorem. Given an Archimedean partially ordered ring (R,P) containing $(\mathbb{Q},\mathbb{Q}_+)$. Set X as above, letting $\Phi : R \to C(X,\mathbb{R})$, $\Phi(a)(\varphi) := \varphi(a)$ be the natural representation. Then the following statements are valid:

i) $X \neq \emptyset$,

ii) the image of Φ is dense in $C(X,\mathbb{R})$,

iii) $\Phi^{-1} C^+(X,\mathbb{R}) = \{a \in R \mid \forall_{r\in\mathbb{Q}, r>0} \; r + a \in P\}$,

iv) $\ker \Phi = \{a \in R \mid \forall_{r\in\mathbb{Q}, r>0} \; r \pm a \in P\}$.

The author's study of the sums of n-th powers in fields as well as the theory of the so-called orderings of higher level, both to be considered as an extension of the Artin-Schreier theory of sums of squares and orders, is essentially based on this theorem. The reader is referred to [B1] and [B2] to see the details. In order to demonstrate the strength of the result, and to display the role of valuation rings, as announced in the introduction, we prove the following surprising statements.

(3.3) Theorem. Let K be a field with char $(K) = 0$. Then the following statements are true:

i) if K is not formally real, then $-1 \in \Sigma K^{2n}$ for every $n \in \mathbb{N}$,

ii) if K is formally real, then

a) $H(K) = \{a \in K \mid \underset{r\in\mathbb{Q}, r>0}{\exists} \; r \pm a \in \Sigma K^{2n}\}$ for every $n \in \mathbb{N}$,

b) $[H(K)^{\times} \cap \Sigma K^2] \subset \bigcap_n \Sigma K^{2n}$.

Remark. Statement i) is due to Joly [J].

Proof. Assume $-1 \notin \Sigma K^{2n}$ for a fixed $n \in \mathbb{N}$. Denote the ring on the right-hand side of ii) a) by B . We proceed as in the proof of (2.16) . Because of

$$\frac{1}{1+a^{2n}}, \frac{a^{2n}}{1+a^{2n}} \in B \quad \text{and} \quad \frac{1}{1+a^{2n}} + \frac{a^{2n}}{1+a^{2n}} = 1$$

we see that, given any maximal ideal $\mathfrak{m}$ of B , $a^{2n} \in B_{\mathfrak{m}}$ or $a^{-2n} \in \mathfrak{m}B_{\mathfrak{m}}$. Hence, the integral closure $\tilde{B}_{\mathfrak{m}}$ of $B_{\mathfrak{m}}$ in K is a valuation ring. Denote the maximal ideal of $\tilde{B}_{\mathfrak{m}}$ by $\tilde{\mathfrak{m}}$. We have $\mathfrak{m}B_{\mathfrak{m}} \subset \tilde{\mathfrak{m}}$. Pick $a \in \tilde{B}_{\mathfrak{m}}$, if a^{2n} were not in $B_{\mathfrak{m}}$, then $a^{-2n} \in \mathfrak{m}B_{\mathfrak{m}}$ and $1 \in \tilde{\mathfrak{m}}$, a contradiction. Hence, $a^{2n} \in B_{\mathfrak{m}}$ for every $a \in \tilde{B}_{\mathfrak{m}}$. Using the identity

$$(3.4) \qquad k!X = \sum_{\ell=0}^{k-1} \pm \binom{k}{1}[(X+\ell)^k - \ell^k]$$

we get $B_{\mathfrak{m}} = \tilde{B}_{\mathfrak{m}}$ because of $\mathbb{Q} \subset B$. Thus, $B_{\mathfrak{m}}$ turns out to be a valuation ring.

We next apply (3.2) to $(B, B \cap \Sigma K^{2n})$ to get $\varphi \in \mathrm{Hom}(B,\mathbb{R})$. Set $\mathfrak{m} = \ker \varphi$, then φ extends to the real place $\hat{\varphi} : B_{\mathfrak{m}} \to \mathbb{R}$. Therefore, by (2.1), K has to be formally real. Consequently, i) is proved. As to ii) , first note $B \subset V$, V a valuation ring if and only if $-1 \notin \Sigma k^{2n}$, k the residue field of V . By i) , this means: $B \subset V$ iff the residue field of V is formally real. Since B is, as a Prüfer ring, the intersection of all its valuation overrings, we get $B = H(K)$ in view of (2.16). As in the remark following (2.17), we obtain $H \cap \Sigma K^{2n} = \Sigma H^{2n}$. The representation space X of (3.2) therefore equals $\mathrm{Hom}(H, \mathbb{R}) \simeq M(K)$. Now, let $\varepsilon \in H^{\times} \cap \Sigma K^2$ be given and set $\hat{\varepsilon} := \Phi(\varepsilon)$, Φ as in (3.2). Then, obviously, $\hat{\varepsilon} > 0$ on X. Since X is compact, we find $r \in \mathbb{Q}$, $r > 0$ with $\hat{\varepsilon} > r$. By (3.2), iii) we get $\varepsilon - r + s \in \Sigma K^{2n}$ for any $s \in \mathbb{Q}$, $s > 0$. In particular, $\varepsilon \in \Sigma K^{2n}$, as to be shown.

Further results concerning the holomorphy ring and sums of n-th powers can be found in [Be 5]. As a first application of (3.3) we show for the rational function field $\mathbb{Q}(X)$:

$$\frac{1+X^2}{2+X^2} \in \Sigma\ \mathbb{Q}(X)^{2n} \quad \text{for every } n \in \mathbb{N}.$$

This follows from ii) a) and b) because of

$$1 \pm \frac{1+X^2}{2+X^2} \in \Sigma\ \mathbb{Q}(X)^2, \quad 2 \pm \frac{2+X^2}{1+X^2} \in \Sigma\ \mathbb{Q}(X)^2.$$

Thus far, no concrete representations of $\frac{1+X^2}{2+X^2}$ as a sum of $2n$-th powers for large n has been found.

As to the next application we consider a number field K. If K is formally real, then $H(K) = K$ because every valuation ring of K with a formally real residue field has to be trivial. Thus ii) b) shows $\Sigma K^2 = \Sigma K^{2n}$ for every n. If K is not formally real, then by i) $-1 \in \Sigma K^{2n}$ and, by using the identity (3.4), we also get $\Sigma K^2 = K = \Sigma K^{2n}$ for every n. Hence, for every number field K and $n \in \mathbb{N}$

$$\Sigma K^2 = \Sigma K^{2n} .$$

This was first proved by Siegel [S, Satz 2] .

4. Strongly anisotropic forms

In this section we start with a field K of arbitrary characteristic. Let $f(X_1,\ldots,X_k)$ be a homogeneous polynomial - a form - of degree d over K ; f is called anisotropic if f has only the trivial zero in K . Given two forms $f(X_1,\ldots,X_k)$, $g(X_1,\ldots,X_\ell)$ of degree d over K , we can form their orthogonal sum

$$(4.1) \qquad (f \perp g)(X_1,\ldots,X_{k+\ell}) := f(X_1,\ldots,X_k) + g(X_{k+1},\ldots,X_{k+\ell}) .$$

In particular, we have the r-fold orthogonal sum $r \times f :=$ $:= f \perp \ldots \perp f$ (r-times) . f is called strongly anisotropic if every multiple $r \times f$, $r \in \mathbb{N}$ is anisotropic; otherwise f is called weakly isotropic. Consider, for example, the case $K := \mathbb{Q}$, $n \geq 2$, $f(X_1,X_2) := X_1^n - 2X_2^n$. Then f is not isotropic but weakly isotropic: $2 \times f$ is isotropic in the usual sense. The notion of a strongly anisotropic form was introduced by Prestel. (See [Pr 1] , [Pr 2] .) This notion comes up naturally in the study of formally real fields for it is easily seen that a field is formally real if and only if the quadratic form $f = X_1^2$ is strongly anisotropic. Moreover, we even have

(4.2) Theorem. K admits a strongly anisotropic form $f \neq 0$ if and only if K is formally real.

Proof. It remains to be shown that K is formally real provided it has a strongly anisotropic form $f(X_1,\dots,X_k)$. First assume $-1 \in \Sigma K^d$, where d is the degree (of homogeneity) of f , say $-1 = a_1^d + \dots + a_r^d$. Then we get $f(1,0,\dots,0) + \sum_1^r f(a_i,0,\dots,0) = 0$ and hence $(r+1) \times f$ is isotropic. Therefore, $-1 \notin \Sigma K^d$. This shows $d = 2n$, $n \in \mathbb{N}$, and then, in view of Theorem (3.3), that K is formally real.

As a consequence of the preceding result, the study of strongly anisotropic forms makes sense only in formally real fields and for forms of degree $2n$. From now on this will be assumed.

Given such a form f , we have the value set

(4.3) $\quad D(f) := \{f(a_1,\dots,a_k) \mid a_1,\dots a_k \in K\}$.

We further set $D_\infty(f) = \bigcup_{r \in \mathbb{N}} D(r \times f)$. Then one readily checks

<u>(4.4) Lemma.</u> Let f be anisotropic, then f is strongly anisotropic if and only if $D_\infty(f) \cap - D_\infty(f) = \{0\}$.

The value set $D_\infty(f)$ has characteristic properties which lead to the following definition: a subset $T \subset K$ is called a <u>ΣK^{2n}-module</u> if T satisfies

(4.5) $\quad T + T \subset T \;,\; K^{2n} T \subset T$

Clearly, $T = K$ and $T = D_\infty(f)$ are examples of ΣK^{2n}-modules.

<u>(4.6) Lemma.</u> Let T be a ΣK^{2n}-module of K . Then the following statements are equivalent :

i) $\quad T = K$,

ii) $\quad T \cap - T \neq \{0\}$.

Proof. Pick $a \in T \cap -T$, $a \neq 0$. From the identity (3.4) we derive $K = \Sigma K^{2n} - \Sigma K^{2n}$. Hence, $K = Ka = (\Sigma K^{2n})a + (\Sigma K^{2n})(-a) \subset T$, i.e., $K = T$.

We continue to study the ΣK^{2n}-module $T \neq K$. By Zorn's lemma we see from (4.6) that every such T is contained in a maximal one S. The maximal ΣK^{2n}-modules $S \neq K$ are called the semiorderings of level n of K. They have the following simple characterization.

(4.7) Theorem.

i) $S \subset K$ is a semiordering of level n if and only if S satisfies:

$$S + S \subset S, \; K^{2n}S \subset S, \; S \cup -S = K, \; S \cap -S = \{0\},$$

ii) if $T \neq K$ is a ΣK^{2n}-module, then

$$T = \cap S,$$

S ranging over all semiorderings of level n, $S \supset T$.

Proof.

i) If S satisfies these axioms, then S is clearly maximal. Conversely, let S be maximal but $a \in K \setminus (S \cup -S)$ might be chosen. Then $T' :=$ $:= S + (\Sigma K^{2n})a$ is a properly larger ΣK^{2n}-module: a contradiction as $T' \neq K$.

ii) Assume the intersection on the right-hand side contains $a \notin T$. Then $T + (\Sigma K^{2n})(-a)$ is a ΣK^{2n}-module $\neq K$ which has to be contained in some of those S's: a contradiction.

Consider for the moment the case $n = 1$. Then a semiordering turns out to be almost an order, only the conditions $1 \in S$, $S \cdot S$ are missing. Because of this similarity the name semiordering was chosen.

There are fields admitting semiorderings S with $1 \in S$ which are not orders. For example, consider the rational function field $K = \mathbb{R}(t)$ and the form

$$f(X_1, X_2, X_3, X_4) = X_1^4 + tX_2^4 + t^2X_3^4 - t^3X_4^4 .$$

Then choose a semiordering $S \supset D_\infty(f)$ and we have $1 \in S$ but S cannot be an order because of $t, t^2, -t^3 \in S$.

The fields with the property that every semiordering S with $1 \in S$ is an order have been characterized in [Pr 1], [B 4] by means of their valuation theory; for example, every number field has this property. The proofs rely heavily on the fact that every semiordering S with $1 \in S$ gives rise to a valuation ring as was first discovered by Prestel [Pr 1].

Given a semiordering S of level n, $1 \in S$, we set

$$A(S) := \{a \in K \mid \exists_{r \in \mathbb{Q}, r>0} \; r \pm a \in S\} ,$$

$$I(S) := \{a \in K \mid \forall_{r \in \mathbb{Q}, r>0} \; r \pm a \in S\} .$$

This is just the construction we applied in the first section to orders. As to the proof of the following theorem, see [Pr 1] for $n = 1$ and [B 4] for the general case.

<u>(4.8)</u> <u>Theorem.</u> ([Pr 1], [B 4]). Under the above assump-

tions the following statements are true:

i) $A(S)$ is a valuation ring with maximal ideal $I(S)$,

ii) $\overline{S} := \{a + I(S) \mid a \in S \cap A(S)\}$ is (even) an Archimedean order of the residue field $A(S)/I(S)$.

In the case $n = 1$, the proof is elementary; whereas in the general situation one needs the Kadison-Dubois theorem. The real worth of this theorem is to be seen in the fact that it leads to a certain local-global theorem for diagonal forms of any even degree. This theorem refers to the property of being weakly isotropic instead of the property of being isotropic as in the classical Hasse-Minkowski theorem. The local fields of the classical situation are replaced by the henselization with respect to non-trivial Krull valuations. As just mentioned, the theorem holds for any even degree form. For the sake of simplicity, however, we give its formulation only in the quadratic case where it is due to Bröcker [Br 1] ; the general case can be found in [B 4] . Note that a quadratic form $f(X_1,\dots,X_k)$ is said to be totally indefinite if $|\mathrm{sgn}_P(f)| < k$ holds for every order P of K .

(4.9) Theorem. A quadratic form f over K is weakly isotropic over K if and only if

i) f is totally indefinite and

ii) f is weakly isotropic over all henselizations with respect to non-trivial Krull valuations.

Of course, one may wonder whether there is a local-global theorem of this type for isotropy. First of all, there is the classical Hasse-Minkowski theorem for global fields. (See, e.g., [O'M] .)

Then, there are fields such that a weakly isotropic form is in fact isotropic. These are just the pythagorean fields. By definition, they are characterized by the property $\Sigma K^2 = K^2$, e.g., $K = \mathbb{R}$. In general, we do not have a local-global theorem for isotropic forms. For example, the famous quadratic form $X_1^2 + X_2^2 + X_3^2 - f(X,Y)X_4^2$ over $\mathbb{R}(X,Y)$, $f(X,Y) = 1 + X^2(X^2-3)Y^2 + X^2Y^4$, (see [C-E-Pf]), is anisotropic but totally indefinite and isotropic over all henselizations. This last statement supplements the results of [C-E-Pf] as was observed by Prestel.

References

[A-S] Artin, E. and Schreier, O.: Algebraische Konstruktion reeller Körper, Abh. Math. Sem. Univ. Hamburg 5 (1927), 85-99.

[Ba] Baer, R.: Über nicht-archimedisch geordnete Körper, Sitz. Ber. der Heidelberger Akad., 8. Abh. (1927), 3-13.

[B 1] Becker, E.: Hereditarily pythagorean fields and orderings of higher level. IMPA Lecture Notes, No. 29 (1978), Rio de Janeiro.

[B 2] Becker, E.: Summen n-ter Potenzen in Körpern, J. reine angew. Mathematik 307/308 (1979), 8-30.

[B 3] Becker, E.: Partial orders on a field and valuation rings, Comm. Alg. 7 (1979), 1933-1976.

[B 4] Becker, E.: Local global theorems for diagonal forms of higher degree, J. reine angew. Math. 318 (1980), 36-50.

[B 5] Becker, E.: The real holomorphy ring and sums of 2n-th powers, these Proceedings.

[B-B] Becker, E. and Bröcker, L.: On the description of the reduced Witt ring, J. Alg. 52 (1978), 328-346.

[B-K] Becker, E. and Köpping, E.: Reduzierte quadratische Formen und Semiordnungen reeller Körper, Abh. Math. Sem. Univ. Hamburg 46 (1977), 143-177.

[Br 1] Bröcker, L.: Zur Theorie der quadratischen Formen über formal reellen Körpern, Math. Ann. 210 (1974), 233-256.

[Br 2] Bröcker, L.: Characterization of fans and hereditarily pythagorean fields, Math. Z. 151 (1976), 149-163.

[Br-M] Brown, R. and Marshall, M.: The reduced theory of quadratic forms, Rocky Mtn. J. Math. 11 (1981), 161-175.

[Bru] Brumfiel, G.: Partially ordered rings and semi-algebraic geometry, Lect. Notes Ser. Lond. Math. Soc. 1979.

[C-E-Pf] Cassels, J.W.S. and Ellison, W.J. and Pfister, A.: On sums of squares and on elliptic curves over function fields, J. Numb. Th. 3 (1971), 125-149.

[C] Chevalley, C.: Introduction to the theory of algebraic functions of one variable, Math. Surveys VI of the Amer. Math. Soc. 1951.

[Dr] Dress, A.: On orderings and valuation of fields, Geometriae Ded. 6 (1977), 259-266.

[D1] Dubois, D.W.: A note on David Harrison's theory of preprimes, Pac. J. Math. 21 (1967), 15-19.

[D2] Dubois, D.W.: Second note on David Harrison's theory of preprimes, Pac. J. Math. 24 (1968), 57-68.

[D3] Dubois, D.W.: Infinite primes and ordered fields, Diss. Math. LXIX (1970).

[D4] Dubois, D.W.: Real commutative algebra I. Places, Revista Matemática Hispana-Americana 39 (1979), 57-65.

[D-E] Dubois, D.W. and Efroymson, G.: Algebraic theory of real varieties, "Studies and Essays" presented to Yu-Why Chen on his 60th birthday (1970), 107-135.

[G] Gilmer, R.: Multiplicative ideal theory, New York 1972.

[Gr] Grothendieck, A.: Eléments de géométrie algébrique, Publ. Math. IHES 20 (1964).

[J] Joly, J.R.: Sommes des puissances d-ièmes dans un anneau commutatif, Acta Arith. 17 (1970), 37-114.

[K] Kadison, R.V.: A representation theorem for commutative topological algebra, Mem. Amer. Math. Soc. 7 (1951).

[Kn] Knebusch, M.: Specialization of quadratic and symmetric bilinear forms, and a norm theorem, Acta Arith. 24 (1973), 279-299.

[L] Lam, T.Y.: The algebraic theory of quadratic forms, Reading 1973.

[Lg] Lang, S.: Algebra, Reading 1965.

[M] Marshall, M.: Abstract Witt rings, Queen's Paper in Pure and Applied Mathematics 57 (1980), Kingston.

[Pf] Pfister, A.: Quadratische Formen in beliebigen Körpern, Invent. Math. 1 (1966), 116-132.

[Pr 1] Prestel, A.: Quadratische Semi-Ordnungen und quadratische Formen, Math. Z. 133 (1973), 319-342.

[Pr 2] Prestel, A.: Lectures on formally real fields, IMPA Lecture Notes No. 22 (1975), Rio de Janeiro.

[S] Siegel, C.L.: Darstellung total positiver Zahlen durch Quadrate, Math. Z. 11 (1921), 246-275.

[Sch 1] Schülting, H.-W.: Über reelle Stellen eines Körpers und ihren Holomorphiering, Ph.D. thesis, Dortmund 1979.

[Sch 2] Schülting, H.-W.: On real places of a field and their holomorphy ring, Comm. Alg., to appear.

Mathematisches Institut
Universität Dortmund
Postfach 500500
4600 DORTMUND 50

An Introduction to Nash Functions

J. Bochnak and Gustave Efroymson

1. Introduction

In this paper, we try to give an introduction to Nash functions. Some of this material appears in [B-E] or in various other papers, e.g., [R]. So the goal here is not originality or even completeness, but to put together some of these results in one place and to at least indicate what we think are some of the simplest proofs. Also, since [B-E] has appeared, there have been some very interesting developments in this area and we wish to mention some of these here.

2. Semi-algebraic Geometry and Nash Functions

Definition: A set S in R^n is called semi-algebraic if it can be written as a finite union of sets of the form $\{x: p_i(x_1,\cdots,x_n) = 0\ ,\ i = 1,\cdots,m\ ,\ p_i(x_1,\cdots,x_n) > 0\ ,\ i = m+1,\cdots,s\ ;$ where the p_i are arbitrary polynomials in $R[x_1,\cdots,x_n]$ and $x = (x_1,\cdots,x_n)\}$.

One would like to consider open semi-algebraic sets in R^n as those defined as above but with only strict inequalities. To do this we need to know that any open semi-algebraic set can be expressed in this form. This is the "Unproved Theorem" of Brumfiel [B], which has been proved and which we will mention later.

It is natural to ask what sort of functions one will allow on semi-algebraic sets. This is a big topic in Brumfiel's book [B], but we will avoid most of this by considering first semi-algebraic functions and then Nash functions.

Definition: A function $f: X \to Y$ from one semi-algebraic set to another is called semi-algebraic if its graph is semi-algebraic.

This naturally leads to the subject of continuous semi-algebraic functions which replace continuous functions and Nash functions which replace analytic functions.

Definition: A function $f: X \to Y$ from one semi-algebraic set to another is called Nash if it is semi-algebraic and real analytic. For X non-singular there is no problem as to what real analytic means since locally X will then be like R^n and we leave the more general case for later.

One of the big advantages of Nash functions over more general real analytic functions is that, since they are algebraic, one can hope to use algebraic methods in dealing with them. But of course one wants to know that they have good properties shared by both polynomials and real analytic functions, and in some cases the best properties of each. This is almost true, but there are still some problems as we will see.

We can summarize the situation now for Nash functions by considering X a non-singular semi-algebraic set and $N(X)$ the ring of Nash functions on X. Then $N(X)$ is a Noetherian ring. The local ring at a point x (in X) has completion which is a power series ring. There is also a connection between the real ideals of $N(X)$ and their zero sets as in the real Nullstellensatz. There is also an approximation theorem which states that a continuous semi-algebraic function on X can be arbitrarily closely approximated by a Nash function on X. This result which, in the case of polynomials, requires X compact shows some of the advantages of Nash functions and semi-algebraic sets. The approximation theorem can be used to prove that one can extend Nash functions from non-singular closed semi-algebraic sets in R^n to all of R^n.

But more basic and necessary for almost all of the above is the Tarski-Seidenberg principle.

Theorem: Let $f: X \to Y$ be a semi-algebraic map (i.e., X and Y are semi-algebraic sets and the graph of f is also semi-algebraic), then the image of f is also semi-algebraic. A better statement might be that if $f: R^n \to R^m$ is a polynomial map and X contained in R^n is semi-algebraic, then $f(X)$ is semi-algebraic also.

This doesn't look much like the Tarski-Seidenberg principle of logic but there is a way of seeing the relationship. To see this, recall the "logic" Tarski-Seidenberg principle. Roughly, this says that if one has a polynomial statement $A(x_1,\cdots,x_n)$, i.e., a Boolean combination of polynomials in $k[x_1,\cdots,x_n]$, and one quantifies A with $\exists$ or $\forall$ then the statement A is true for one real closed field k containing the coefficients iff it is true for all real closed fields containing k. These two versions of Tarski-Seidenberg have a "common ancestor":

Theorem: Let k be a real closed field and let $A(x_1,\ldots,x_n)$ be a polynomial relation in $k[x_1,\ldots,x_n]$. Then there exists a polynomial relation $B(x_1,\ldots,x_n)$ in $k[x_1,\ldots,x_n]$ such that $(\exists x_1,\cdots,x_n \in L : C(x_1,\cdots,x_n,y_1,\cdots,y_m))$ is true for L iff the statement $B(y_1,\cdots,y_m)$ is true for L. Since this holds for any L which contains the y_i s, we find that the truth of A depends only on the truth of B which holds independent of which real closed filed L containing k we choose. So choose $L = k$ and the result follows.

Now for an application of Tarski-Seidenberg we consider first the slick proof of Hilbert's 17th problem. For this, let $K = R(X_1,\cdots,X_n)$ and consider an element $f(X_1,\cdots,X_n) = p(X_1,\cdots,X_n)/q(X_1,\cdots,X_n)$ of K where p and q are polynomials. We wish to prove that if $f(a_1,\cdots,a_n)$ is $\geqslant 0$ whenever it is defined for real a, then f can be expressed as a sum of squares in K. The proof is based on the not too difficult result that if f can't be so expressed, then K can be ordered so that f is negative. We now choose L to be a real closure of K so ordered. Now consider the statement: For all $a_1,\cdots,a_n$ in $k \supset R$, $q(a_1,\cdots,a_n) \neq 0$ implies $p(a_1,\cdots,a_n)$ $q(a_1,\cdots,a_n) \geqslant 0$. By our hypothesis, this statement is true for $k = R$. So by Tarski-Seidenberg, it is also true for $k = L$. But $R(X_1,\cdots,X_n)$ is in L and, in particular, so are $X_1, \cdots, X_n$. But $q(X_1, \cdots, X_n) \neq 0$, so we find $p(X_1, \cdots, X_n)/q(X_1, \cdots, X_n) \geq 0$ which is a contradiction! This proof appears in Cohen, [C], which also contains a beautiful proof of Tarski-Seidenberg.

Of course, the above argument can easily be adapted to give a proof of the real Nullstellensatz. For this, one easily reduces to the case of prime ideal P in $R[X_1, \cdots, X_n]$. Then if $P = (f_1, \cdots, f_s)$, one wants to show that for any g in $R[X_1, \cdots, X_n]$, that g is in P iff g vanishes on the real zero set of P when P is a "real" prime ideal. This means that $R[X_1, \cdots, X_n]/P$ can be ordered. So we let L be a real closure of the quotient field of $R[X_1, \cdots, X_n]/P$ with this ordering. Then we consider the statement: For all $a_1, \cdots, a_n$ in k, $f_1(a_1, \cdots, a_n) = 0, \cdots, f_s(a_1, \cdots, a_n) = 0$ implies $g(a_1, \cdots, a_n) = 0$. Since this statement is true for $k = R$, it is true for $k = L$. We apply the statement when $a_i = \overline{x}_i$, the image of X_i in L. But, by definition, all $f_i(\overline{x}_1, \cdots, \overline{x}_n) = 0$, so $g(\overline{x}_1, \cdots, \overline{x}_n) = 0$ which is the same as saying that g is in P.

Our real goal is to try and carry the above proof over to the case where we consider the Nash ring $N(R^n)$. It looks deceptively easy, but there are problems. So consider the case of a real prime ideal P in $N(R^n)$. As above we let L = a real closure of $N(R^n)/P$. Then we wish to show that g in $N(R^n)$, is in P iff g vanishes on the real zeros of P. We let ψ be the induced homomorpsism from $N(R^n)$ to L. In trying to carry through the proof above, we see that we must find a way to define g on L. This can be done using as usual the Tarski-Seidenberg principle. For the Nash function g will have a semi-algebraic graph which will then be defined by a polynomial statement: $(A(x_1, \cdots, x_n, z)$ *iff* $g(x_1, \cdots, x_n) = z$.) But then one can use A to define a semi-algebraic set in L^{n+1} which will also be the graph of a function $g_L : L^n \to L$. It is easy to see that we will have our Nullstellensatz once we prove the following theorem.

Substitution Theorem: With the above notation, $g_L(\psi X_1, \cdots, \psi X_n) = \psi g$. There is a very nice proof of this by M. Coste, [C_2]. We give here a proof which is a combination of his proof and the original proof. The main advantage (if it is an advantage) of this proof over Coste's is that it avoids the use of the Artin-Mazur definition of Nash functions. The proof still uses the basic ideas in the Artin-Mazur paper of n normalization and Zariski's main theorem. We still need the Mostowski separation theorem, which will be proved later in this paper. To state the separation theorem requires that we define an extension of the polynomial ring on R which consists of adjoining the square root of any polynomial p which is positive on U and then considering the extension of the ring so obtained which we get by again adjoining square roots of those elements of this ring which are positive on U. We call this ring $Q(U)$ and note that it is clearly contained in N(U). The separation theorem states that if S_1 and S_2 are disjoint closed semi-algebraic sets in U, then there exists an element of Q(U) which is positive on S_1 and negative on S_2. It is also almost immediate that the substitution theorem applies to Q(U).

We would also like to use the fact that the ring N(U) of Nash functions on U is integrally closed. This follows from the usual argument for complex analytic functions. Let $V \subset R^n$ be a non-singular algebraic set and $U \subset V$ an open connected semi-algebraic set. Let $\bar{x}_i$ be the image of x_i in $\Gamma[V] = R[X_1, \cdots, X_n]/(ideal\ (V))$. We also identify $\bar{x}_i$ with $x_i | V$.

Proposition: Let $f: U \to R$ be a Nash function. Then there is an algebraic set W in R^{n+1+s} and Nash functions $g_i: U \to R$, $i = 1, \cdots, n+1+s$ with $g_i = \bar{x}_i$, $i = 1, \cdots, n$, $g_{n+1} = f$, such that if $g: U \to R^{n+1+s}$ is defined by $g(x) = (g_1(x), \cdots, g_{n+1+s}(x))$ which is the same as $(x_1, \cdots, x_n, f(x), g_{n+2}(x), \cdots, g_{n+1+s}(x))$, then g(U) is a connected component of $W \cap (U \times R^{s+1})$.

<u>Proof</u>. Let $P_f = \{p \text{ in } \Gamma[V][Z]: p(x,f(x)) = 0 \text{ for all } x \text{ in } U\}$. The set of zeros of P_f is the Zariski closure of the graph of f in $V \times R$. Let $D = \Gamma[V][Z]/P_f$. Take the normalization B of D in its field of fractions $D_{(0)}$. Choose $t_1, \cdots, t_s$ in $\Gamma[V][Z]_{(0)}$ whose image in $D_{(0)}$ generates B over D. Now $t_i(x,f(x))$ will define a Nash function since it is integral over the ring of Nash functions. We denote this function by g_{n+1+i}, $i = 1, \cdots, s$. Now consider B as a quotient ring of $R[X_1, \cdots, X_n, Z, T_1, \cdots, T_s] = R[X,Z,T]$ by an ideal J of R[X,Z,T] and D as a quotient of R[X,Z] by P_f. The natural inclusions $R[X,Z] \subset R[X,Z,T]$ and $P_f \subset J$ induce the projection $\pi: (\mathbb{C}^{n+1+s}, W_2) \to (\mathbb{C}^{n+1}, W_1)$, where $\pi(x,z,t) = (x,z)$ and

$$W_1 = \{(x,z) \text{ in } V_{\mathbb{C}} \times \mathbb{C}: P_{\mathbb{C}}(x,z) = 0 \text{ for all } p \text{ in } P_f\}.$$

$$W_2 = \{(x,z,t) \text{ in } \mathbb{C}^{n+1+s}: r_{\mathbb{C}}(x,z,t) = 0 \text{ for all } r \text{ in } J\}.$$

By construction, g(U) is contained in a connected component of $W \cap (U \times R^{s+1})$, but $W = W_2 \cap R^{n+1+s}$. However, g(U) must be the whole component because Zariski's main theorem insures that through any point of $W \cap (V \times R^{s+1})$, there is exactly one branch of W_2.

<u>Substitution Theorem</u>: Let $\psi: N(U) \to L$ be a homomorpsism, $\psi(1) = 1$, L a real closed field, $R \subset L$. Then

(i) $\psi(\bar{x}) = (\psi(\bar{x}_1), \cdots, \psi(\bar{x}_n))$ is in U_L and

(ii) for all f in N(U), one has $\psi(f) = f_L(\psi(x))$.

<u>Proof</u>: (i) If we assume that U is described as the set of points in R defined by strict inequalities, the proof is easy. Shiota (see [B-E]) has shown that this restriction is not necessary. So suppose that $U = \{x \in R: h_1(x) > 0, \cdots, h_s(x) > 0\}$. Then consider the function:

$$h(x) = \prod(|h_i(x)| + h_i(x)).$$

It is then clear that $h(x)$ is positive on U and vanishes off U. So $h(x) > 0$ will describe U and also will describe U_L. But it is easily seen that $h(\psi x) = \psi h(x)$ Since $h(x)$ is a square in $N(U)$, it follows that $\psi h(x)$ is a square and so positive and this implies ψx is in U_L. For the general case, we let X be the smallest real algebraic set in R^n which contains U. We can assume that X is irreducible. Now let $D = X \setminus U$. We work by induction on $dim(X)$. We wish to construct a semi-algebraic function h on R^n which is Nash on $R^n \setminus D$ and vanishes on D. Let $X = g^{-1}(0)$ for some polynomial $g(x)$. By the "unproved theorem", we can take $D = \{x \text{ in } R^n : g(x) = 0, p_1(x) \geq 0, \cdots, p_t(x) \geq 0\}$, for polynomials g, $p_1, \cdots, p_t$. We can assume that X is not contaned in any $p_i^{-1}(0)$ and $\mathrm{nf}\{p_i(x): x \in d\} = 0$, for all i. Now let $D = D \cap \cup p_i^{-1}(0)$, $K = D \setminus D'$, $K' = X \setminus D$. Then $dim(D') < m$. Moreover, K and K' are disjoint. By induction, there exists g_1 semi-algebraic on R^n such that $g_1 \geq 0$ and $g_1^{-1}(0) = D$. Also g_1 is Nash on $R^n \setminus D$. Now consider the sets $L = \{(x,y)$ in $R^n xR : x \in K, yg_1(x) = 1\}$ $L' = (x,y)$ in $R^n xR : x \in K', yg_1(x) = 1\}$. Then L and L' are semi-algebraic, closed and disjoint. So there exists a function, F of the type of the separation theorem so that $F(L) > 0$ and $F(L) < 0$. Next consider the function $G(x) = (g_1(x) + F^2(x,1/g_1(x))^{1/2} - F(x,1/g_1(x))$. Finally let $h = g^N G(x)$ for N a sufficiently large integer. This h can be used as above to show that $\phi(x)$ is in U_L.

(ii) Let f be in $N(U)$ and let W and $g = (g_1,\ldots,g_{n+1+s})$ be as in Propositon 1. Since $g(U)$ is a connected component of $W \cap \pi^{-1}(U)$ where $\pi : R^{n+1+s} \to R^n$ is the natural projection, there is an element h in $Q(U\times R\times R^s)$ such that $h(g(U)) > 0$ and $h(W\cap\pi^{-1}(U)\setminus g(U)) > 0$.

Let K be a real closed field containing R. We apply Tarski's principle to the formula

$$F_K: \text{(For all } (x,z) \text{ in } U_K \times K^{n+1+s},\ z \text{ in } W_K \cap \pi_K^{-1}(U_K),$$

$$\pi_K(z) = x,\ h_K(z) > 0 \text{ implies that } z = g_K(x)).$$

Now F_K is true for $K = R$, so it is true for any real closed field $K \supset R$, and in particular for $K = L$. Let's take $z = \psi(g) = (\psi(g_1),\cdots,\psi(g_{n+1+s})) = (\psi(x),\psi(f),\psi(g_{n+2}),\cdots,\psi(g_{n+s+1}))$ and $x = \psi(x)$. Obviously $\pi_L(\psi(g)) = (\psi(g_1),\cdots,\psi(g_n)) = \psi(x)$ and $\psi(g)$ is in $W_L \cap \pi_L^{-1}(U_L)$ for if λ is a polynomial in n+1+s variables with $\lambda^{-1}(0) = W$, then $\lambda_L(\psi(g)) = \psi(\lambda \cdot g) = \psi(0)$ so $\psi(g)$ is in W_L. Finally it has been noted that since h is in $Q(U \times R \times R^s)$ that $h_L(\psi(g)) = \psi(h \cdot g)$. From this it follows that $h_L(\psi(g)) = \psi(h \cdot g) > 0$ since $h \cdot g$ is strictly positive on U. Hence $\psi(g) = g_L(\psi(\bar{x}))$ and in particular $\psi(f) = \psi(g_{n+1}) = g_{n+1L}(\psi(\bar{x})) = f_L(\psi\bar{x}))$.

We would like to give a sketch of the proof of the result which Coste calls the generalized Thom lemma. this first appeared in $[E_1]$, with improved versions in $[C_1]$ and $[C_2]$. This version is due to Houdebine $[C_2]$.

<u>Theorem</u>: Let $p_i(x_1,\cdots,x_n)$, $i = 1,\cdots,m$ be polynomials in n variables. Then there exist more polynomials p_i, $i = m+1,\cdots,q$ so that if we let $S(e_1,\cdots,e_q) = \{x \text{ in } R^n \text{ such that sign}(p_i) = e_i\}$, then

(i) all $S(e_1,\cdots,e_q)$ are connected

(ii) $S(e_1,\cdots,e_q)$ is contained in the closure of $S(e_1',\cdots,e_q')$ if and only if $e_i \neq 0$ implies $e_1' = e_i$. Here $e_i = +, -,$ or 0.

<u>Proof:</u> First reduce to the case where all the p_i are monic in some variable x_n. Then we work by induction since the theorem is fairly obvious for n = 0 or 1. We also need to consider the hyper-resultants of two polynomials p and q in $R[x_1,\cdots,x_n]$. We consider the factorization $p = \prod_{i=1}^{d} (x_n - a_i(x_1,\cdots,x_{n-1}))$ and then consider the various $s_k(q(a_1),\cdots,q(a_d))$ the kth

symmetric polynomial of the "roots" of p evaluated by q. We take all the p_i and all $\partial^k p_i/\partial x_n^k$ and then all hyper-resultants of pairs of these polynomials. Then we use induction to construct a list of polynomials in $x_1,\cdots,x_{n-1}$ which has the required property on R^{n-1}. The first m polynomials are the ones in $x_1,\cdots,x_n$ while the polynomials p_i, $i > m$, are only polynomials in $x_1,\cdots,x_{n-1}$. By the properties of the hyper-resultants, above any $S(e_{m+1},\cdots,e_q)$, the roots of any of the p_i's have constant multiplicity and either coincide or do not intersect. Now consider $S(e_1,\cdots,e_q)$ which is not contained in the closure of $S(e_1',\cdots,e_q')$. If $S(e_{m+1},\cdots,e_q)$ is not contained in the closure of $S(e_{m+1}',\cdots,e_q')$, then we are done by induction. So suppose that $S(e_{m+1},\cdots,e_q)$ is contained in the closure of $S(e_{m+1}',\cdots,e_q')$. Now consider the fiber in $S(e_1,\cdots,e_q)$ of some x in $S(e_{m+1},\cdots,e_q)$ which will be either empty, a point, or an interval (possibly infinite at one or both ends). The same holds for the primes. Since the region $S(e_1',\cdots,e_q')$ is connected and since $p_1,\cdots,p_m$ are all monic, the closure of $S(e_1',\cdots,e_q')$, will intersect the fiber of any point above $S(e_{m+1}',\cdots,e_q')$. To finish, we need only show if the closure of $S(e_1',\cdots,e_q')$ intersects $S(e_1,\cdots,e_q)$, then the first contains the second. If $S(e_1,\cdots,e_q)$ has one point fiber, this is obvious. But if the fiber is an interval, then none of the p_i, $i = 1,\cdots,q$ can change sign there and so for an interior point of the interval, the sign will be constant and so the inequality will hold on $S(e_1,\cdots,e_q)$, which is what we want to show.

Now the separation theorem can be easily proved as in $[E_1]$. For this let A and B be disjoint closed semi-algebraic sets in R^n. Then apply the generalized Thom lemma to the polynomials defining A and B to get $p_1,\cdots,p_q$ in $R[x_1,\cdots,x_n]$ as above. Now fix one of the $S(e_1,\cdots,e_q)$ in A. Then B is a union of some of the $S(e_1',\cdots,e_q')$ and for each of these, there exists some p_i which is positive on $S(e_1,\cdots,e_q)$ and negative on $S(e_1',\cdots,e_q')$ or vice versa since neither is contained in the closure of the other. Actually it is possible there is say p_1 which is positive on the first set and 0 on the second and p_2 which is positive on the second and 0 on the first, but then $p_1 - p_2$ will do. Then taking the sum of the $|p_l| - p_l$, we obtain a function $h(x)$, which is zero on $S(e_1,\cdots,e_q)$ and positive on B. Since h is positive on B and B is closed, by the usual Tarski-Seidenberg argument, there exists some $\epsilon(x)$ of the form $1/(C+r^{2m})$ for large enough constant C and large integer m so that $h(x) > \epsilon(x)$ on B. Then we can replace $|p_l|$ by $(p_l^2 + \epsilon'(x))^{1/2}$ where $\epsilon'(x)$ is a function similar to $\epsilon(x)$ above. Then the new function $\sum[(p_l^2 + \epsilon'^2)^{1/2} - p_l - \epsilon']$ which is Nash of the required type will be $<$ *some* $-\epsilon''(x)$ on $S(e_1,\cdots,e_q)$ and $> \epsilon''(x)$ on B. Then repeating the procedure for all the $S(e_1,\cdots,e_q)$ whose union is A, we are done.

Also, as Coste has pointed out, the unproved proposition is an easy consequence of the generalized Thom lemma. This goes as follows. Let U be an open semi-algebraic set so U is open but the defining statement may include some equalities. So consider the polynomials used to define U and apply the generalized Thom lemma to find $p_1,\cdots,p_m$ with the properties given above. Then, $U = \bigcup_l S_l(e_1,\cdots,e_m)$. Now if $S_l(e_1,\cdots,e_m)$ has $e_{l_1} = 0,\cdots,e_{l_k} = 0$, we consider the set $T_l = \{x : p_l(x) = e_l \text{ for } l \neq \text{any } e_{l_\ell}\}$. For example, if S_l is $p_1 > 0,\cdots,p_s > 0, p_{s+1} = 0,\cdots,p_m = 0$, then T_l is $p_1 > 0,\cdots,p_s > 0$. Now consider $V = \bigcup_l T_l$. We claim that $V = U$. It is clear that $V \supset U$. So suppose we take one of the T_l's. Then T_l is a union of S_l and sets of the form $W_l = S(e_1,\cdots,e_m)$ where more of the e_l are + or −, than in S_l. So the closure of any W_l contains S_l. So closure $(W_l) \cap U \neq \emptyset$. But since U is open, $W_l \cap U \neq \emptyset$. But by the partition property, $W_l \subset U$. Q.E.D.

3. Applications and New Results.

There are many applications of Nash functions. Some of these were discussed at the conference and so presumably will be covered in other papers in this volume. We only briefly mention some of these. There is first the result of Mahe [Ma1], [Ma2] on separation of components of real algebraic varieties by signatures of quadratic forms. His proof makes use of the separation theorem, plus, of course, many techniques which have nothing to do with Nash functions. Then there is the important result of Tougeron [T] which may be stated as follows: Let $\Omega \subset R^n$ be an open semi-algebraic subset and let $f : \Omega \to R^p$ be a proper Nash map. Then the image $f^*(C(R^p))$ is closed in $C^\infty(\Omega)$, here $C^\infty(R^p)$ is the space of C^∞ functions on R^p and $f^*(\psi) = \psi \circ f$ for $\psi \in C(R^p)$. Then there is the solution of the Nash conjecture by Tognoli presented in [Tog1], and generalizations by Bennedetti and Tognoli reported in [Tog2], and in very great generality by Akbulut and King [A-K]. These really are quite different in nature than the results presented in this paper since they deal with the question of what differentiable "manifolds" (they may be singular) are diffeomorphic to real algebraic manifolds.

There is a very interesting question about Nash varieties which was not discussed at the conference and which has been recently solved by Shiota. This is the question of deciding when two Nash manifolds are Nash isomorphic. The natural question is whether two diffeomorphic Nash varieties are Nash isomorphic. The answer turns out to be "not necessarily". This is "because a Nash isomorphism will carry over to the boundary". In his paper [S], Shiota constructs non-singular algebraic sets X and X′ which are C^∞ diffeomorphic but not Nash isomorphic. This is quite amazing!

Other interesting results on Nash functions include questions about the ring of Nash functions on a Nash manifold. There is the theorem of Bochnak, Kucharz, and Shiota that connects $H^1(X, Z_2)$ with the class group of X, for X a Nash manifold, [B-K-S].

There are also questions about the relation between analytic components and Nash components of algebraic varieties. Do they always coincide? Mostowski, [M], using his separation theorem, showed that disjoint analytic components are also Nash components. Efroymson, $[E_2]$, has shown that for the case of the plane, where the interesting case is for real curves, the two coincide.

Then there is the extension problem for Nash functions. Mostowski conjectured that the following is true and even had a proof, unfortunately not correct.

Theorem: Let h be a Nash function on R^n. Let U be a semi-algebraic open neighborhood of $X = h^{-1}(0)$ and $f: U \to R$ be Nash. Then there exists $g: R^n \to R$ which is Nash so $g = f$ on X. This has now been proved by Efroymson, $[E_3]$.

A natural extension of the above question is whether one can extend a function f merely defined on $h^{-1}(0)$. If $h^{-1}(0)$ is non-singular, there is no problem, but in the singular case, there is the question of what a non-singular function on $X = h^{-1}(0)$ means. So suppose we consider $X \subset R^n$. Then define $f: X \to R$ to be Nash if for every P in X, there exists a neighborhood U_p of P and $f_p: U_p \to R$, f_p Nash on U_p so that $f = f_p$ on $U_p \cap X$. Now the question of extension can be posed. In general, there is no extension as in the example of the Whitney umbrella $(x^2 + y^2)z = x^3$, on which there is a function which can't be extended. The problem seems to be the non-coherence of the Whitney umbrella. However, for curves in R^n, such extensions can be made. See $[E_4]$ for details. Also for certain other varieties such as the cone $x^2 + y^2 = z^2$, it is not hard to show that the methods for curves applies to show all Nash functions can be extended.

Bibliography

[A-K] S. Akbulut, H. King, Real Algebraic Structures on Topological Spaces, Publ. Math. No. 53, I.H.E.S. 79-162, 1981.

[B] G. Brumfiel, Partially Ordered Rings and Semi-algebraic Geometry, Cambridge University Press, London Mathematical Society Lecture Notes, Series 37, 1979.

[B-E] J. Bochnak, G. Efroymson, Real algebraic geometry and the Hilbert 17th problem, Math. Ann. 251, 213-241 (1980).

[B-K-S] J. Bochnak, W. Kucharz, M. Shiota, The divisor class groups of some rings of global real anaytic, Nash or rational regular functions, preprint.

[C_1] M. Coste, M.F. Coste-Roy, Topologies for real algebraic geometry, in Topos methods in geometry, Aarhus Universitet, pub. no. 30, 1979.

[C_2] M. Coste, preprint.

[Co] P. Cohen, Decision procedure for real and P-adic fields, Comm. Pure and Appl. Math. 22, 131-151 (1969).

[E_1] G. Efroymson, Substitution in Nash functions, Pac. J. of Math. 54, 101-112 (1976).

[E_2] G. Efroymson, Nash rings on planar domains, Trans. A.M.S. 249, 435-445 (1979).

[E_3] G. Efroymson, Extension of Nash functions, preprint.

[E_4] G. Efroymson, Extension of Nash functions on real curves, in preparation.

[Ma1] L. Mahé, Separation des composantes réelles par les signatures d'espaces quadratiques, Comptes Rend. Acad. Sc. t. 292, 769-771 (1981).

[Ma2] L. Mahé, Signatures et composantes connexes, (to appear in Math. Annalen).

[M] T. Mostowski, Some properties of the ring of Nash functions, Ann. Sc. Norm. Sup. Pisa C., Sci. III, 243-266 (1976).

[R] J. J. Risler, Sur l'anneau des fonctions de Nash globales, Ann. Sci. Ecole

Norm sup. 8, 365-378 (1975).

[S] M. Shiota, Classification of Nash manifolds, preprint.

[Tog1] A. Tognoli, Algebraic Geometry and Nash Functions, Inst. Naz. dialta Math., Institutiones Mathematicae, v. iii, Academic Press, 1978.

[Tog2] A. Tognoli, Algebraic approximation of manifolds and spaces, Sem. Bourbaki vol. 1979/80, Expose 548, Lecture Notes in Math no. 842, Springer-Verlag, 1981.

[T] J. A. Tougeron, Fonctions composees differentiables: cas algébrique, Ann. Inst. Fourier, 30 (4), 51-74 (1980)

Gustave Efroymson

Department of Mathematics and Statistics

University of New Mexico

Albuquerque, New Mexico 87131

and

J. Bochnak

Vrije Universiteit

Subfaculteit Wiskunde

Amsterdam, The Netherlands

REAL VALUATION RINGS AND IDEALS

G. W. Brumfiel

Introduction

This paper is based on lectures given at Rennes, before and during the conference on Real Algebraic Geometry and Quadratic Forms, in May 1981. However, I learned many things in Rennes, and I have included some of them in this revised version of the lectures.

My own interest originally in semi-algebraic geometry was that it provided a very constructive, finitistic description of many parts of topology of interest to algebraic topologists. This includes smooth manifold theory via non-singular semi-algebraic manifolds and Nash functions. One can also study more functions, semi-algebraic functions on semi-algebraic sets, or fewer functions, rational regular functions on semi-algebraic sets, providing a kind of subdivision of real semi-algebraic geometry into three branches (or categories) analogous to differential topology, combinatorial topology, and algebraic geometry. I have realized that the abstract study of real commutative algebra is quite relevant for the delicate rational function theory of semi-algebraic sets, but is not so efficient for the description of topological phenomena, that is, semi-algebraic functions, where one is better off dealing directly with sets of points, rather than rings and ideals. These lectures on real valuation rings and ideals definitely belong to the study of rational function theory and abstract real commutative algebra.

One of my long term goals in this subject is to describe many real algebra-geometric phenomena in terms of rings and ideals which satisfy some kind of "reality hypotheses". I suppose the best possible situation would be to have a category of "good" rings, so that the allowable ideals were simply those for which the residue rings were "good". This works perfectly for real prime ideals, which correspond in this way to integral domains with formally real field of fractions. But it now seems very unlikely to me that all the interesting reality hypotheses on ideals can be captured with one definition, or that the definitions should only depend on the residue ring. In these notes, I first study real valuation rings in formally real fields. Then I study arbitrary intersections of ideals in real valuation rings with other subrings of the field (for example, affine coordinate rings of real varieties). These ideals in the subring cannot be characterized by internal properties of the residue rings alone, but seem to be very interesting geometrically. The new results are from the Stanford thesis of my student, Robert O. Robson.

The paper begins with some preliminary results which are known to many. After studying the work of M. Coste and M. F. Coste-Roy and after very enlightening discussions with L. Mahé at Rennes, I have elected to emphasize the concept of a "precone" in a ring, rather than a partial order. In a field, the two notions coincide, but in a ring precones fit in better with the Coste and Coste-Roy theory of the real spectrum. Also, the important "Positivstellensatz" of Stengle (characterizing functions which are non-negative on a semi-algebraic set) is very naturally a consequence of considerations with precones, using some old ideas of Prestel. This was explained to me by Mahé, and I have included this proof of Stengle's Theorem in the paper, in the section on real function fields.

The proofs of two other main theorems were greatly clarified by discussions at Rennes with Houdebine and Rolland. These are the Real Place Existence Theorem (Theorem 2) and a finiteness theorem of Robson (Theorem 11) about intersections of real valuation ideals in Noetherian domains. In both cases, the point of the new proofs is to exploit the fact that the real valuation rings in a totally ordered field are themselves well-ordered by inclusion. This fact makes certain aspects of the study of real valuation rings easier than classical (Krull) valuation theory. But I think there is still much work to be done and many results to be discovered before real valuation theory will be fully appreciated.

<u>Definitions</u>

A is always a commutative ring with 1.

A <u>precone</u> in A is a subset $\mathfrak{P} \subset A$ such that (1) $\mathfrak{P} + \mathfrak{P} \subset \mathfrak{P}$ and $\mathfrak{P} \cdot \mathfrak{P} \subset \mathfrak{P}$ (2) $a^2 \in \mathfrak{P}$, all $a \in A$, and (3) $-1 \notin \mathfrak{P}$.

A precone $\mathfrak{P} \subset A$ is a <u>preorder</u>, or <u>partial order</u>, if $\mathfrak{P} \cap -\mathfrak{P} = (0)$. Given a preorder, we obtain a partial ordering of A by $a \leq b$ if and only if $b - a \in \mathfrak{P}$. We refer to $(A,\mathfrak{P})$ as a <u>partially ordered ring</u>. A preorder is a <u>total order</u> if $\mathfrak{P} \cup -\mathfrak{P} = A$. Note that if A is a field, any precone is a preorder.

If $\mathfrak{P} \subset A$ is a precone, an ideal $I \subset A$ is $\mathfrak{P}$<u>-convex</u> if $\Sigma p_i \in I$, all $p_i \in \mathfrak{P}$, implies all $p_i \in I$. Equivalently, if $\mathfrak{P}/I \subset A/I$ is the set of cosets of the form $p + I$, $p \in \mathfrak{P}$, then I is $\mathfrak{P}$-convex exactly when $\mathfrak{P}/I$ is a preorder in A/I. Similarly, a precone $\mathfrak{P} \subset A$ is a preorder exactly when (0) is a $\mathfrak{P}$-convex ideal.

Preliminaries on Convex Ideals

Let $\mathfrak{P} \subset A$ be a precone.

I1. Arbitrary intersections of $\mathfrak{P}$-convex ideals are $\mathfrak{P}$-convex. In particular, any ideal $I \subset A$ is contained in a smallest $\mathfrak{P}$-convex ideal, which we denote $H(I,\mathfrak{P})$, the $\mathfrak{P}$-hull of I. (Of course, perhaps $H(I,\mathfrak{P}) = A$.)

I2. If a union $I = \cup I_\alpha$ of ideals is an ideal and if each I_α is $\mathfrak{P}$-convex, then I is $\mathfrak{P}$-convex. In particular, by Zorn's lemma, any proper $\mathfrak{P}$-convex ideal is contained in a maximal (proper) $\mathfrak{P}$-convex ideal ($\neq$ convex maximal ideal in general).

I3. If I is $\mathfrak{P}$-convex, then $\sqrt{I}$ is $\mathfrak{P}$-convex.

I4. If I is $\mathfrak{P}$-convex and radical $(I = \sqrt{I})$, then the quotient ideals $(I : x) = \{y \in A | yx \in I\}$ are $\mathfrak{P}$-convex.

I5. Maximal $\mathfrak{P}$-convex ideals are prime. (Proof: $P = \sqrt{P}$ by I3. If $ab \in P$, $a \notin P$, then $P \subsetneq (P : b) = A$ by I4, so $b \in P$.)

I6. If $I \subset A$ is any ideal, then

$$\sqrt{H(I,\mathfrak{P})} = \bigcap_{\substack{P = \text{convex prime} \\ P \supset I}} P$$

(Proof: The key step is showing that if $f \notin \sqrt{H(I,\mathfrak{P})}$, then there is a convex prime P which contains I but no power of f. Choose a convex ideal Q maximal among those with these properties. Then $Q = \sqrt{Q}$ and $Q = (Q : f^n)$, all n, by I3 and I4. If $ab \in Q$, $a \notin Q$, then $Q \subsetneq (Q : b)$, so some $f^n \in (Q : b)$ which is the same as $b \in (Q : f^n) = Q$. Thus Q is prime.)

I7. If $I \subset A$ is any ideal, then

$$\sqrt{H(I,\mathfrak{P})} = \{c \in A \mid c^{2s} + p \in I, \text{ some } p \in \mathfrak{P},\ s \geq 1\}.$$

(Proof: The only hard part is showing the right-hand side is closed under sums. Then it is easily seen to be a convex, radical ideal containing I and contained in $\sqrt{H(I,\mathfrak{P})}$, hence it must coincide with $\sqrt{H(I,\mathfrak{P})}$. Suppose $c^{2s} + p' = a$, $d^{2r} + q' = b$, $p',q' \in \mathfrak{P}$, $a,b \in I$. We may assume $s = r$. Then $((c+d)^2 + (c-d)^2)^{2s} = (2c^2 + 2d^2)^{2s} = c^{2s}p'' + d^{2s}q''$ with $p'',q'' \in \mathfrak{P}$, hence $((c+d)^2 + (c-d)^2)^{2s} + p'p'' + q'q'' = ap'' + bq'' \in I$, which has the form $(c+d)^{4s} + p \in I$, with $p \in \mathfrak{P}$).

I8. Let $S \subset A$ be a multiplicative set such that $S \cap \mathfrak{P} \cap -\mathfrak{P} = \emptyset$. Form the localization A_S and define $\mathfrak{P}_S = \{[a/s] \mid ast^2 \in \mathfrak{P} \text{ some } t \in S\}$. Then $\mathfrak{P}_S \subset A_S$ is a precone. If $I \subset A$ is $\mathfrak{P}$-convex, then $IA_S \subset A_S$ is $\mathfrak{P}_S$-convex. (Examples of such S are $S(1) = \{1+p \mid p \in \mathfrak{P}\}$, complements of $\mathfrak{P}$-convex prime ideals, and any S if $\mathfrak{P}$ is a preorder.)

I9. If A is an integral domain, $\mathfrak{P} \subset A$ a precone, then in the field of fractions of A we have

$$(A_{S(1)}, \mathfrak{P}_{S(1)}) = \bigcap_{\substack{Q = \text{maximal} \\ \mathfrak{P}\text{-convex ideal}}} (A_{(Q)}, \mathfrak{P}_{(Q)}).$$

If $I \subset A$ is any ideal, then $IA_{S(1)} = \cap IA_{(Q)}$. (Proof: Suppose $x/b \in \cap IA_{(Q)}$. Write $x/b = x_Q/b_Q$, with $x_Q \in I$ and $b_Q \notin Q$. Then by I2, $1 \in \sqrt{H(\{b_Q\},\mathfrak{P})}$ since no proper convex ideal contains all the b_Q. By I7, we get $1 \leq \Sigma a_{Q_i} b_{Q_i}$ for suitable $a_{Q_i} \in A$, and $x/b = (a_{Q_i} x_{Q_i})/(a_{Q_i} b_{Q_i}) = (\Sigma a_{Q_i} x_{Q_i})/(\Sigma a_{Q_i} b_{Q_i}) \in IA_{S(1)}$. The proof that $\mathfrak{P}_{S(1)} = \cap \mathfrak{P}_{(Q)}$ is almost identical.)

Preliminaries on Precones and Preorders

We fix a ring A and consider some constructions with the sets of all precones and preorders on A. These sets are inductively ordered by inclusion, $\mathfrak{P}_1 \subset \mathfrak{P}_2$. We say $\mathfrak{P}_2$ refines $\mathfrak{P}_1$, or that $\mathfrak{P}_1$ is weaker than $\mathfrak{P}_2$. If $\mathfrak{P} \subset A$ is a subset and $g_i \in A$ are elements, we set $\mathfrak{P}[g_i] = \{\Sigma p_I g_I | p_I \in \mathfrak{P}, g_I = g_{i_1} \cdot \ldots \cdot g_{i_r}\}$.

O1. An arbitrary intersection of precones (respectively preorders) is a precone (respectively preorder). The weakest possible precone is $\mathfrak{P}_W = \{\Sigma a_i^2 | a_i \in A\}$, and $\mathfrak{P}_W$ is a precone exactly when -1 is not a sum of squares in A. In this case, by I7, $H(0), \mathfrak{P}) \subset A$ is a proper $\mathfrak{P}$-convex ideal, hence there exist $\mathfrak{P}$-convex prime ideals. Every precone (respectively preorder) admits maximal precone (respectively maximal preorder) refinements.

O2. If $\mathfrak{P} \subset A$ is a precone, $g \in A$, then $\mathfrak{P}[g]$ is a precone if and only if $(1+p') + (1+p'')g \neq 0$ for all $p', p'' \in \mathfrak{P}$. Also, $\mathfrak{P}[g]$ is a precone if and only if $\mathfrak{P} \subset A$ extends to a precone in the ring $A[T]/(T^2 - g)$. The weakest such extension intersects A exactly in $\mathfrak{P}[g]$.

If $\mathfrak{P} \subset A$ is a preorder, $g \in A$, and if $p' + p''g \neq 0$ for all $p', p'' \in \mathfrak{P}$ with $p'' \neq 0$, then $\mathfrak{P}$ extends to a preorder on $A[T]/(T^2 - g)$, which intersects A exactly in $\mathfrak{P}[g]$. Conversely, if $\mathfrak{P}[g]$ is a preorder on A and if A has no nilpotent elements or if g is not a zero divisor, then $\mathfrak{P}$ extends to a preorder on $A[T]/(T^2 - g)$.

(Proof: The statements about precones are trivially checked if the condition $(1+p') + (1+p'')g \neq 0$ is replaced by $1+p+qg \neq 0$, all $p, q \in \mathfrak{P}$. But if one has $1+p+qg = 0$, then also $0 = (1+p+qg)(1+p+g) = 1+p^2+2p+qg^2 + (1+p)(1+q)g = (1+p') + (1+p'')g$. The statements about preorders are slightly tricky to prove in general. The only case we will use is when A is a field and we want to adjoin a square root, preserving a partial ordering. This classical result is quite easy.)

03. If $\mathfrak{P} \subset A$ is a precone, $g \in A$, then either $\mathfrak{P}[g]$ or $\mathfrak{P}[-g]$ is a precone. In particular, if $\mathfrak{P}$ is a maximal precone, then $\mathfrak{P} \cup -\mathfrak{P} = A$. If $\mathfrak{P} \cup -\mathfrak{P} = A$, then $\mathfrak{P} \cap -\mathfrak{P} = I$ is a $\mathfrak{P}$-convex ideal of A and $\mathfrak{P}/I \subset A/I$ is a total order. In the case that $\mathfrak{P}$ is a maximal precone, $\mathfrak{P} \cap -\mathfrak{P}$ is a prime ideal of A. (So $\mathfrak{P}$ is a <u>prime precone</u>, a point of the real spectrum of M. Coste and M. F. Coste-Ray.)

(Proof: If $\mathfrak{P}[g]$ and $\mathfrak{P}[-g]$ are not precones, then we get $1+p+qg = 0$ and $1+p'-q'g = 0$ with $p,p',q,q' \in \mathfrak{P}$. Then $(1+p)(1+p') = -qq'g^2$, or $1+p+p'+pp'+qq'g^2 = 0$, a contradiction. The only other non-obvious statement is that $\mathfrak{P} \cap -\mathfrak{P}$ is a prime ideal if $\mathfrak{P}$ is a maximal precone. Suppose $ab \in \mathfrak{P} \cap -\mathfrak{P}$, with $a \notin \mathfrak{P} \cap -\mathfrak{P}$. We may assume $a \notin \mathfrak{P}$. Since $\mathfrak{P}$ is maximal, $\mathfrak{P}[a]$ is not a precone, so $1+p+qa = 0$, for some $p,q \in \mathfrak{P}$. Then $b+pb+qab = 0$. Now observe that $b \in \mathfrak{P}$ implies $b \in -\mathfrak{P}$ and $b \in -\mathfrak{P}$ implies $b \in \mathfrak{P}$, so, in fact, $b \in \mathfrak{P} \cap -\mathfrak{P}$.)

04. If $\mathfrak{P}_1$ and $\mathfrak{P}_2$ are two precones in A and if a prime ideal $P \subset A$ is $\mathfrak{P}_1 \cap \mathfrak{P}_2$-convex, then P is either $\mathfrak{P}_1$-convex or $\mathfrak{P}_2$-convex.

(Proof: If not, then $x+p \in P$ for suitable $x,p \in \mathfrak{P}_1$, $x \notin P$ and $y+q \in P$ for suitable $y,q \in \mathfrak{P}_2$, $y \notin P$. Then P contains $(x+p)^2y^2 + x^2(y+q)^2$ $= x^2y^2 + (2px+p^2)y^2 + x^2(y+q)^2 = x^2y^2 + (x+p)^2y^2 + x^2(2qy+y^2)$. But we see this element has the form $x^2y^2 + r$ with $r \in \mathfrak{P}_1 \cap \mathfrak{P}_2$, so $x^2y^2 \in P$, which contradicts $x,y \notin P$.)

O5. If A is an integral domain, $\mathfrak{P} \subset A$ a preorder, and $g \in A$, then either $\mathfrak{P}[g]$ or $\mathfrak{P}[-g]$ is a preorder on A. In particular, any maximal preorder on an integral domain is a total order. If K is a field, $\mathfrak{P} \subset K$ a preorder, $g,-g \notin \mathfrak{P}$, then both $\mathfrak{P}[g]$ and $\mathfrak{P}[-g]$ are preorders and $\mathfrak{P} = \mathfrak{P}[g] \cap \mathfrak{P}[-g]$. In particular, if $A \subset K$ is a subring and $P \subset A$ is a prime $\mathfrak{P} \cap A$-convex ideal, then there exist total order refinements $\hat{\mathfrak{P}}$ of $\mathfrak{P}$ in K such that P is still $\hat{\mathfrak{P}} \cap A$-convex.

(Proof: If $\mathfrak{P}[g]$, $\mathfrak{P}[-g]$ are both not preorders, then $0 = p+qg = p'-q'g$ for elements $p,p',q,q' \in \mathfrak{P}$, $q,q' \neq 0$, by O2. Thus $pp' = -qq'g^2$. But this can hold only if $pp' = qq'g^2 = 0$ and since A is an integral domain, this is impossible. In the field case, if $p+qg = 0$, then $g = -p/q = -pq/q^2 \in -\mathfrak{P}$, so either $g \in -\mathfrak{P}$ or $\mathfrak{P}[g]$ is a preorder. If $f = p+qg = p'-q'g$, with $p,q,p',q' \in \mathfrak{P}$, then $(q+q')f = pq'+p'q$, so either $q = q' = 0$ or $f = (pq'+p'q)/(q+q')$. In either case, $f \in \mathfrak{P}$. The last statement now follows from O4 by an application of Zorn's lemma.)

Real Valuation Rings

Let $(K,\mathfrak{P})$ be a partially ordered field. We denote valuation rings in K by A_v, with corresponding valuation $v\colon K^* \to \Gamma_v$ = totally ordered abelian group. The maximal ideal is $P_v \subset A_v$ and the residue field is $\Delta_v = A_v/P_v$, with $p_v\colon K \to \Delta_v, \infty$ the associated place.

Definition 1. A_v is $\mathfrak{P}$-real if the following equivalent conditions hold:

(i) A_v is a $\mathfrak{P}$-convex subring of K.

(ii) $P_v \subset A_v$ is a $\mathfrak{P} \cap A_v$-convex ideal.

(iii) Δ_v can be ordered such that $p_v(x) \geq 0$ in Δ_v, for all $x \in \mathfrak{P} \cap A_v$.

(iv) If $x,y \in K$, $0 \leq x \leq y$, then $v(x) \geq v(y)$ in Γ_v.

If K is a field and $A_v \subset K$ any valuation ring such that the residue field $\Delta_v = A_v/P_v$ is formally real, then K is formally real and the sums of squares in A_v coincide with the elements of A_v which are sums of squares in K. The proof is easy: if $\Sigma\alpha_i^2 \in A_v$, $\alpha_i \in K$, then for some index, say $i = 1$, $\Sigma\alpha_i^2 = \alpha_1^2(1 + \Sigma(\alpha_i/\alpha_1)^2)$, with all $\alpha_i/\alpha_1 \in A_v$. Now if $\alpha_1 \in A_v$, we have our sum of squares in A_v and if $\alpha_1 \notin A_v$, multiply by $(1/\alpha_1)^2 \in P_v$ and reduce to 0 in Δ_v, contradicting reality of Δ_v. As a consequence , we have that for any subring $A \subset A_v$, the center $P = P_v \cap A$ is $\mathfrak{P}_w \cap A$-convex, where $\mathfrak{P}_w \subset K$ denotes the preorder of sums of squares in K. In fact, an old theorem of Baer and Krull asserts that any total order on Δ_v is induced by some total order on K for which $P_v \subset A_v$ is convex. On the other hand, examples can be given of preorders on valuation rings A_v for which P_v is convex but which do not extend to preorders on K for which P_v is convex. This phenomenon does not seem very interesting geometrically, so we insist in our study of reality and valuation rings that we begin with preorders on fields K.

Theorem 2 (Real Place Existence Theorem). Let $(K,\mathfrak{P})$ be a partially ordered field, $A \subset K$ a subring, $P \subset A$ a $\mathfrak{P} \cap A$-convex prime ideal. Then for each total order refinement $\hat{\mathfrak{P}}$ of $\mathfrak{P}$ for which P is still $\hat{\mathfrak{P}} \cap A$-convex, there exists a unique valuation ring $A_v \subset K$ such that (1) A_v is $\hat{\mathfrak{P}}$-real, (2) $A \subset A_v$ and $P = P_v \cap A$, and (3) the induced total order on Δ_v is Archimedean over the field of fractions of A/P in Δ_v.

Proof: First note that the existence of such $\hat{\mathfrak{P}}$ is just preliminary result O5. The proof we give now of the theorem was suggested by Houdebine. With K totally ordered, the set of valuation rings in K which are real is itself totally ordered by inclusion and corresponds to certain Dedekind cuts of K, namely those cuts t which satisfy $2t = t$, $t^2 = t$. Any subring of K determines such a cut--the elements bounded in absolute value by an element of the subring. In our situation, we form the local ring $A_{(P)}$ and take the induced cut of $(K,\hat{\mathfrak{P}})$. This gives a $\hat{\mathfrak{P}}$-real valuation ring A_v containing $A_{(P)}$. If $x \in P$, then $|x| \leq 1$, so $1 \leq |1/x|$. If $|1/x| \leq |y|$, $y \in A_{(P)}$, then y is a unit in $A_{(P)}$ since $PA_{(P)}$ is convex by I8. Thus $|1/y| \leq |x|$, which contradicts $x \in P$, so we conclude $|1/x| \notin A_v$, that is, $x \in P_v$. Conversely, if $x \in P_v \cap A$, then $x \in P$, since otherwise $1/x \in A_{(P)} \subset A_v$. We have now established $P = P_v \cap A$, and (1) and (2) of the theorem.

Consider the extension of totally ordered fields $A_{(P)}/PA_{(P)} \subset A_v/P_v$, say $F \subset E$. It is now a fairly elementary fact, argued using cuts, that E is Archimedean over F exactly when E admits no non-trivial $\hat{\mathfrak{P}}$-real places which are trivial on F. But this is precisely the same as the assertion that $A_v \subset K$ is the smallest $\hat{\mathfrak{P}}$-real valuation ring containing $A_{(P)}$. Thus we have (3) of the theorem, as well as the uniqueness of A_v.

<u>Proposition 3</u>. Let $(K,\mathfrak{P})$ be a partially ordered field, $A \subset K$ a subring. Then the following subrings of K coincide, which we denote by $\overline{A}$ and call the <u>semi-integral closure</u> of A in $(K,\mathfrak{P})$.

(i) $\displaystyle\bigcap_{\substack{A_v = \mathfrak{P}\text{-real val. ring} \\ A_v \supset A}} A_v$

(ii) $\{x \in K \mid x^{2n} + a_1x^{2n-1} + \cdots + a_{2n} \leq 0,\ \text{some } a_i \in A,\ n \geq 1\}$

(iii) $\{x \in K \mid a' < x < a'',\ \text{some } a',a'' \in A\}$

(iv) $\{x \in K \mid -a < x < a, \text{ some } a \in A\}$

(v) $\{x \in K \mid x^2 \leq a^2, \text{ some } a \in A\}$

(vi) $\{x \in K \mid$ for all total order refinements $\hat{\mathfrak{P}}$ of $\mathfrak{P}$, $|x|$ is bounded rel $\hat{\mathfrak{P}}$ by an element of $A\}$

(vii) $A\{\frac{1}{1+p} \mid p \in \mathfrak{P}\}$

Proof: We will show (i) $\subset$ (ii) $\subset$ (iii) $\subset$ (iv) $\subset$ (v) $\subset$ (vi) $\subset$ (i) and then show (vii) = (iii).

(i) $\subset$ (ii) If $x \in K$ does not satisfy such a semi-integral inequality as in (ii), then $H((1/x^2)A[1/x^2], \mathfrak{P}) \subset A[1/x^2]$ is proper. Otherwise, by I7 there would be an inequality $1 \leq a_1/x^2 + \cdots + a_n/x^{2n}$, $a_i \in A$, which we could multiply by x^{2n}. By I2 and I5, we can find a $\mathfrak{P}$-convex prime in $A[1/x^2]$ containing $1/x^2$, and then by the Real Place Existence Theorem, we can find a $\mathfrak{P}$-real valuation ring $A_v \subset K$ with $A[1/x^2] \subset A_v$, $1/x^2 \in P_v$. Thus $x^2 \notin A_v$ hence $x \notin A_v$.

(ii) $\subset$ (iii) In my book, I prove by induction on degree that for the general monic even degree polynomial $f(T) = T^{2n} + a_1T^{2n-1} + \cdots + a_{2n}$, there are universal polynomials $\beta^+(a_1 \cdots a_{2n})$ and $\beta^-(a_1 \cdots a_{2n}) \in Q[a_1 \cdots a_{2n}]$ and $h_i^+(T), h_i^-(T) \in Q[a_1 \cdots a_{2n}, T]$ such that

$$\beta^-(a_1 \cdots a_{2n}) - T + \Sigma h_i^-(T)^2 = f(T) = T - \beta^+(a_1 \cdots a_{2n}) + \Sigma h_i^+(T)^2 .$$

We then set $T = x$ and see that if $f(x) \leq 0$, then

$$\beta^-(a_1 \cdots a_{2n}) \leq x \leq \beta^+(a_1 \cdots a_{2n}) .$$

<u>(iii) ⊂ (iv)</u> We have $0 \le (b - \frac{1}{2})^2$ so $b \le b^2 + 1$. If $a' < x < a''$, then $-a < x < a$, with $a = 1 + (a')^2 + (a'')^2$.

<u>(iv) ⊂ (v)</u> If $-a < x < a$, then $a^2 - x^2 = (a-x)(a+x) \ge 0$.

<u>(v) ⊂ (vi)</u> Obvious.

<u>(vi) ⊂ (i)</u> If an element x is not in some $\mathfrak{P}$-real valuation ring $A_v \supset A$, then $1/x \in P_v$. By 05, refine $\mathfrak{P}$ to a total order $\hat{\mathfrak{P}}$ keeping A_v $\hat{\mathfrak{P}}$-real. If $|x| \le |a|$ (rel $\hat{\mathfrak{P}}$), with $a \in A$, then $1 \le |a|/|x| \in P_v$, which contradicts convexity of P_v. Thus $|x|$ is not $\hat{\mathfrak{P}}$-bounded by an element of A.

<u>(vii) ⊂ (iii)</u> If $p \in \mathfrak{P}$, then $0 < \frac{1}{1+p} < 1$.

<u>(iii) ⊂ (vii)</u> If $a' < x < a''$, then $0 < y = x-a' < a'' - a'$. It suffices to prove $y \in$ (vii). But $p = \frac{a'' - a'}{y} - 1 \ge 0$, so $\frac{1}{1+p} = \frac{y}{a'' - a'}$ and $y = (a'' - a')/1+p$.

Semi-integrally closed rings are rather large, not as well-behaved as integral closures. For example, if A is a finitely generated domain of transcendence degree greater than 1 over some ground field, then the semi-integral closure is generally not Noetherian. This is a consequence of the following lemma, together with the fact that valuation rings are mostly not Noetherian.

<u>Lemma 4</u>. Let $(K,\mathfrak{P})$ be a partially ordered field, $A \subset K$ a subring.

(a) If $S \subset A$ is a multiplicative set, then $(\overline{A_S}) = (\overline{A})_S$.

(b) If A is semi-integrally closed, then all ideals $I \subset A$ are $\mathfrak{P}$-convex.

(c) If A is a local, semi-integrally closed subring, then A is a real valuation ring.

Proof:

(a) One can use any of the characterizations (ii)-(vii) of semi-integral closures in Proposition 3.

(b) If $0 \leq x \leq y$, $x,y \in A$, $y \in I$, then $0 \leq x/y \leq 1$, so $x/y \in A$ and $x = (x/y)y \in I$.

(c) Suppose $x \in K$, $1/x \notin A$. We will show $x^2 \in A$ which implies $x \in A$ (semi-integrally closed implies integrally closed). Since $\frac{1}{1+x^2} \in A$, we know $1 - \frac{1}{1+x^2} = \frac{x^2}{1+x^2} \in A$. If $\frac{1}{1+x^2} \in P$, the maximal ideal of A, it would follow that $1 - \frac{1}{1+x^2}$ is a unit in the local ring A. That is, $\frac{1+x^2}{x^2} = \frac{1}{x^2} + 1 \in A$ which gives $\frac{1}{x^2} \in A$, contradicting $1/x \notin A$. Therefore, $\frac{1}{1+x^2}$ is a unit in A, so $1+x^2 \in A$ and $x^2 \in A$. This shows A is a valuation ring, which is $\mathfrak{P}$-real by part (b).

Corollary 5. Any localization at a prime ideal of a semi-integrally closed domain is a real valuation ring. The $\mathfrak{P}$-real valuation rings containing a subring A of a partially ordered field $(K,\mathfrak{P})$ correspond bijectively with the localizations of $\overline{A}$ at prime ideals $\overline{P} \subset \overline{A}$, that is, with points of $\mathrm{Spec}(\overline{A})$.

Proof: The first statement follows Lemma 4(a), (c). If A_V is any $\mathfrak{P}$-real valuation ring containing A, let $\overline{P} = P_V \cap \overline{A}$. Then $\overline{A}_{(\overline{P})} \subset A_V$. But $\overline{A}_{(\overline{P})}$ is also a real valuation ring and no element of $\overline{P}\,\overline{A}_{(\overline{P})}$ is invertible in A_V. Thus $\overline{A}_{(\overline{P})} = A_V$.

Given a subring A of a partially ordered field $(K,\mathfrak{P})$, it is of interest to investigate the maximal and minimal primes of $\overline{A}$.

Proposition 6.

(a) $$\overline{A} = \bigcap_{\text{maximal } \overline{P}} \overline{A}_{(\overline{P})}.$$

(b) The prime $\overline{P} \subset \overline{A}$ is maximal if and only if for all total order refinements $\hat{\mathfrak{P}}$ of $\mathfrak{P}$ such that $\overline{A}_{(\overline{P})}$ is $\hat{\mathfrak{P}}$-real we have (1) $P = \overline{P} \cap A$ is a $\hat{\mathfrak{P}} \cap A$-maximal convex ideal and (2) the induced total order on the residue field $\overline{A}_{(\overline{P})}/\overline{P}\,\overline{A}_{(\overline{P})}$ is Archimedean over the subfield $A_{(P)}/P\,A_{(P)}$.

Proof:

(a) This is obvious since the minimal elements among the $\mathfrak{P}$-real valuation rings containing A are the $\overline{A}_{(\overline{P})}$, with $\overline{P}$ maximal.

(b) The key point is similar to part of the proof of the Real Place Existence Theorem. Once K is totally ordered by $\hat{\mathfrak{P}}$, there is only one minimal $\hat{\mathfrak{P}}$-real valuation ring containing $A_{(P)}$ and the residue field is Archimedean over $A_{(P)}/P\,A_{(P)}$. If $\overline{P} \subset \overline{A}$ is maximal, this has to be $\overline{A}_{(\overline{P})}$. If $P \subset A$ were not $\hat{\mathfrak{P}} \cap A$-maximal convex, we could find a real valuation ring in $\overline{A}_{(\overline{P})}/\overline{P}\,\overline{A}_{(\overline{P})}$, with non-trivial center on A/P, which would correspond to a prime in $\overline{A}$ properly containing $\overline{P}$. Conversely, if $\overline{P} \subsetneq \overline{Q}$, choose a total order $\hat{\mathfrak{P}}$ so that $\overline{A}_{(\overline{Q})}$ is $\hat{\mathfrak{P}}$-real. Then $\overline{A}_{(\overline{P})}$ is also $\hat{\mathfrak{P}}$-real and either (1) fails or $\overline{A}_{(\overline{P})}/\overline{P}\,\overline{A}_{(\overline{P})}$ is not Archimedean over $A_{(P)}/P\,A_{(P)}$, since $\overline{A}_{(\overline{Q})} \subsetneq \overline{A}_{(\overline{P})}$, so (2) fails.

The hypotheses of (b) simplify greatly if the subring $A = k \subset K$ is a field which inherits a total order from $\mathfrak{P}$. Then (1) is automatic and (2) says that any total order on $\overline{\Delta} = \overline{A}_{(\overline{P})}/\overline{P}\,\overline{A}_{(\overline{P})}$ compatible with $\mathfrak{P}$ must be Archimedean over k.

Proposition 7. The prime $\overline{P} \subset \overline{A}$ is minimal if and only if the value group $\Gamma = K^*/\overline{A}^*_{(\overline{P})}$ associated to the valuation ring $\overline{A}_{(\overline{P})} \subset K$ is an Archimedean ordered group, that is, a subgroup of $(\mathbb{R},+)$, the additive real numbers, as ordered group.

Proof: This is immediate from valuation theory since $\overline{A}_{(\overline{P})}$ has a unique non-zero prime, which means Γ has rank 1.

Included in the above rank 1 valuations are, of course, any $\mathfrak{P}$-real discrete rank 1 valuations $v\colon\ K^* \to \mathbb{Z}$, non-negative on A. It is amusing that when $A = \mathbb{Q}$, the rationals, then the maximal primes of $\overline{A}$ correspond to places $p\colon\ K \to (\mathbb{R},\infty)$ and the minimal primes of $\overline{A}$ correspond to valuations $v\colon\ K^* \to (\mathbb{R},+)$, both times with values in $\mathbb{R}$, but different sorts of objects.

Real Valuation Ideals

As always, $(K,\mathfrak{P})$ is a partially ordered field and $A \subset K$ a subring. By a $\mathfrak{P}$-real valuation ideal, we mean the contraction to A of an ideal in some $\mathfrak{P}$-real valuation ring containing A. We will study the class of ideals of A which are intersections of $\mathfrak{P}$-real valuation ideals. The results below are due to my student, R. Robson, at Stanford.

Since the intersection of a real valuation ring in K with the field of fractions of A is still a real valuation ring, we see that we lose no generality by assuming that our field K is the field of fractions of A when studying real valuation ideals. Note that we observed in Lemma 4 that all ideals in $\mathfrak{P}$-real valuation rings are $\mathfrak{P}$-convex. Thus, intersections of $\mathfrak{P}$-real valuation ideals in A are certainly $\mathfrak{P}$-convex. We will call this class of ideals of A, $\mathfrak{P}$-<u>completely convex ideals</u> ($\mathfrak{P}$-c.c. ideals). The Real Place Existence Theorem shows that prime $\mathfrak{P}$-convex ideals are $\mathfrak{P}$-real valuation ideals, hence all radical $\mathfrak{P}$-convex ideals belong to our class of $\mathfrak{P}$-c.c. ideals. Obviously, there is a smallest $\mathfrak{P}$-c.c. ideal containing any ideal I, which we denote by $CH(I,\mathfrak{P})$, the <u>complete hull of</u> I, <u>rel</u> $\mathfrak{P}$.

We begin our study of $\mathfrak{P}$-c.c. ideals with the following real analogue of a result in Zariski-Samuel, vol. II, in an appendix on complete ideals.

<u>Proposition 8</u>. Let $(K,\mathfrak{P})$ be a partially ordered field, $A \subset K$ a subring, $I \subset K$ any A-submodule. Then the following $\bar{A}$-submodules of K coincide:

(i) $\displaystyle\bigcap_{\substack{A_v = \mathfrak{P}\text{-real val. ring} \\ A_v \supset A}} IA_v$ (which is a module over $\bar{A} = \cap A_v$)

(ii) $\{x \in K \mid x^{2n} + a_1 x^{2n-1} + \cdots + a_{2n} \leq 0, \text{ for some } a_j \in I^j\}$

(iii) $\{x \in K \mid$ for all total order refinements $\hat{\mathfrak{P}}$ of $\mathfrak{P}$, $|x|$ is bounded rel $\hat{\mathfrak{P}}$ by an element of $I\}$

(iv) $I\bar{A}$.

Moreover, if A is a local ring, with $\mathfrak{P} \cap A$-convex maximal ideal $P \subset A$, then

$$I\bar{A} = \bigcap_{\substack{A_v = \mathfrak{P}\text{-real val. ring} \\ A_v \supset A \\ P_v \cap A = P}} IA_v$$

Proof: We show (i) $\subset$ (ii) $\subset$ (iii) $\subset$ (i) = (iv).

(i) $\subset$ (ii) Let $x \in \cap IA_v$ and consider the elements $\{a/x\}$, $a \in I$, and the ring $A\{a/x\}$. For every $\mathfrak{P}$-real A_v containing $A\{a/x\}$, we have $x = \Sigma a_i y_i$, $a_i \in I$, $y_i \in A_v$. We may assume all $a_i/a_1 \in A_v$, so $x/a_1 \in A_v$. Thus by the Real Place Existence Theorem, the $\{a/x\}$ cannot all be contained in any proper $\mathfrak{P}$-convex ideal of $A\{a/x\}$. Thus by I7, there is an inequality $1 \leq \Sigma a_j/x^j$, $a_j \in I^j$ which gives $x^{2n} \leq \Sigma a_j x^{2n-j}$ for suitable n.

(ii) $\subset$ (iii) Assume $x^{2n} \leq b_1 x^{2n-1} + \cdots + b_{2n}$. Fix the total order refinement $\hat{\mathfrak{P}} \supset \mathfrak{P}$, and write $b_i = \Sigma(c_{i1} \cdots c_{ii}) \in I^i$. Then working rel $\hat{\mathfrak{P}}$,

$$x^{2n} \leq \sum \left| \frac{c_{i1} \cdots c_{ii}}{x^i} \right| x^{2n} \leq 2n \max_{i,j} \left| \frac{c_{ij}}{x} \right|^i x^{2n}$$

so

$$1 \leq \max_{i,j} \left| \frac{2n\, c_{ij}}{x} \right| ,$$

which gives $|x| \leq |2n\, c_{ij}|$ for some $c_{ij} \in I$.

(iii) $\subset$ (i) Given $\mathfrak{P}$-real A_v, we know there is a total order refinement $\hat{\mathfrak{P}}$ of $\mathfrak{P}$ such that A_v is $\hat{\mathfrak{P}}$-real. Suppose $|x| < |a|$ (rel $\hat{\mathfrak{P}}$), $x \in K$, $a \in I$. We want $x \in IA_v$. But $|x/a| \leq 1$, so $x/a \in A_v$ and $x = a(x/a) \in IA_v$.

(i) = (iv) Obviously (iv) $\subset$ (i). But from Corollary 5, the localizations $\bar{A}_{(\bar{P})}$ give all the $\mathfrak{P}$-real valuation rings $A_v \supset A$. Thus

$$\cap IA_v = \bigcap_{\bar{P} \in \operatorname{spec}(\bar{A})} I\bar{A}_{(\bar{P})} .$$

But from I9 and the fact that $\bar{A} = \bar{A}_{S(1)}$ since $\bar{A}$ is semi-integrally closed, we get $\cap I\bar{A}_{(\bar{P})} = I\bar{A}$.

Finally, assume A is a local ring and suppose $x \notin I\bar{A}$. By refining the order $\mathfrak{P}$, we may assume $1-b \in \mathfrak{P}$ for all $b \in P$, since $P \subset A$ is $\mathfrak{P} \cap A$-convex. A real valuation ring for this refined order will also be $\mathfrak{P}$-real. We claim $H(P, \{a/x\}, \mathfrak{P}) \subset A\{a/x\}$ is proper, where, as before, a runs over all elements of I. Otherwise, $1 \leq b + a_1/x + \cdots + a_n/x^n$, where $b \in P$, $a_j \in I^j$, hence $x^{2n} \leq \Sigma a_j' x^{2n-j}$, where $a_j' = a_j/(1-b) \in I^j$, since $1-b$ is a positive unit in A. Now a choice of a real valuation ring A_v with $P \subset P_v$ and $\{a/x\} \subset P_v$ gives $x \notin IA_v$ and $P_v \cap A = P$.

Note that if $I = A$, we just obtain $\bar{A}$, the semi-integral closure, in Proposition 8. In general, we denote by $\bar{I}$, the $\bar{A}$-module defined by the Proposition. Note that from (ii) or (iii), $\bar{I}$ does not depend on the ring A, as long as I is an A-module. In particular, we can replace A by $\mathbb{Z}$ and conclude $\bar{I} = \cap IA_v$, the intersection taken over all $\mathfrak{P}$-real valuation rings A_v, not just those containing A.

<u>Corollary 9</u>. If $I \subset A$ is an ideal, then $CH(I, \mathfrak{P}) = I\bar{A} \cap A = \{x \in A \mid x^{2n} + a_1x^{2n-1} + \cdots + a_{2n} \leq 0$, some $a_j \in I^j\}$.

<u>Proof</u>: The smallest $\mathfrak{P}$-c.c. ideal containing I is obviously $(\cap IA_v) \cap A = \bar{I} \cap A$, which then has the <u>single formula</u> characterization from Proposition 8 (ii).

We collect a few other pleasant properties of $\mathfrak{P}$-c.c. ideals.

Lemma 10.

(a) If $I \subset A$ is $\mathfrak{P}$-c.c., $x \in A$, then $(I : x)$ is $\mathfrak{P}$-c.c. In particular, associated primes of I are $\mathfrak{P}$-convex.

(b) If $S \subset A$ is any multiplicative set and $I \subset A$ is $\mathfrak{P}$-c.c., then $IA_S \subset A_S$ is $\mathfrak{P}$-c.c.

(c) If $I \subset A$ is any ideal, then the complete hulls of powers of I satisfy $CH(I^n)CH(I^m) \subseteq CH(I^{n+m})$. Thus there is an associated graded ring

$$\bigoplus_{n \geq 0} CH(I^n)/CH(I^{n+1})$$

Proof:

(a) If $y \in (I : x)\bar{A} \cap A$, then $yx \in I\bar{A} \cap A = I$, so $y \in (I : x)$, which shows $(I : x) = (I : x)\bar{A} \cap A$.

(b) The semi-integral closure of A_S is $(\bar{A})_S$ by Lemma 4. If $y \in I(\bar{A})_S \cap A_S$, then for some $s \in S$, $ys \in I\bar{A} \cap A = I$ so $y = ys/s \in IA_S$, which shows $IA_S = I\bar{A}_S \cap A_S$.

(c) Clearly $(I^n\bar{A} \cap A)(I^m\bar{A} \cap A) \subseteq I^{n+m}\bar{A} \cap A$.

Next, define a $\mathfrak{P}$-c.c. ideal to be $\mathfrak{P}$-irreducibly completely convex ($\mathfrak{P}$-i.c.c.) if whenever $I = I_1 \cap I_2$, with I_j also $\mathfrak{P}$-c.c., then either $I = I_1$ or $I = I_2$. The following very nice finiteness theorem is due to Robson.

Theorem 11. If $I \subset A$ is a $\mathfrak{P}$-i.c.c. ideal, then I is a real valuation ideal. Consequently, if A is a Noetherian domain, then all $\mathfrak{P}$-c.c. ideals are finite intersections of $\mathfrak{P}$-real valuation ideals.

Proof: The second statement is immediate from the first since the usual argument gives that in the Noetherian case all $\mathfrak{P}$-c.c. ideals are finite intersections of $\mathfrak{P}$-i.c.c. ideals.

The proof we give here of the first statement is based on the method of Houdebine used earlier for the Real Place Existence Theorem. Namely, it suffices to argue that there exists a total order refinement $\hat{\mathfrak{P}}$ of $\mathfrak{P}$ such that I is still $\hat{\mathfrak{P}}$-c.c. Because then there is a smallest $\hat{\mathfrak{P}}$-real valuation ring A_{v_o}, containing A, so $I = \bigcap_{\hat{\mathfrak{P}}\text{-real } A_v} IA_v \cap A = IA_{v_o} \cap A$.

Consider $g \notin \mathfrak{P} \cup -\mathfrak{P}$. Then by O5, $\mathfrak{P} = \mathfrak{P}[g] \cap \mathfrak{P}[-g]$. Any $\mathfrak{P}$-real valuation ring $A_v \subset K$ is either $\mathfrak{P}[g]$-real or $\mathfrak{P}[-g]$-real by O4. Therefore,

$$I = \bigcap IA_v \cap A = \left(\bigcap_{A_v = \mathfrak{P}[g]\text{real}} IA_v \cap A\right) \cap \left(\bigcap_{A_w = \mathfrak{P}[-g]\text{real}} IA_w \cap A\right).$$

Since I is irreducible, we conclude $I = \bigcap_{A_v = \mathfrak{P}[g]\text{real}} IA_v \cap A$ or $I = \bigcap_{A_w = \mathfrak{P}[-g]\text{real}} IA_w \cap A$. That is, I is either $\mathfrak{P}[g]$-c.c. or $\mathfrak{P}[-g]$-c.c. Of course, when we refine the partial order, I remains irreducible too.

Now we want to use Zorn's lemma. Let $\mathfrak{P}_\alpha$ be a totally ordered (by refinement) chain of partial orders on K for which I is $\mathfrak{P}$-c.c. and consider $\mathfrak{P}' = \bigcup_\alpha \mathfrak{P}_\alpha$. The semi-integral closure $\overline{I}'$ of I with respect to $\mathfrak{P}'$ is seen by Proposition 8(ii) to coincide with $\bigcup_\alpha \overline{I}_\alpha$, where $\overline{I}_\alpha$ is the $\mathfrak{P}_\alpha$-semi-integral closure of I. Therefore, I is $\mathfrak{P}'$-c.c., so Zorn's lemma and the previous paragraph complete the proof of the theorem.

Corollary 12. If $I \subset A$ is a $\mathfrak{P}$-i.c.c. ideal in a Noetherian domain A, then the associated primes of I form a chain $P_1 \subset P_2 \subset \cdots \subset P_r$. In fact, all the P_i come from primes in one valuation ring.

Proof: We know $I = IA_{v_0} \cap A$ for some $\mathfrak{P}$-real valuation ring $A_{v_0} \supset A$. The P_j have the form $P_j = (I : x_j) \subset A$. But from Lemma 10, $(I : x_j)A_{v_0} \cap A = (I : x_j)$. Now $(I : x_j)A_{v_0}$ may not be prime in A_{v_0}, but any ideal of A_{v_0} maximal among those with the property that they contract to P_j in A will be prime. The corollary is then proved since the ideals in any valuation ring are totally ordered by inclusion.

Real Function Fields

We now assume $A = R[x_1 \cdots x_n]$ is a finitely generated integral domain over a real closed field R, with formally real field of fractions K. Any partial order $\mathfrak{P}$ on A extends to a partial order $\mathfrak{P}_*$ on K by localization, $\mathfrak{P}_* = \{f/g \mid fgh^2 \in \mathfrak{P}, \text{ some } h \neq 0\}$. In general, $\mathfrak{P} \neq \mathfrak{P}_* \cap A$.

If $\{g_i\}$ are finitely many elements of A, let the smallest possible precone containing $\{g_i\}$ be denoted by $\mathfrak{P}_w[g_i]$. If this is a preorder, denote the induced preorder on K by $\mathfrak{P}_w(g_i)$. Thus we distinguish sums of squares in A and K by writing $\mathfrak{P}_w[1]$ and $\mathfrak{P}_w(1)$.

We define the following subsets of R^n.

X = the real points of the irreducible variety defined by A.

X_0 = the set of simple points in X.

$U\{g_i\} = \{x \in X \mid g_i(x) > 0 \text{ all } i\}$.

$W\{g_i\} = \{x \in X \mid g_i(x) \geq 0 \text{ all } i\}$.

If $I \subset A$ is an ideal set $X(I) = \{x \in X \mid f(x) = 0, \text{ all } f \in I\}$. The following results constitute a rapid course in semi-algebraic geometry.

A) $\mathfrak{P}_w[g_i]$ is a precone if and only if $W\{g_i\} \neq \emptyset$. If all $g_i \neq 0$ in A, then $\mathfrak{P}_w[g_i]$ is a preorder if and only if $U\{g_i\} \cap X_0 \neq \emptyset$. X_0 is an R-manifold of dimension equal to the transcendence degree of A over R and no non-zero element of A vanishes on any open subset of X_0.

B) If $\mathfrak{P}_W[g_i]$ is a precone, then a prime ideal $P \subset A$ is $\mathfrak{P}_W[g_i]$-convex if and only if $W\{g_i\}$ contains an open set of simple points of the variety $X(P)$. In particular, the maximal $\mathfrak{P}_W[g_i]$-convex ideals of A correspond to the points of $W\{g_i\}$. Given an ideal $I \subset A$ and $f \in A$, then $f(x) = 0$ all $x \in X(I) \cap W\{g_i\}$ if and only if there is a formula $f^{2s} + p \in I$, some $p \in \mathfrak{P}_W[g_i]$, $s \geq 1$.

C) If $\mathfrak{P}_W[g_i]$ is a precone and $f \in A$, then $f(x) > 0$ all $x \in W\{g_i\}$ if and only if there is a formula $f(1+p) = 1+q$, with $p,q \in \mathfrak{P}_W[g_i]$. Also, $f(x) \geq 0$ all $x \in W\{g_i\}$ if and only if there is a formula $f(f^{2s}+p) = q$, $p,q \in \mathfrak{P}_W[g_i]$, some $s \geq 1$.

D) If $\mathfrak{P}_W[g_i]$ is a preorder and all $g_i \neq 0$, then a prime ideal $P \subset A$ is $\mathfrak{P}_W(g_i) \cap A$-convex if and only if the closure $\overline{U\{g_i\} \cap X_o}$ contains an open set of simple points of $X(P)$. In particular, the maximal $\mathfrak{P}_W(g_i) \cap A$-convex ideals of A correspond to the points of $\overline{U\{g_i\} \cap X_o}$.

E) If $\mathfrak{P}_W[g_i]$ is a preorder, all $g_i \neq 0$, and $f \in A$, then $f \in \mathfrak{P}_W(g_i) \cap A$ if and only if $f(x) \geq 0$ for all $x \in \overline{U\{g_i\} \cap X_o}$.

It is not our intention to prove all these results here. They can all be interpreted as versions of either the Real Nullstellensatz or Hilbert's 17th Problem. The fastest proofs combine the Tarski-Seidenberg principle with the elementary preliminary results on real commutative algebra given at the beginning of this article, especially I5, I6, I7, O2, and O3.

For example, let us look at C), Stengle's theorem. Suppose $f(x) > 0$ all $x \in W\{g_i\}$. Then $\mathfrak{P}_W[g_i,-f]$ cannot be a precone on A. Because if it were, we could extend it to a maximal precone. By O3, this would give a homomorphism φ: $A \to L$, with L a totally ordered field, with $\varphi(g_i) \geq 0$

and $\varphi(f) \leq 0$. Then Tarski's Principle would give $x \in W\{g_i\}$ with $f(x) \leq 0$. So we conclude from O2 that there is a formula $(1+p)f = 1+q$, $p,q \in \mathfrak{P}_W[g_i]$. In the case that $f(x) \geq 0$ all $x \in W\{g_i\}$, we replace A by the localization $A[1/f]$ to get the desired formula from the first case, at least if $\mathfrak{P}_W[g_i]$ is a precone in $A[1/f]$. But by I8, this will be the case unless $f^s \in \mathfrak{P} \cap -\mathfrak{P}$, some s, in which case $f^{2s} \in -\mathfrak{P}$ is already a relation of the desired sort. $(f(f^{2s} + (-f^{2s})) = 0.)$ This elegant proof of Stengle's theorem was shown to me by L. Mahé, combining an old theorem of Prestel with ideas of M. Coste and M. F. Coste-Roy.

I don't think result D) is quite as easy to deduce from the Tarski Principle as A), B), and C). The reason is, statements about all $f \in \mathfrak{P}_W(g_i) \cap A$ are not so obviously statements about only finitely many elements of A. However, E) can be proved first and then one can exploit the theorem that any closed semi-algebraic set is a finite union of sets of the form $W\{h_i\}$. In my book, I gave a long proof of D) without using this last fact. I still don't know any really quick proofs of D).

In any event, our intention in stating results A) - E) is just to remind the reader of our geometric motivation for studying partial orders on rings and fields. We now return to our main theme and study real valuations of real function fields.

The simplest valuations of function fields K over a ground field k are the "prime divisors". These are discrete rank one valuations $v: K^* \to \mathbb{Z}$, whose valuation ring is the localization of a finitely generated integrally closed domain A' over k with fraction field K, at a minimal prime of A'. The residue field is also a function field K', with $\text{tr. deg.}(K'/k) = \text{tr. deg.}(K/k) - 1$. In fact, any valuation of K whose residue field K' has transcendence degree one less than that of K, is such a prime divisor.

One can actually see the picture here. A' corresponds to an affine variety X' of dimension equal to tr. deg.$(K/k) = r$ and the minimal prime corresponds to a subvariety V' of dimension $r-1$.

Each function $f \in A'$ has associated an integer $v(f) \geq 0$, the order to which f vanishes generically across V'. This v extends to a valuation of K by $v(f/g) = v(f) - v(g)$.

It is a classical result that if A is any finitely generated domain over a field with quotient field K and $P \subset A$ is any prime, then there is a prime divisor v of K with valuation ring $A_v \supset A$ such that $P_v \cap A = P$. This refines the place existence theorem by providing a very nice kind of place with preassigned center $P \subset A$.

Here is roughly the geometry behind the classical proof, although the algebra can be done directly without any reference to geometry. (The key algebraic step is the Krull Principal Ideal Theorem which implies the irreducible components of varieties defined by a principal ideal all have codimension 1.) The prime $P \subset A$ corresponds to a subvariety $V \subset X$ of codimension perhaps greater than 1. If we blow up X along V, then take integral closure, we get a normal variety X birationally equivalent to X such that the inverse image of V now has irreducible components of codimension 1. (Points on varieties here have coordinates in an algebraically closed field.) Any of these components which maps onto V provides a hypersurface $V' \subset X'$ and a prime divisor of K with center P on A.

Now, this proof does not go through in the real case either algebraically or geometrically. Geometrically, the real points over V of the blown up variety X' over X do not necessarily increase in dimension because there may not be enough real tangent lines to X at points of V. Algebraically,

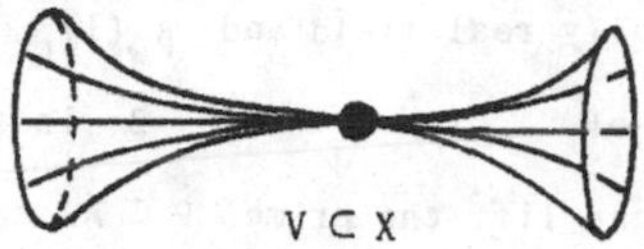

the problem is that principal ideas are generally not convex, and one has no control over their zeros.

I conjecture that this elementary proof can be pushed through for certain orders on function fields by iterating the blowing up process. Somehow the singularity should get better and eventually a nice real hypersurface produced over V. I give an example after the next theorem of this iteration of elementary blow-ups.

In any event, if we are willing to quote the powerful resolution of singularities theorem, we can obtain the desired result for certain preorders on function fields. The proof below owes much to a conversation with Efroymson and Tognoli. The result was also observed by Risler for real varieties.

Theorem 13. Let $A = R[x_1 \cdots x_n]$ be a finite domain over a real closed field R, with fraction field K. Let $\mathfrak{P}_w(g_{ij})$, $1 \leq i \leq m$, be finitely generated preorders on K and let $\mathfrak{P} = \bigcap_{i=1}^{m} \mathfrak{P}_w(g_{ij})$. Then for any $\mathfrak{P} \cap A$-convex prime $P \subset A$ there exists a $\mathfrak{P}$-real prime divisor of K with center P on A.

<u>Proof</u>: First, P will be $\mathfrak{P}_w(g_{ij}) \cap$ A-convex for some i by 04. Thus, we may assume $\mathfrak{P} = \mathfrak{P}_w(g_j)$. Secondly, we want to reduce to the case of the weak preorder $\mathfrak{P} = \mathfrak{P}_w(1)$ on K. To do this, we construct an algebraic extension field $L \supset K$ by adding square roots of enough of the g_j until all g_j are squares in L. The fact that $\mathfrak{P}_w(g_j)$ is a preorder on K guarantees that L is still a formally real field and $\mathfrak{P}_w(1)_L \cap K = \mathfrak{P}_w(g_j)_K$. Let B denote the integral closure of A in L. Then B is still of finite type over R. Moreover, we can lift the prime $P \subset A$ to a $\mathfrak{P}_w(1) \cap$ B-convex prime $Q \subset B$, for example by using the Real Place Existence Theorem for $A \subset L$. Now any $\mathfrak{P}_w(1)$-real prime divisor of L with center Q on B will contract to a $\mathfrak{P}_w(g_i)$-real prime divisor of K with center P on A, since the residue field of the valuation ring in L is algebraic over the residue field of the contracted valuation ring in K, hence they have the same transcendence degree over R. This achieves the reduction to $\mathfrak{P} = \mathfrak{P}_w(1) \subset K$.

Let X be the real affine variety associated to A, $V \subset X$ the subvariety corresponding to P. From statement D) above the closure $\overline{X}_0$ contains an open set of simple points of V, since $P \subset A$ is assumed $\mathfrak{P}_w(1) \cap$ A-convex.

Let π: $X' \to X$ be an affine desingularization of X, that is, a birational equivalence of real affine varieties with X' non-singular. The main point is that this can be done so that $\pi(X') = \overline{X}_0$, the closure of the simple points. (The degenerate points $X - \overline{X}_0$ will never be in $\pi(X')$.) The technique is to first resolve the singularities of some projective closure of X, then restrict to appropriate affine pieces, as in [G. Efroymson, <u>Local Reality on Algebraic Varieties</u>, J. Algebra 29 (1974), 133-142]. Then for some subvariety $V' \subset X'$ we will have $\pi(V') \subset V \cap \overline{X}_0$ and $\pi(V')$ will contain an open set of simple points of V. Namely, V' will be a suitable component of $\pi^{-1}(V')$, but we don't yet know the dimension of V'. However, since X' is non-singular, there are plenty of real tangent lines to X' through points of V'. Generically, V' is a submanifold of the manifold X', with a nice normal bundle. Blowing up X' along V' will produce a variety X" with a real hypersurface V" over V'. Generically, V" is the real projective space bundle associated to the real normal bundle of V' in X'. Finally, we normalize to get a real hypersurface V''' in a normal variety X''' over V" in X". Now $V''' \subset X'''$ defines our $\mathfrak{P}_W(1)$-real prime divisor of K. The center on A is our original prime P because the projection of V''' to X lies in $V \cap \overline{X}_0$ and contains at least an open set of simple points of V.

Example 14. Let $A = R[x,y]$, $\mathfrak{P} = \mathfrak{P}_W(x,y,x^2-y) \subset K = R(x,y)$, $P = (x,y)$. Now $X = X_0 = \mathbb{R}^2$ and $V = \{(0,0)\}$, but note that there is only one real tangent direction to R^2 at the origin which lies in $W\{x,y,x^2-y\}$. When we blow up

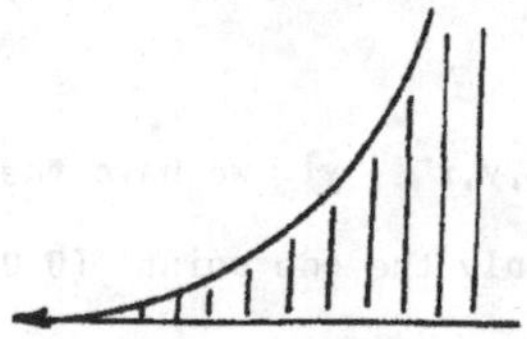

the origin π: $X' \to X$, then $\pi^{-1}(0,0) \subset X'$ is a circle, but only one point of this circle is in the closure of the set where the functions x,y, and x^2-y are positive. The exact picture near the point is the following. An affine part of X' identifies with the surface $wx = y$ in space, with the w-axis lying above (0,0). In the plane region $W\{x,y,x^2-y\}$ we have the curves $y = cx^2$, $0 \le c \le 1$, $0 \le x$, but these curves all have the same tangent at the

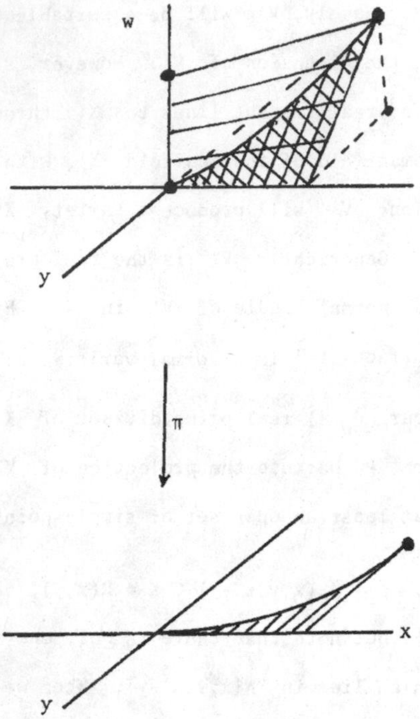

origin. On the surface $wx = y$ above $W\{x,y,x^2 - y\}$ we have the set $\{(w,x,y)\,|\,wx = y,\ 0 \leq x,\ 0 \leq y \leq x^2\}$ with only the one point $(0,0,0)$ over $(0,0)$. But over the curve $y = cx^2$ lies the curve $w = cx$, $y = cx^2$, and these have <u>distinct</u> tangents as c varies, $0 \leq c \leq 1$. Thus if we <u>iterate</u>, and blow up the surface $wx = y$ at the origin, we produce a third surface so that a whole <u>interval</u> of points lies above $(0,0)$ and belongs to the closure of the set of points where $0 < x$ and $0 < y < x^2$. This gives us a real prime divisor. Of course, in the proof of Theorem 13, we have avoided this specific consideration by adjoining square roots of x, y, and x^2-y. But the resulting surface $R[\sqrt{x}, \sqrt{y}, \sqrt{x^2-y}]$ probably is not so nice, and we have only hidden the difficulty by quoting the resolution of singularities theorem.

Note also that if we replace $0 \leq y \leq x^2$ by $0 \leq y \leq x^n$, then we would be required to iterate blowing up n times before separating the tangents.

Example 15. Next suppose we partially order $R(x,y)$ with $0 < x$ and $0 < y < x^n$ for all $n \geq 1$, that is, we take $\mathfrak{P} = \mathfrak{P}_w(x,y,\ x^n - y | n \geq 1)$. The prime ideal $(x,y) \subset R[x,y]$ is certainly $\mathfrak{P}$-convex, but now it is easy to see that (x,y) cannot be the center of any $\mathfrak{P}$-real prime divisor of $R(x,y)$. Namely, we would then have $v\colon\ R(x,y)^* \to \mathbb{Z}$, a valuation compatible with $\mathfrak{P}$, with $v(x) > 0$ and $v(y) > 0$. But $0 \leq y \leq x^n$ requires $v(y) \geq nv(x)$ for all n, which is impossible. Thus the finiteness assumption on $\mathfrak{P}$ in Theorem 13 is quite important.

Let us return briefly to Propositions 6 and 7 where we characterized maximal and minimal primes in semi-integral closures. Suppose the field K is a function field $K = R(x_1 \cdots x_n)$ and $A = R[x_1 \cdots x_n]$. When does a real valuation ring $A_v \subset K$ correspond to a maximal prime $\overline{P} \subset \overline{A}$, that is, $A_v = \overline{A}_{(\overline{P})}$, $P_v = \overline{P}\,\overline{A}_{(\overline{P})}$? It is clear from Proposition 6 that a sufficient condition is that the center $P = P_v \cap A$ correspond to a point on the variety X associated to A and that the residue field A_v/P_v is R. This will be the case for any real discrete rank r valuation on K, where $r = \text{tr. deg.}(K/R)$ However, there are many other (complicated) examples of minimal valuation rings and in general the maximal spectrum of $\overline{A}$ seems of limited geometric interest.

From Proposition 7 we see that any real prime divisor of K corresponds to a minimal prime of $\overline{A}$. A classical result about function fields is that the integral closure B' of a finitely generated domain B over a field is the intersection of its localizations at minimal primes $B' = \bigcap_{\min P'} B'_{(P')}$.

As stated earlier, all these $B'_{(p')}$ are prime divisors of the field of fractions of B. We now establish a real analogue of this result.

Theorem 16. Let $A = R[x_1 \cdots x_n]$ be a domain with fraction field K and let $\mathfrak{P} = \bigcap_i \mathfrak{P}(g_{ij}) \subset K$ be a preorder obtained as a finite intersection of finite refinements of the weak preorder. Then

$$\bar{A} = \bigcap_{\substack{A_v = \mathfrak{P}\text{-real prime div.} \\ A_v \supset A}} A_v .$$

More generally, if $I \subset K$ is a finitely generated A-module, then

$$I\bar{A} = \bigcap_{\substack{A_v = \mathfrak{P}\text{-real prime div.} \\ A_v \supset A}} IA_v .$$

Finally, if $P \subset A$ is a $\mathfrak{P} \cap A$-convex prime ideal and $I \subset K$ is a finitely generated $A_{(P)}$ module, then

$$\overline{IA_{(P)}} = \bigcap_{\substack{A_v = \mathfrak{P}\text{-real prime div.} \\ A_v \supset A \\ P_v \cap A = P}} IA_v .$$

Proof: The first statement is just the second with $I = A$. So let $(a_1,\ldots,a_k)$ generate I as a module. Suppose $y \notin I\bar{A} = \bigcap_{\text{all } A_v} IA_v$, say $y \notin IA_v$. Then $y/a_i \notin A_v$, $1 \leq i \leq k$, hence $a_i/y \in P_v$. Consider $A' = A[a_i/y]$, still a finite domain over R, with $A' \subset A_v$ and $a_i/y \in P' = P_v \cap A'$, $1 \leq i \leq k$. Apply Theorem 13 to get a $\mathfrak{P}$-real prime, divisor $A_{v'} \supset A'$ with center $P_{v'} \cap A' = P'$. Then $a/y \in P' \subset P_{v'}$ all $a \in I$, hence $y \notin IA_{v'}$.

The last statement of Theorem 16 follows by the same proof, using the last statement of Proposition 8.

The point of Theorem 16 is that it explains how semi-integral closures can be computed using only the simplest possible real valuation rings (under suitable hypotheses). Recall a $\mathfrak{P}$-real prime divisor containing A has the form $A'_{(P')}$ where A' is an integrally closed, finitely generated domain containing A, and $P' \subset A'$ is a $\mathfrak{P} \cap A'$ convex minimal prime ideal. The <u>only</u> ideals in the discrete valuation ring $A'_{(P')}$ are powers of the maximal ideal, which is principal,

$$(P')^n A'_{(P')} = \{\alpha \in K \mid v'(\alpha) \geq n\},$$

where $v'\colon K^* \to \mathbb{Z}$ is the associated valuation. Geometrically, A' corresponds to a variety X' over X, $P' \subset A'$ corresponds to a $\mathfrak{P}$-real hypersurface $V' \subset X'$ and the valuation v' measures the order to which functions vanish across V'. Thus under the finiteness hypotheses of Theorem 16 we get a good geometric feeling for $\mathfrak{P}$-c.c. ideals $I \subset A$, namely, they are defined by specifying the orders to which functions f on X should vanish across all possible such $V' \subset X'$ over X. Note the contracted ideals $IA'_{(P')} \cap A = \{g \in A \mid v'(g) \geq \min v'(f),\ f \in I\}$ are primary with radical $P = P' \cap A$, since the ideals $(P')^n A'_{(P')}$ are primary with radical P'.

A classical result about complete ideals (arbitrary intersections of valuation ideals) in a Noetherian domain is that they can be expressed as finite intersections of ideals associated to discrete rank 1 valuations, even prime divisors in the function field case. In particular, this provides a primary decomposition theorem. In our real case, we have Robson's Theorem 11 to the effect that every $\mathfrak{P}$-c.c. ideal in a Noetherian domain is a finite intersection of $\mathfrak{P}$-real valuation ideals and we have Theorem 16, which provides a decomposition into very nice primary $\mathfrak{P}$-real valuation ideals, but without a finiteness statement. We conjecture that under the finiteness hypotheses on $\mathfrak{P}$ of Theorem 16 it is, in fact, true that every $\mathfrak{P}$-c.c. ideal I is a finite intersection of ideals $IA_v \cap A$, where $A_v \supset A$ is a $\mathfrak{P}$-real prime divisor.

Example 17 (Robson). Let $K = R(x,y)$, $v\colon\ R(x,y)^* \to \mathbb{Z}\times\mathbb{Z}$ the valuation $v(f(x,y)) = (n,m)$ if $f(x,y) = y^n(x^m r(x) + ys(x,y)) \in R[x,y]$ with $r(0) \neq 0$. $\mathbb{Z}\times\mathbb{Z}$ is ordered by $(n,m) > (n',m')$ if $n > n'$ or $n = n'$ and $m > m'$. The associated places are the y-adic place on $R(x,y)$ and the x-adic place on $R(x)$, that is, $R(x,y) \to R(x), \infty \to R, \infty$. In particular, the valuation v is $\mathfrak{P}$-real for the preorder $\mathfrak{P} = \mathfrak{P}_w(x,y,\ x^n - y \mid n \geq 1)$ on $R(x,y)$. We claim that the ideal $(xy,y^2) = \{f(x,y) \in R[x,y] \mid v(f(x,y)) \geq (1,1)\}$ is a $\mathfrak{P}$-i.c.c. ideal. The associated primes are $(y) \subset (x,y)$ (see Corollary 12). The point is, in any ideal decomposition $(xy,y^2) = I_1 \cap I_2$, with $I_1, I_2 \neq (xy,y^2)$, then at least one of I_1, I_2 contains some power x^n and the other contains y. $\mathfrak{P}$-convexity would then put y in both ideals since $0 \leq y \leq x^n$, all n.

This example shows that it is impossible to get a finite $\mathfrak{P}$-c.c. primary decomposition without some hypothesis on $\mathfrak{P}$. However, if in this example we replace $\mathfrak{P}$ by $\mathfrak{P}' = \mathfrak{P}_w(x,y,\ x^j - y \mid j \leq n)$, then $(xy,y^2) = (y) \cap (x^{2n}, xy, y^2)$ is a $\mathfrak{P}'$-c.c. decomposition. In particular, $\mathfrak{P}$-real valuation ideals are certainly not always irreducible. In fact, our conjecture just above Example 17 is that under the finiteness hypothesis on $\mathfrak{P}$, the irreducible $\mathfrak{P}$-real valuation ideals are exactly those corresponding to $\mathfrak{P}$-real prime divisors.

We will state two other geometric characterizations of the semi-integral closure $\overline{A}$, where $A = R[x_1 \cdots x_n]$ and $\mathfrak{P} = \bigcap_i \mathfrak{P}_w(g_{ij}) \subset K = R(x_1 \cdots x_n)$ satisfies our usual finiteness hypothesis. Let X_0 be the simple points of the variety associated to A and let $W = \bigcup_i \overline{U\{g_{ij}\} \cap X_0} \subset \overline{X}_0$. Then by statement E) above, $f \in \mathfrak{P} \cap A$ if and only if $f|_W \geq 0$. Each rational function $f/g \in K$ defines an R-valued function $f/g\colon W - V(g) \to R$ where $V(g)$ denotes the zeros of g.

Proposition 18. $\overline{A} = \{f/g \in K \mid |f/g|$ is bounded on every bounded subset of $W - V(g)\}$.

Proof: If $f/g \in \overline{A}$, then there is an element $a \in A$ with $-a < f/g < a$ (rel $\mathfrak{P}$) by Proposition 8. This puts the desired bounds on the function f/g.

Conversely, if $f/g \notin \overline{A}$, then $f/g \notin A'_{(P')}$ for some $\mathfrak{P}$-real prime divisor containing A by Theorem 16. This means g vanishes to higher order than f across some hypersurface $V' \subset X'$, where X' is some normal variety with a projection $\pi\colon X' \to W$. This yields a little bounded piece of $W - V(g)$, for example, an arc of a curve, on which f/g is not bounded.

The second characterization was suggested in a paper of Schülting, who refers to some work of Bröcker. Let us say $f/g \in K$ is properly birationally regular on $W \subset \overline{X}_0$ if there is a non-singular affine variety X' and a map $\pi\colon X' \to X$ which is a birational equivalence such that $\pi\colon W' \to W$ is surjective and proper, where $W' = \pi^{-1}(W) \subset X'$, and such that $f/g = f'/1+p'$, $f', p' \in A'$, $p' \in \mathfrak{P} \cap A'$, where A' is the affine coordinate ring of X'. In particular, the function $f/g = f'/1+p'\colon W' \to R$ is defined on all of W', since $p'|_{W'} \geq 0$.

Now it is the case that by using Proposition 18, and the resolution of singularities theorem to construct suitable affine desingularizations, one can prove the following.

Proposition 19. $\bar{A} = \{f/g \in K \mid f/g$ is properly birationally regular on $W\}$.

Real Curves

We will now discuss my student Robson's results on the local structure of a semi-algebraic subset of a real curve near a point. From the point of view of semi-algebraic, or topological classification, curves are not interesting. At a point there is some finite number of intervals coming in--that is all. The

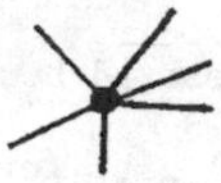

interest is in the more delicate function theory of the affine coordinate ring or rational function field. (Of course, the same applies in the classical case, curves over $\mathbb{C}$.)

Even the more delicate local structure of real curves is rather thoroughly understood within the clasical theory. At a point P, which we assume is the origin, of a curve $C \subset R^n$, there are a finite number of branches. Each branch α is given by a collection of algebraic-analytic real power series for the coordinates

$$\alpha \begin{cases} x_1 = a_{10}t^e + a_{11}t^{e+1} + a_{12}t^{e+2} + \cdots \\ x_2 = a_{20}t^e + a_{21}t^{e+1} + a_{22}t^{e+2} + \cdots \\ \vdots \\ x_n = a_{n0}t^e + a_{n1}t^{e+1} + a_{n2}t^{e+2} + \cdots . \end{cases}$$

(It should be remarked that a linear change of coordinates may be necessary before all leading terms of the $x_i(t)$ have the same degree.) We will refer to the lowest possible exponent e of a parameter t of a given branch as the <u>order</u> of the branch. The sum of the orders e_i along all real branches α_i we will call the <u>real multiplicity</u> of the curve C at P, denoted $m_P(C) = \sum_{\alpha_i} e_i$. More generally, if $W \subset C$ is a semi-algebraic subset containing P, $m_P(W)$ will denote the sum of the orders of branches with germ in W.

This last sentence refers to the following. If R is a real closed field which contains an element whose powers form a null-sequence (a <u>microbe</u> in the papers of Bukowski and Dubois), then each power series associated to a branch actually converges to an element of R for each t in some interval $(-\varepsilon,\varepsilon)$. The $x_i(t)$ thus actually define a function $\alpha\colon (-\varepsilon,\varepsilon) \to C$, $\alpha(0) = P$. If there are r branches, we get 2r half-arcs by restricting t to $[0,\xi)$ and $(-\varepsilon,0]$. If the order e is odd, these half-arcs emanate from P along opposite tangent rays, but if e is even, then both arcs have the same tangent ray and double back like a cusp. When we say a branch has germ in W, we mean at least one of these half-arcs lies in W, for sufficiently small ε.

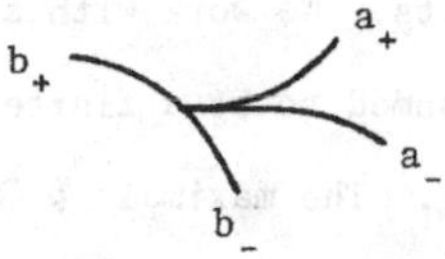

Fields which contain microbes include real closures of all finitely generated fields over Q. (Archimedean is certainly not required.) But no matter how big the original ground field, the equations which define the curve have only finitely many coefficients. Thus we can work over the real closure of this coefficient field. Tarski's Principle, cleverly used, then gives infinitely differentiable, semi-algebraic functions $\alpha\colon (-\varepsilon,\varepsilon) \to C$ for each branch, even over the larger field where convergence may be meaningless. The images of these functions represent all points on C near P.

Now, the classical theory in the real case is usually embedded in the complex theory. Our point of view will be to work directly inside the real affine coordinate ring localized at P and the real function field, with various preorders. We want t› see how the number of branches, their orders, and other data emerges from real commutative algebra alone. One reason this is interesting is that instead of looking at the entire curve C near P, we can look at semi-algebraic subsets of C. Now some of the half-arcs may be lost. This geometric step is performed algebraically by replacing the weak preorder $\mathfrak{P}_w(1)$ by other preorders $\mathfrak{P}$ in the function field. This changes the maximal convex ideals (points), semi-integral closures, and complete hulls of ideals. In fact, our main tool will be the graded ring associated to complete hulls of the powers of the maximal ideal of the local ring at P.

So, to begin, let $A = R[x_1 \cdots x_n]$ be a domain with $\text{tr.deg.}(A/R) = 1$, $K = R(x_1 \cdots x_n)$, which we assume is a real field. Let $C \subset R^n$ be the variety associated to A, $C_o \subset C$ the simple points. We work with a preorder $\mathfrak{P} = \bigcap_i \mathfrak{P}_w(g_{ij}) \subset K$, which, as usual, is assumed to be a finite intersection of finite refinements of the weak preorder. The maximal $\mathfrak{P} \cap A$-convex ideals correspond to the points of $W = \bigcup_i \overline{U\{g_{ij}\} \cap C_o}$. We assume $P = (0,\ldots,0) \in W$.

Form the local ring $A_{(P)} = R[x_1 \cdots x_n]_{(x_1 \cdots x_n)}$ which has a unique non-zero prime ideal, which by slight abuse of notation we will call P. The integral closure B of $A_{(P)}$ in K is a Dedekind domain with finitely many prime ideals, which we denote $P_1,\ldots,P_r$. The localizations $B_{(P_i)}$ are discrete valuation rings and, in fact, give all the non-trivial valuation rings in K which contain $A_{(P)}$. Since P is $\mathfrak{P} \cap A_{(P)}$-convex, it is the center of some $\mathfrak{P}$-real valuation ring in K, so we conclude at least one $P_k \subset B$ is $\mathfrak{P} \cap B$-convex.

Each $\mathfrak{P}$-real $B_{(P_k)}$ has residue field $B_{(P_k)}/P_kB_{(P_k)} = R$. In fact, for $n \geq 0$, $P_k^nB_{(P_k)}/P_k^{n+1}B_{(P_k)} \simeq R$, with generator t^n, where $(t) = P_kB_{(P_k)}$. In the usual way, each real P_k gives a power series embedding $\alpha_k\colon B_{(P_k)} \to R[[t]]$

Lemma 20. The prime $P_k \subset B$ is $\mathfrak{P}$-convex if and only if at least one of the two arc germs of the function $\alpha_k\colon (-\varepsilon,\varepsilon) \to C$ defined by the power series $x_i(t) = \alpha_k(x_i)$, $1 \leq i \leq n$, has image in $W \subset C$.

Proof: Recall $\mathfrak{P} = \bigcap_i \mathfrak{P}_w(g_{ij})$, so if P_k is $\mathfrak{P}$-convex, it is $\mathfrak{P}_w(g_{ij})$-convex for some i. Recall $W = \bigcup_i \overline{U\{g_{ij}\} \cap C_o}$, so to prove that an arc of the curve belongs to W, it is sufficient to show that the functions $g_{ij}(x_1(t),\ldots,x_n(t))$ are non-negative on $(-\varepsilon,0]$ or $[0,\varepsilon)$. This depends only on the lowest degree term of $g_{ij}(x_1(t),\ldots,x_n(t)) = at^m + (\text{higher})$, and obviously holds if m is odd (in which case only one of the half-arcs belongs to W) or if m is even and $a > 0$ (in which case both half-arcs belong to W). But if m is even, then $g_{ij}(x_1(t)\ldots x_n(t))/t^m$ belongs to $\mathfrak{P}_w(g_{ij})$ and is a unit in $B_{(P_k)}$. The coefficient a is just the value of g_{ij}/t^m in the residue field R, which is positive since P_k is $\mathfrak{P}_w(g_{ij})$-convex.

Conversely, for each prime $P_k \subset B$, the residue field $\Delta_k = B_{(P_k)}/P_kB_{(P_k)}$ is R or $R[\sqrt{-1}]$. We always get a power series embedding $\alpha_k\colon B_{(P_k)} \to \Delta_k[[t]]$. If for each real value $t \in [0,\varepsilon)$, $\alpha_k(t) = (x_1(t),\ldots,x_n(t))$ belongs to $C \subset R^n$, then all the coefficients of the $x_i(t)$ are real. (The derivatives at $t = 0$ of the $x_i(t)$ will be real.) Since every element of B is a quotient of elements of A, we see that the power series expansion of any element of B has real coefficients, that is, $\Delta_k = R$.

Suppose now for $t \in [0,\varepsilon)$, we have $\alpha_k(t) \in W = \bigcup_i \overline{U\{g_{ij}\} \cap X_0}$. Then for some i and possibly a smaller ε, we know $\alpha_k(t) \in \overline{U\{g_{ij}\} \cap X_0}$, all $t \in [0,\varepsilon)$. Define a total order $\hat{\mathfrak{P}}$ on B by $f \in \hat{\mathfrak{P}}$ if the leading coefficient of $\alpha_k(f)$ is positive. Then $P_k \subset B$ is clearly $\hat{\mathfrak{P}} \cap B$-convex and also $g_{ij} \in \hat{\mathfrak{P}}$, so P_k is $\mathfrak{P}_w(g_{ij}) \cap B$-convex.

We now study the $\mathfrak{P}$-complete hulls of powers of P in $A_{(P)}$. From the last part of Proposition 8, we know $P^m \overline{A_{(P)}} = \bigcap_{\mathfrak{P}\text{-convex } P_k} P^m B_{(P_k)}$. In the Dedekind domain B, let $PB = P_1^{e_1} \cdot \ldots \cdot P_r^{e_r}$, so $P^m B = P_1^{me_1} \cdot \ldots \cdot P_r^{me_r}$. If $v_k\colon K^* \to \mathbb{Z}$ is the valuation associated to $B_{(P_k)}$, then $e_k = \min\{v_k(f) \mid f \in P = (x_1 \ldots x_n)\}$, so e_k is the order of the branch $\alpha_k\colon A_{(P)} \to R[[t]]$ at P, that is, the least power of t occuring among the power series $\alpha_k(x_i)$, $1 \le i \le n$. We conclude

<u>Lemma 21.</u> $P^m \overline{A_{(P)}} \cap B = \prod_{\mathfrak{P}\text{-convex } P_k} P_k^{me_k} = \{f \in B \mid v_k(f) \ge me_k\}$. Thus $CH(P^m, \mathfrak{P}) = \{f \in A \mid v_k(f) \ge me_k, \text{ all } \mathfrak{P}\text{-convex } P_k\}$.

The next remark is that the <u>conductor</u> of B in $A_{(P)}$ is non-zero. That is, there is $h \in A_{(P)}$, $h \neq 0$, with $hB \subset A_{(P)}$. Since $v_k(hg) = v_k(h) + v_k(g)$ and since we can obviously find elements g in the UFD, B with any prescribed $v_k(g)$, we conclude that for any <u>sufficiently large</u> prescribed integers b_k, we can find $f = hg \in A_{(P)}$ with $v_k(f) = b_k$, all k.

Let $G_m = CH(P^m, \mathfrak{P})/CH(P^{m+1}, \mathfrak{P})$ the terms in the graded ring $G_* = \bigoplus_{m \ge 0} G_m$ naturally associated to the point $P \in W \subset C$.

Proposition 22. (Robson) For m sufficiently large,

$$m_P(W) = \sum_{\mathfrak{P}\text{-convex } P_k} e_k = \dim_R(G_m) .$$

Proof: We know we can produce elements $f \in CH(P^m, \mathfrak{P})$ so that $(v_k(f))$ represents any preassigned tuple (b_k), if $b_k \geq me_k$, m sufficiently large. Since elements with different values are linearly independent, this proves $\dim_R(G_m) \geq \Sigma e_k$.

For the other direction, we have an injection

$$G_m = \frac{P^m \overline{A_{(P)}} \cap A}{P^{m+1} \overline{A_{(P)}} \cap A} \longrightarrow \frac{P^m \overline{A_{(P)}} \cap B}{P^{m+1} \overline{A_{(P)}} \cap B} .$$

But from Lemma 21, the right-hand side is $\Pi P_k^{me_k} / \Pi P_k^{(m+1)e_k}$, which has R-dimension Σe_k, so $\dim_R(G_m) \leq \Sigma e_k$.

We now want to indicate how the functions $f \in A_{(P)}$ which are non-negative near P show up in the graded ring $G_* = \bigoplus_{m \geq 0} G_m$, and determine the number of branches, their orders, and for each whether one arc or two arcs are present in W. This is also part of Robson's thesis.

The first step is to refine the order $\mathfrak{P}$, in order to deal with infinitesimal behavior of functions near P. Set $\mathfrak{P}_\varepsilon = \bigcap_i \mathfrak{P}_W(g_{ij}, \delta^2 - \Sigma x_k^2 | \delta > 0)$. In other words, $f \in \mathfrak{P}_\varepsilon \cap A_{(P)}$ if and only if f is non-negative on some $W \cap B_\delta$, where B_δ is the ball of radius δ centered at P. Although $\mathfrak{P}_\varepsilon$ is an infinite refinement of $\mathfrak{P}$, it is still defined by elementary conditions over R, so has much better properties than an order like the one considered in Example 15. In fact, one of the valuation rings $B_{(P_k)}$ over $A_{(P)}$ is $\mathfrak{P}_\varepsilon$-real exactly when it is $\mathfrak{P}$-real, so the graded ring G_* does not change.

Suppose we have k-branches $\alpha_1,\ldots,\alpha_k$ in W, where $\alpha_i: A_{(P)} \to R[[t_i]]$ is a power series representation of $A_{(P)}$. If the branch α_i has order e_i, then the R-vector space G_m has basis corresponding to the powers $\{t_i^{me_i+j-1}\}$, $1 \le i \le k$, $1 \le j \le e_i$. That is, for m sufficiently large and for any linear combination of these powers of t_i, there is a function $f \in A_{(P)}$ so that $\alpha_1(f) + \cdots + \alpha_k(f) \in \oplus R[[t_i]]$ is congruent to this linear combination modulo higher power of the t_i. We actually think of a finite sum $\Sigma a_{ij}t_i^{me_i+j-1}$ as being a function germ on the curve W near P, where each t_i is non-zero on the ith branch and zero on all other branches. Then f agrees with such a function germ, modulo higher powers of the t_i, hence the question of whether f is non-negative near P is entirely determined by the coefficients a_{ij}. In fact, just as in the proof of Lemma 20, on the ith branch the function f is non-negative near P if its leading coefficient a_{ij} belongs to an even power of t_i and is positive, or belongs to an odd power of t_i and has the same sign as t_i on the one half-arc of the branch in W. We may as well assume t_i positive on all such single half-arc branches.

What this shows is that if we set $\mathfrak{P}_m = \text{image}(\mathfrak{P}_\varepsilon \cap CH(P^m,\mathfrak{P})) \subset G_m = CH(P^m,\mathfrak{P})/CH(P^{m+1},\mathfrak{P})$, then $\mathfrak{P}_m$ is a convex cone in the R-vector space G_m defined by a conjunction over i of disjunctions of conditions of the form "some $a_{ij'} = 0$, some $a_{ij''} > 0$". It turns out that the linear geometry of $\mathfrak{P}_m \subset G_m$ for large even and odd m together determines the number of branches, their individual multiplicities, and, for each, whether one arc or two arcs lie in W. For example, if m is even, then the first coefficients a_{io} all belong to even powers $t_i^{me_i}$, and we see that the interior of $\mathfrak{P}_m \subset G_m$, m even, is one component of the complement of k hyperplanes, where k is the number of branches. If m is odd, then the first coefficients a_{io} belonging to branches of odd order e_i with two arcs in W must vanish.

Thus the dimension of $\mathfrak{P}_m \subset G_m$, m odd, determines the number of such branches. In this subspace of G_m, the interior of $\mathfrak{P}_m$ is bounded by k-ℓ hyperplanes, where ℓ is the number of branches of order 1, with two arcs in W. One next looks at the structure of $\mathfrak{P}_m$ in the intersection of these k hyperplanes of G_m, m even, or k-ℓ hyperplanes of a subspace of G_m, m odd, to make further deductions, and then iterates this process. It is a little complicated, but in the end all branch data is deducible.

Let us look at pictures of $\mathfrak{P}_{odd}$ and $\mathfrak{P}_{even}$ for points of multiplicity 1 and 2.

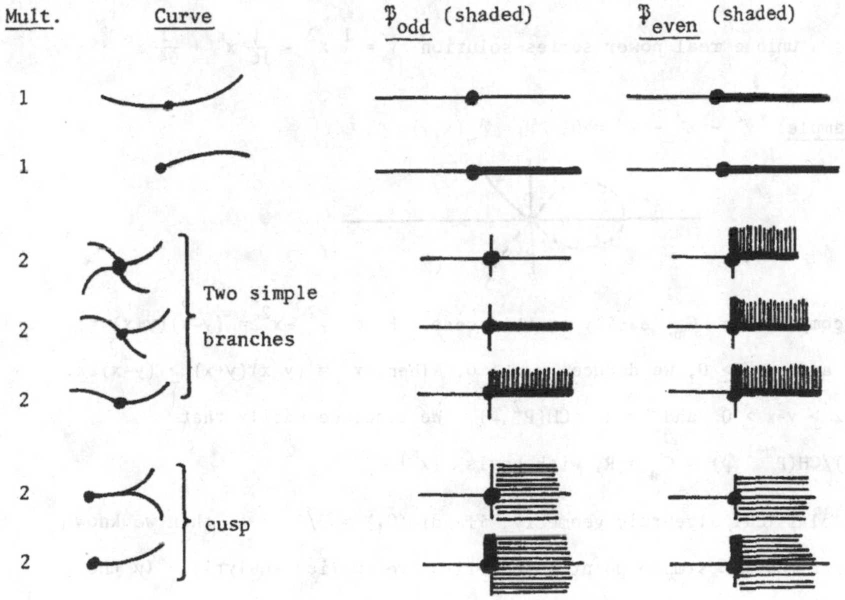

Here, $\mathfrak{P}_{odd}$ alone is sufficient to distinguish all possibilities, but already for points of multiplicity 3, it is necessary in some cases to look at both $\mathfrak{P}_{odd}$ and $\mathfrak{P}_{even}$. There are ten possible branch structures with multiplicity 3, and after that it really gets complicated.

If $m_P(W) = 1$, then $\dim_R(G_m) = 1$ for all $m \geq 0$. (The proof of Proposition 22 shows $\dim_R(G_m) \leq m_P(W)$, all m.) In this case, the curve must be parametrized near P in W by exactly one algebraic-analytic power series and P is a simple point from the point of view of Nash functions. The point P need not be an algebraic simple point.

<u>Example</u>: $y^3 + 2x^2y - x^4 = 0$, $\mathfrak{P} = \mathfrak{P}_W(1)$.

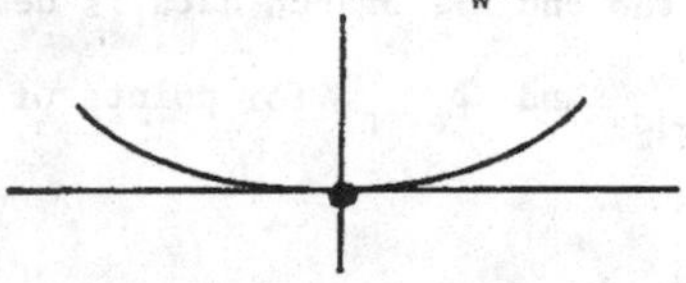

There is a unique real power series solution $y = \frac{1}{2}x^2 - \frac{1}{16}x^4 + \frac{1}{64}x^6 \cdots$.

<u>Example</u>: $y^2 - x^2 - x^3 = 0$, $\mathfrak{P} = \mathfrak{P}_W(x,y)$.

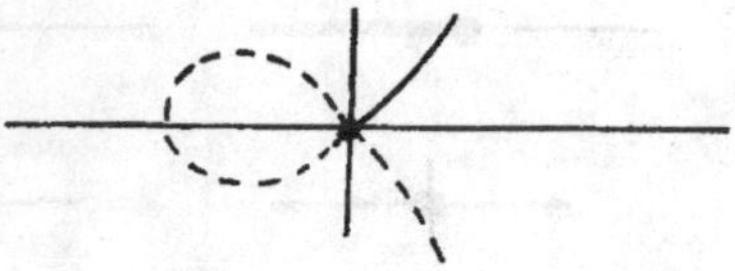

We can compute the G_m easily in this case. From $y^2 - x^2 = (y-x)(y+x) = x^3 \geq 0$ and $x,y \geq 0$, we deduce $y-x \geq 0$. Then $x^3 = (y-x)(y+x) \geq (y-x)2x$, so $x^2/2 \geq y-x \geq 0$ and $y-x \quad CH(P^2,\mathfrak{P})$. We conclude easily that $CH(P^m,\mathfrak{P})/CH(P^{m+1},\mathfrak{P}) = G_m \simeq R$, with basis $\{x^m\}$.

In classical algebraic geometry, if $\dim(G_1) = P/P^2 = 1$, then we know P is an algebraic simple point--one derivative implies analytic. In the real case this is not true and our graded rings associated with completely convex hulls provide a measure of degree of differentiability of real semi-algebraic functions.

Example: $y^3 - x^{100} = 0$, $\mathfrak{P} = \mathfrak{P}_w(1)$.

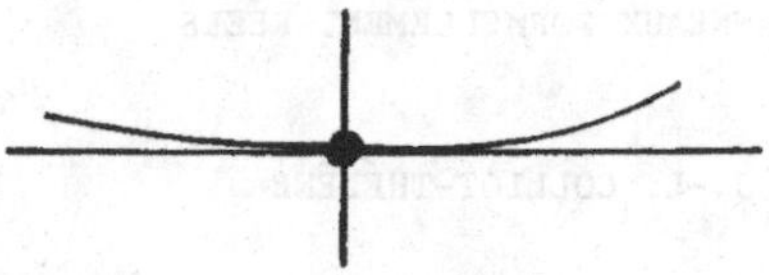

In the germ $0 \leq y^2 \leq x^{66}$, so $y \in CH(P^{33},\mathfrak{P})$. We have $G_j = R$, with basis $\{x^j\}$ for $0 \leq j \leq 32$, but $G_{33} = R \oplus R$ with basis $\{x^{33},y\}$. In fact, $G_j = R \oplus R$ with basis $\{x^j, x^{j-33}y\}$ for $33 \leq j < 66$ and $G_j = R \oplus R \oplus R$ with basis $\{x^j, x^{j-33}y, x^{j-66}y^2\}$ for $j \geq 66$, so the multiplicity is 3. Of course, we knew this from the parametrization $x = t^3$, $y = t^{100}$.

The theory of completely convex hulls and semi-integral closures for real curves is simple because <u>all</u> the valuation rings in a function field of dimension 1 are so easily described. In higher dimensions, there are many challenging problems concerning applications of real valuation rings to the study of the function theory of a real variety. The graded rings and completions of local rings associated to completely convex hulls of powers of the maximal ideal look very useful. For example, there certainly should exist a nice charatcerization of Nash simple points. Also, we wonder when these graded rings and completions are Noetherian.

Departement of Mathematics
Stanford University
Stanford, California 94305
U.S.A.

VARIANTES DU NULLSTELLENSATZ RÉEL
ET ANNEAUX FORMELLEMENT RÉELS

J.-L. COLLIOT-THÉLÈNE

Dans la première partie de ce texte, on voit comment la conjonction du théorème d'homomorphisme de Lang [Lg] et d'un lemme très simple d'algèbre commutative observé par M.-F. Roy [Cl] permet d'obtenir les divers Nullstellensätze et Positivstellensätze dans le cas des fonctions polynomiales sur les variétés algébriques. Ces théorèmes, qu'on peut considérer comme des variantes précises de la solution d'Artin du $17^{\text{ème}}$ problème de Hilbert, sous la forme que lui a donnée Lang, ont été obtenus par Dubois, Risler et Stengle. Plusieurs auteurs, à la suite de Stengle [S], ont insisté sur la simplicité de l'algèbre commutative requise pour ces variantes : Prestel [P], Lorenz [Lo1,Lo2], Bröcker [Br], Bröcker-Dress-Scharlau [BDS]. L'approche ici décrite, connue de nombreux spécialistes, semble la plus courte.

Le lemme d'algèbre commutative est aussi utilisé dans le seconde partie du texte, où l'on met en regard les mérites des deux définitions d'anneau commutatif réel (l'une schématique, l'autre birationnelle), qui ont un sens pour les anneaux de type géométrique.

Ce texte ne prétend à aucune originalité, sauf peut-être pédagogique. Pour un historique de ces questions, ainsi qu'une abondante bibliographie, on consultera l'article de T. Y. Lam [L].

I. Préliminaires.

LEMME 1. (cf. [Cl]) : *Soit* A *un anneau (commutatif unitaire). Les énoncés suivants sont équivalents :*

i) -1 *est une somme de carrés dans* A ;

ii) *Pour tout idéal premier* p *de* A, -1 *est une somme de carrés dans le corps des fractions* A(p) *de* A/p *(i.e. le corps* A(p) *n'est formellement réel pour aucun* p*).*

Démonstration de (ii) ⟹ (i). Notons $\boxed{A} \subset A$ l'ensemble des sommes de carrés dans A. Si -1 n'est pas une somme de carrés dans A, l'ensemble multiplicatif $S = \{1+x \mid x \in \boxed{A}\}$ ne contient pas 0. Ainsi l'anneau $B = S^{-1}A$ n'est pas l'anneau nul. Soit donc m un idéal maximal de B, et soit $p = i^{-1}(m)$, où i désigne l'application naturelle $A \longrightarrow S^{-1}A$. Le corps des fractions de A/p est inclus dans le corps B/m. Par hypothèse, -1 est une somme de carrés dans le premier corps : il l'est donc aussi dans B/m. Il y a donc un élément de m de la forme $1 + \sum x_i^2$, avec chaque x_i dans B, soit de la forme a_i/s_i, avec $a_i \in A$ et $s_i \in S$. L'élément

$$t = \prod_i s_i^2 + \sum_i a_i^2 \prod_{j \neq i} s_j^2 \in A$$

est donc dans p. Comme chaque s_i est le la forme $1 + u_i$ avec $u_i \in \boxed{A}$, il en est de même de $\prod_i s_i^2$, donc aussi de t. Ainsi $p \cap S \neq \emptyset$, ce qui est absurde ([Bo], chap. II, § 2, Prop. 11) : -1 est donc une somme de carrés dans A. ∎

On a la variante et généralisation suivante (cf. [BDS]) :

LEMME 1 bis : *Soit* A *un anneau (commutatif unitaire). Soit* $T \subset A$ *tel que* $1 \in T$ *et* $T.T \subset T$. *Les énoncés suivants sont équivalents :*

i) *On peut écrire* -1 *sous la forme* $\sum_i a_i x_i^2$, *avec* $a_i \in T$ *et* $x_i \in A$;

ii) *Pour tout idéal premier* p *de* A, *il existe dans le corps des fractions* A(p) *de* A/p *une représentation de* -1 *sous la forme* $\sum_i a_i x_i^2$, *avec* $x_i \in A(p)$ *et* a_i *la classe dans* A/p *d'un élément de* $T \subset A$.

La démonstration est identique : on considère le sous-ensemble S de A constitué des éléments de la forme $1 + \sum_i a_i x_i^2$ avec $a_i \in T$ et $x_i \in A$.

THEOREME 1 (Artin-Lang [Lg]) : *Soit* k *un corps réel clos, et soit* A *une* k*-algèbre de type fini intègre. Si* -1 *n'est pas une somme de carrés dans le corps des fractions* de A *(i.e. si ce corps est ordonnable), alors il existe un* k*-homomorphisme (de* k*-algèbres) de* A *dans* k.

On a la variante et généralisation suivante (cf. [L], p. 51) :

THEOREME 1 bis : *Soit* (k,p) *un corps ordonné* (p = *l'ensemble des éléments positifs de* k *pour l'ordre donné). Soit* A *une* k*-algèbre de type fini intègre. S'il existe sur le corps des fractions de* A *un ordre qui prolonge celui de* k*, alors il existe un* k*-homomorphisme (de* k*-algèbres) de* A *dans la clôture réelle de* (k,p).

II. Nullstellensätze et Positivstellensätze.

Dans les deux propositions suivantes, A désigne une k-algèbre de type fini quelconque (non nécessairement intègre, ni même réduite). On note V = SpecA la k-variété algébrique affine qu'elle définit. Pour L un surcorps de k, on note $V(L) = \mathrm{Hom}_{k\text{-alg}}(A,L)$ l'ensemble des L-points de V.

PROPOSITION 1 : *Soit* k *un corps réel clos, et* A *et* V *comme ci-dessus. Les conditions suivantes sont équivalentes :*

i) -1 *est une somme de carrés dans* A ;

ii) V(k) *est vide.*

Démonstration de (ii) ⟹ (i) : Si -1 n'est pas une somme de carrés dans A, le lemme montre l'existence d'un idéal premier $\mathfrak{p}$, tel que le corps des fractions de $A/\mathfrak{p}$ soit ordonnable. Le théorème 1 assure alors l'existence d'un k-homomorphisme de k-algèbres de la k-algèbre de type fini intègre $A/\mathfrak{p}$ dans k. Composant avec $A \to A/\mathfrak{p}$, on obtient un point de V(k). ∎

Soient $\{P_i\}_{i\in I}$, $\{Q_j\}_{j\in J}$, $\{R_h\}_{h\in H}$ des familles finies (éventuellement vides) d'éléments de A. Soit Q le sous-monoïde multiplicatif de A engendré par 1 et les Q_j. Soit S le sous-semi-anneau de A engendré par les carrés de A, les P_i et les Q_j. Soit $\mathfrak{J}$ l'idéal de A engendré par les R_h.

PROPOSITION 2 : *Soient* k *un corps réel clos, et* A, V, $\{P_i\}$,$\{Q_j\}$,$\{R_h\}$, Q, S, $\mathfrak{J}$ *comme ci-dessus. Soit* f *dans* A, *et soit* Ω *le sous-ensemble semi-algébrique de* V(k) *défini par :*

$$(M \in V(k)) \quad \forall\, i \in I \quad P_i(M) \geq 0 \; ; \; \forall\, j \in J \quad Q_j(M) > 0 \; ; \; \forall\, h \in H \quad R_h(M) = 0 .$$

a) f *est strictement positif sur* Ω *si et seulement s'il existe* s *et* s' *dans* S, q *dans* Q *et* u *dans* $\mathfrak{J}$ *avec :*

$$sf = q + s' + u.$$

b) f *est positif ou nul sur* Ω *si et seulement s'il existe un entier* $n \geq 0$, s *et* s' *dans* S, q *dans* Q *et* u *dans* $\mathfrak{J}$ *avec :*

$$sf = qf^{2n} + s' + u.$$

c) f *est nul sur* Ω *si et seulement s'il existe un entier* $n \geq 0$, s *dans* S, q *dans* Q *et* u *dans* $\mathfrak{J}$ *avec :*

$$qf^{2n} + s + u = 0.$$

Démonstration : Notons $A_o = A$, puis $A_1 = A_o/\mathfrak{J}$, puis $A_2 = Q^{-1}A$, puis $A_3 = A_2[T_j]_{j\in J} / (T_j^2 - Q_j)_{j\in J}$, puis $A_4 = A_3[U_i]_{i\in I} / (U_i^2 - P_i)_{i\in I}$, où les T_j et les U_i sont des variables.

De façon générale, pour A un anneau commutatif unitaire, $T_1,\ldots,T_n$ des variables, et $\alpha_1,\ldots,\alpha_n$ des éléments de A, l'anneau $B = A[T_1,\ldots,T_n] / (T_1^2-\alpha_1,\ldots,T_n^2-\alpha_n)$ est un A-module libre de rang 2^n, de base les éléments :

$$v_{e_1\ldots e_n} = T_1^{e_1}\ldots T_n^{e_n} \qquad \text{avec} \qquad 0 \leq e_i \leq 1 .$$

La coordonnée selon $v_{o\ldots o}$, qui définit une rétraction de l'homomorphisme naturel $A \longrightarrow B$, sera notée $Tr_{B/A} : B \longrightarrow A$ (elle ne dépend pas de l'ordre des T_i).

Par ailleurs, il est clair que si A est une k-algèbre de type fini, il en est de même de chacun des anneaux A_i définis ci-dessus.

Pour démontrer l'implication non triviale de l'énoncé a), introduisons la k-algèbre de type fini $A_5 = A_4[X]/X^2+f$. L'hypothèse implique que cette k-algèbre n'admet pas de k-homomorphisme dans k. La proposition 1 permet d'écrire -1 comme une somme de carrés dans A_5. En appliquant à une telle égalité successivement Tr_{A_5/A_4}, puis Tr_{A_4/A_3}, puis Tr_{A_3/A_2}, puis en chassant les dénominateurs, enfin en remontant dans A_o, on obtient une égalité du type annoncé.

Pour démontrer l'implication non triviale de l'énoncé b), introduisons $A_5 = A_4[1/f]$, puis $A_6 = A_5[X]/X^2+f$. L'hypothèse implique que -1 s'écrit comme une somme de carrés dans A_6. On applique ici Tr_{A_6/A_5}, on chasse ensuite les dénominateurs pour revenir dans A_4, et on continue ensuite comme ci-dessus.

Dans le cas c), on introduit $A_5 = A_4[1/f]$. L'hypothèse implique que -1 est une somme de carrés dans A_5. On chasse les dénominateurs pour revenir dans A_4, et on continue comme ci-dessus. ∎

Remarques.

R1 : Si l'on multiplie l'égalité donnée en b) par f, et l'on fait passer le terme fu dans l'autre membre, on obtient une version plus connue du Positivstellensatz.

R2 : Le lemme 1 bis et le théorème 1 bis permettent de généraliser la proposition 2 au cas où k est un corps ordonné, et où l'on se donne des conditions de positivité ou de nullité sur les points à coordonnées dans la clôture réelle de k d'un ensemble semi-algébrique défini par des équations à coefficients dans k. On obtient alors des égalités entre fonctions, à coefficients dans k, dès que f est dans A, i.e. elle-même à coefficients dans k.

R3 : La proposition 2 admet des variantes "formelles" du type suivant : soit A un anneau, et soit f dans A ; si pour tout homomorphisme de A dans un corps réel clos, l'image de f est strictement positive, alors il y a dans A une égalité du type $1 + \sum x_i^2 = (\sum y_j^2)f$. Ces variantes formelles résultent uniquement du lemme 1 (ou du lemme 1 bis). Comme me l'a signalé L. Mahé, on peut d'ailleurs commencer par établir ces variantes formelles, qui sont donc très simples, puis utiliser par exemple le théorème de Tarski-Seidenberg pour déduire la proposition 2. Cette méthode a l'avantage, comme l'a remarqué M. Coste [C2] de s'appliquer à divers anneaux de fonctions de Nash, grâce au théorème de substitution de Bochnak et Efroymson : l'algèbre commutative simple utilisée dans la proposition 2 est naturellement présente dans la démonstration des Positiv- et Nullstellensätze pour les fonctions de

Nash donnée par Bochnak et Efroymson [BE].

R4 : M. Coste a donné lors de la conférence un exemple de Nullstellensatz réel qui ne semble pas se déduire simplement de la proposition 2.

III. Qu'est-ce qu'un anneau réel ?

On trouve dans la littérature plusieurs notions d'anneau (formellement) réel. L'une d'entre elles demande que tout élément de l'anneau A de la forme $1 + \sum_i x_i^2$ $(x_i \in A)$ soit inversible. Cette condition n'est guère raisonnable pour les anneaux de type géométrique (de type fini sur un corps). Une fois cette définition écartée, il reste encore deux définitions raisonnables : ou bien l'on demande que -1 ne soit pas une somme de carrés dans l'anneau A, ou bien on demande qu'aucune somme de carrés non triviale ne soit nulle.

PROPOSITION 3 : *Soit* A *un anneau commutatif unitaire. Les conditions suivantes sont équivalentes :*

(i) -1 *n'est pas une somme de carrés dans* A ;

(ii) *le spectre réel* $\mathrm{Spec}_R A$ *de* A *est non vide ;*

(iii) *il existe un homomorphisme de* A *dans un corps réel clos ;*

(iv) *il existe un idéal premier* $\mathfrak{p}$ *de* A *tel que le corps des fractions* $A(\mathfrak{p})$ *de* $A/\mathfrak{p}$ *soit ordonnable ;*

(v) *(supposant de plus* A *de type fini sur un corps réel clos* k*). Il existe un* k*-homomorphisme de* A *dans* k, *i.e.* $V(k) \neq \emptyset$, *avec* $V = \mathrm{Spec}A$;

(vi) *l'anneau* A *possède une signature, i.e. un homomorphisme surjectif de l'anneau de Witt* W(A) *vers* **Z** ;

(vii) *l'anneau* A *possède un préordre, i.e. une partie* $S \subset A$ *stable par addition, multiplication, contenant les carrés de* A, *et ne contenant pas* (-1).

Dans la proposition suivante, nous nous limitons à un cas très géométrique, renvoyant à [CDLR], Theorem 7.3 (avec A. Wadsworth) pour un énoncé général.

PROPOSITION 4 : *Soit* A *une* k*-algèbre de type fini intègre sur un corps* k *réel clos.*

Soit V = SpecA. *Les conditions suivantes sont équivalentes :*

(i) *Une égalité* $\sum_i a_i^2 = 0$ *n'est possible dans* A *que si chaque* a_i *est nul ;*

(ii) -1 *n'est pas une somme de carrés dans le corps des fractions de* A *;*

(iii) *tout ouvert de Zariski non vide* U *de* V *possède un* k*-point (i.e.* $U(k) \neq \emptyset$*) ;*

(iv) V *possède un* k*-point lisse ;*

(v) *le corps des fractions de* A *possède une* k*-place à valeurs dans* k.

<u>Démonstration de la proposition</u> 3 (ou références) : Tout d'abord, les données suivantes sont clairement équivalentes :

a) un idéal premier $\mathfrak{p}$ de A, et un ordre du corps des fractions de $A/\mathfrak{p}$;

b) un homomorphisme de A dans un corps réel clos ;

c) un point du spectre réel de A ;

d) une partie S de A satisfaisant les conditions de (vii), et telle que de plus : $S \cup -S = A$ et $S \cap -S$ est un idéal premier $\mathfrak{p}$ de A (S n'est autre que l'ensemble des éléments de A dont la réduction dans $A/\mathfrak{p}$ est positive ou nulle pour l'ordre de $A(\mathfrak{p})$).

Ceci assure l'équivalence de (ii), (iii) et (iv). L'équivalence de (i) et (iv) fait l'objet du lemme 1, celle de (i) et (v) est la proposition 1. L'implication (iii) $\Longrightarrow$ (vi) est claire, et sa réciproque est une conséquence immédiate du théorème de factorisation des signatures de M. Knebusch [K2], pp. 245-246, lui-même tiré de théorèmes de Dress (passage au cas d'un anneau local) et de Kanzaki-Kitamura (cas d'un anneau local). Vu d) ci-dessus, l'implication (iv) $\Longrightarrow$ (vii) est claire. Sa réciproque est due à A. Prestel ([P], lemme 1.4) qui a montré qu'un préordre S <u>maximal</u> est toujours du type décrit en d) ci-dessus. ∎

Pour démontrer la proposition 4, nous utiliserons le

LEMME 2 : *Soit* A *un anneau local régulier. Si* -1 *est une somme de* m *carrés dans le corps des fractions de* A, *c'est aussi une somme de* m *carrés dans le corps résiduel de* A.

Démonstration : Ceci est clair si A est un anneau de valuation discrète, i.e. si la dimension dimA de A est 1. Pour obtenir le cas général, on raisonne alors par récurrence sur la dimension de A. Soit n = dimA > 1, et soit t, dans l'idéal maximal de A, un paramètre régulier, et p = (t) l'idéal premier qu'il engendre. Alors A_p est un anneau de valuation discrète dont le corps résiduel est le corps des fractions de l'anneau local régulier A/p, dont la dimension est (n-1). ∎

Démonstration de la proposition 4 (ou références) : Les énoncés (i) et (ii) sont identiques. L'implication (v) ⟹ (i), qui est facile, est l'un des premiers énoncés de [Lg] : sa réciproque est aussi dans [Lg], c'est d'ailleurs une version raffinée de cette réciproque qui donne le théorème 1. A l'usage du lecteur qui veut voir le théorème 1 comme une application immédiate du théorème de Tarski-Seidenberg, notons (cf. [Kl]) que la récurrence utilisée dans le lemme 2 permet aussi, une fois donné un k-point lisse de V, de construire une k-place (non unique) du corps des fractions de A à valeurs dans k. Ainsi : (iv) ⟹ (v).

L'implication (ii) ⟹ (iii) résulte du théorème 1 et du fait que les ouverts du type SpecA[1/f] forment une base de la topologie de SpecA.

L'implication (iii) ⟹ (iv) résulte du fait que V possède un ouvert non vide lisse sur k : l'ouvert complémentaire du fermé défini par les mineurs principaux de la matrice jacobienne associée à une présentation de A comme quotient d'une algèbre de polynômes.

L'implication (iv) ⟹ (ii) résulte du lemme 2 : soit P un point k-rationnel lisse de V, et soit A_P l'anneau local de P sur V. C'est un anneau local régulier ; si -1 était une somme de carrés dans le corps des fractions de A, qui est aussi celui de A_P, -1 serait une somme de carrés dans le corps résiduel k de A_P, ce qui est absurde. ∎

Si l'on se donne une k-algèbre de type fini intègre (k réel clos) A, telle que la k-variété V = SpecA soit <u>lisse</u> sur k, les énoncés (vi) de la proposition 3 et (iv) de la proposition 4 coïncident. Cette réconciliation des deux définitions possibles de réalité formelle a lieu dans un cadre plus large :

PROPOSITION 5 (cf. [Ba], [BDS]) : *Soit A un anneau (noethérien) régulier intègre. Si -1 est une somme de carrés dans le corps des fractions de A, c'est une somme de carrés dans A.*

Démonstration : Si -1 n'est pas une somme de carrés dans A, il existe (Lemme 1) un idéal premier p de A tel que -1 n'est pas une somme de carrés dans le corps des fractions $A(p)$ de A/p. Mais $A(p)$ est le corps résiduel de l'anneau local régulier A_p, dont le corps des fractions coïncide avec celui de A. Le lemme 2 montre que -1 ne peut donc pas être une somme de carrés dans le corps des fractions de A. ∎

Remarque sur la proposition 3 (communiquée par M. Coste) : On a vu dans la démonstration une liste (a,b,c,d) d'objets équivalents à la donnée d'un point du spectre réel de A. On a vu aussi que l'existence d'une telle donnée, pour un anneau A, équivaut à l'existence d'une signature de l'anneau A. Mais la donnée d'un point du spectre réel n'est pas en général équivalente à la donnée d'une signature. C'est le cas si A est un corps, mais c'est déjà faux pour A un anneau de valuation discrète, comme on va le voir. Pour A un anneau quelconque, on dispose d'une application évidente $f : \mathrm{Spec}_R A \longrightarrow \mathrm{Sign}\, A$ du spectre réel de A dans l'ensemble des signatures de A. Pour A un anneau local (ou même semi-local connexe), le procédé de Kanzaki-Kitamura (cf. Knebusch [K2]) permet de définir une application $g : \mathrm{Sign}A \longrightarrow \mathrm{Spec}_R A$, dont f est une rétraction : $f \circ g = \mathrm{id}_{\mathrm{Sign}A}$. Soit alors $A = \mathbb{R}[X]_{(X)}$ le localisé de la droite affine sur les réels au point $X = 0$. Considérons les deux points du spectre réel de A donnés, l'un par la réduction modulo X, et l'ordre de $\mathbb{R}$, l'autre par l'inclusion de A dans son corps des fractions $\mathbb{R}(X)$, muni de l'ordre pour lequel X est infiniment petit positif (par rapport à l'ordre de $\mathbb{R}$). On vérifie facilement que les deux signatures associées coïncident sur les unités de A, donc aussi sur W(A) : l'application f n'est pas injective, g identifie SignA à un sous-ensemble propre de $\mathrm{Spec}_R A$.

REFERENCES

[Ba] R. BAEZA : *Über die Stufe von Dedekind-Ringen*, Archiv der Math. 33 (1979) p. 226-231.

[Bo] N. BOURBAKI : *Algèbre commutative*, Chap. II, Hermann, Paris (1961).

[B-E] J. BOCHNAK, G. EFROYMSON : *Real Algebraic Geometry and the 17th Hilbert Problem*, Math. Ann. 251 (1980) p. 213-241.

[Br] L. BRÖCKER : *Positivbereiche in kommutativen Ringen*, erscheint in Abh. Math. Sem. Univ. Hamburg.

[BDS] L. BRÖCKER, A. DRESS, R. SCHARLAU : *An (almost) trivial local-global principle for the representation of -1 as a sum of squares in an arbitrary commutative ring* (Vorabdruck).

[CDLR] M. D. CHOI, Z. D. DAI, T. Y. LAM, B. REZNICK : *The Pythagoras Number of Some Affine Algebras and Local Algebras* (Preprint).

[C1] M.-F. COSTE-ROY : Thèse, Université de Paris-Nord (1980).

[C2] M. COSTE : *Ensembles semi-algébriques et fonctions de Nash*, Prépublication de l'Université de Paris-Nord (1981).

[K1] M. KNEBUSCH : *Specialization of quadratic and symmetric bilinear forms, and a norm theorem*, Acta Arithmetica 24 (1973) p. 279-299.

[K2] M. KNEBUSCH : *Symmetric bilinear forms over algebraic varieties*, in Conference on Quadratic Forms, Queen's papers in pure and applied Mathematics n° 46 (1977) p. 103-283.

[L] T. Y. LAM : *The theory of ordered fields, in Ring theory and algebra* III, Lecture notes in pure and applied mathematics, Vol. 55, Marcel Dekker (1980).

[Lg] S. LANG : *The theory of real places*, Annals of Mathematics, Vol. 57 (1953) p. 378-391.

[Lo1] F. LORENZ : *Quadratische Formen und die Artin-Schreiersche Theorie der formal reellen Körper*, Bull. Soc. Math. France, Mémoire 48 (1977) p. 61-73.

[Lo2] F. LORENZ : *Einige Bemerkungen zu einem Satz von Sylvester*, Vorabdruck (1978).

[P] A. PRESTEL : *Lectures on formally real fields*, IMPA Lecture Notes, Rio de Rio de Janeiro (1975).

[S] G. STENGLE : *A Nullstellensatz and a Positivstellensatz in Semialgebraic Geometry*, Math. Ann. 207 (1974) p. 87-97.

C.N.R.S. Mathématiques
Bâtiment 425
Université de Paris-Sud
91405 - ORSAY
FRANCE

ENSEMBLES SEMI-ALGEBRIQUES

par

Michel COSTE

(Université de Rennes I et Université de Niamey)

I - INTRODUCTION.

DEFINITION 1.1 : *Les ensembles semi-algébriques de* $\mathbb{R}^n$ *forment la plus petite collection de parties de* $\mathbb{R}^n$ *contenant toutes les parties du genre* $\{(x_1,\ldots,x_n) \in \mathbb{R}^n | P(\underline{x}) > 0\}$ *et stable par intersection finie, union finie et passage au complémentaire. De façon équivalente, si on appelle condition de signe sur le polynôme* P *une des conditions* $P(\underline{x}) > 0$, $P(x) = 0$ *ou* $P(\underline{x}) < 0$, *un semi-algébrique est donné par une combinaison booléenne (obtenue par disjonction, conjonction et négation) de conditions de signe sur un nombre fini de polynômes.*

Exemples :

1) Toute variété algébrique réelle est un semi-algébrique.

2) $\{(x,y) \in \mathbb{R}^2 | x^2 + y^2 \leq 1$ et $(x \leq 0$ ou $y \leq 0)\}$ est un semi-algébrique

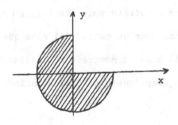

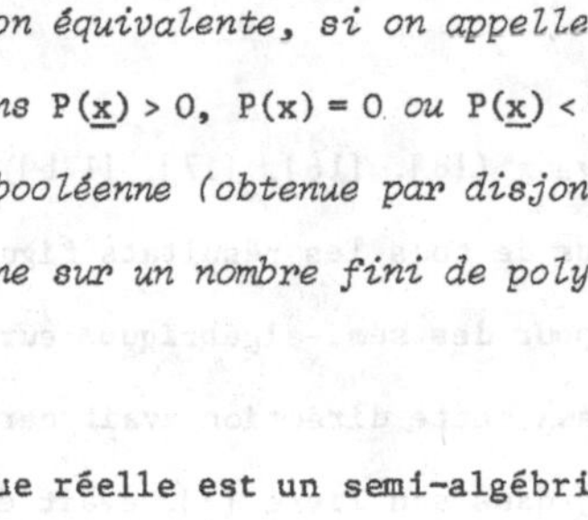

3) $\{(x,y) \in \mathbb{R}^2 \mid y = e^x\}$ n'est pas semi-algébrique.

4) $\{(x,y) \in \mathbb{R}^2 \mid \exists\ n \in \mathbb{N}\quad y = nx\}$ n'est pas semi-algébrique.

5) Les semi-algébriques de $\mathbb{R}$ sont les réunions finies de points et d'intervalles quelconques (ouverts ou fermés ou ouverts d'un côté et fermés de l'autre, à bornes finies ou non).

Cet article fait le tour des propriétés topologiques de base des ensembles semi-algébriques réels. Tous les résultats énoncés ici sont déjà connus (nombre fini de composantes connexes, triangulation, trivialité locale des morphismes semi-algébriques, structure conique locale, etc...). Ces résultats sont plus ou moins éparpillés dans un certain nombre d'articles (voir la bibliographie). Par ailleurs, les semi-algébriques sont assez souvent abordés dans la littérature comme cas particuliers de semi-analytiques (par exemple chez Lojasiewicz [19]) ou avec des méthodes non élémentaires (voir la démonstration de la structure conique locale chez Milnor, théorème 2.10 de [20]).

Dans une série de travaux ([8], [16], [17], [17b]), H. Delfs et M. Knebusch ont donné des démonstrations de tous les résultats figurant dans ce papier, par des méthodes élémentaires, et pour des semi-algébriques sur un corps réel clos quelconque. Un des premiers pas dans cette direction avait certainement été l'article de Cohen [4] ; Brumfiel aussi, dans son livre [3], avait établi de cette manière quelques résultats pour les semi-algébriques sur un corps réel clos quelconque. Delfs et Knebusch utilisent leurs résultats pour développer une cohomologie des ensembles semi-algébriques qui redonne pour un corps réel clos quelconque les résultats que l'on a sur $\mathbb{R}$. Leur travail a aussi montré que l'utilisation conséquente de méthodes élémentaires n'amène pas à compliquer les démonstrations, bien au contraire.

Dans la présentation que je donne ici, j'ai suivi cette voie, mais seulement à moitié : les méthodes de démonstration sont élémentaires, mais on ne s'intéresse qu'aux semi-algébriques réels. Je veux simplement signaler ici que la difficulté principale que l'on rencontre pour le passage à un corps réel clos quelconque est le fait que $\mathbb{R}$ est le seul corps réel clos localement connexe. Il est donc nécessaire de modifier convenablement la notion de connexité : on peut dire qu'un semi-algébrique A est "semi-algébriquement connexe" s'il n'est pas réunion disjointe de deux semi-algébriques non vides ouverts dans A [3], ou utiliser une notion de connexité par arc semi-algébrique [16][17], ou passer au spectre réel [6] ; ces trois méthodes donnent le même résultat.

Il doit être clair que ce texte ne prétend à aucune originalité. Sa seule ambition est de regrouper un ensemble de résultats de base sur les semi-algébriques réels, ceci dans un espace restreint, et en redémontrant autant que possible tous les résultats utilisés. Les preuves sont agencées de manière à montrer que tout repose sur deux outils de base : le "saucissonnage" de Cohen (section II) et le lemme de Thom (section III).

Je remercie M. Knebusch et L. Mahé pour leurs remarques qui m'ont permis de préciser certains points obscurs des démonstrations.

II - LE "SAUCISSONNAGE" DE COHEN.

Le théorème 2.3 ci-dessous est vraiment la base de tous les raisonnements par induction qui suivront. La démarche suivie est celle de Cohen [4] , revue par Brumfiel [3].

DEFINITION 2.1 : *Soit* S *un semi-algébrique de* $\mathbb{R}^n$. *Une fonction* $f : S \longrightarrow \mathbb{R}$ *est dite semi-algébrique si, pour tout semi-algébrique* $T \subset \mathbb{R}^{p+1}$, *l'ensemble* $\{(\underline{x},\underline{t}) \in \mathbb{R}^{n+p} | \underline{x} \in S$ *et* $(\underline{t},f(\underline{x})) \in T\}$ *est un semi-algébrique de* $\mathbb{R}^{n+p}$.

On définit d'habitude une fonction semi-algébrique comme une fonction dont le graphe est semi-algébrique. Nous verrons plus loin que les deux définitions sont équivalentes. La définition 2.1 permet d'obtenir immédiatement quelques propriétés des fonctions semi-algébriques :

PROPOSITION 2.2 : *La différence et le produit de deux fonctions semi-algébriques sont semi-algébriques. Les fonctions semi-algébriques forment un anneau. Par ailleurs, si* f *est semi-algébrique ainsi que* $g_1,\ldots,g_n$ *, la fonction composée* $f(g_1,\ldots,g_n)$ *est encore semi-algébrique.*

THEOREME 2.3 : *Soit* $P(\underline{x},y)$ *(avec* $\underline{x}=(x_1,\ldots,x_n)$*) un polynôme. Il existe une partition de* $\mathbb{R}^n$ *en semi-algébrique* $A_1,\ldots,A_m$ *tels que pour chaque* $i=1,\ldots,m$

- ou bien $P(\underline{x},y)$ *a un signe constant* (>0, <0 *ou* $=0$) *pour tout* $\underline{x}$ *de* A_i *et tout* y *de* $\mathbb{R}$ *;*

- ou bien, il existe un nombre fini de fonctions semi-algébriques continues $\xi_1<\ldots<\xi_{\ell_i}$ *de* A_i *dans* $\mathbb{R}$ *telles que* $\{\xi_1(\underline{x}),\ldots,\xi_{\ell_i}(\underline{x})\}$ *est l'ensemble des zéros de* $P(\underline{x},y)$ *pour tout* $\underline{x}$ *de* A_i *, et que le signe de* $P(\underline{x},y)$ *ne dépend que des signes de* $y-\xi_j(\underline{x})$ *pour* $j=1,\ldots,\ell_i$.

Ceci revient à dire que l'on a découpé le cylindre de base A_i (voici le saucisson !) en tranches, au moyen des fonctions ξ_j , de telle façon que P soit nul le long de ces coupes et de signe constant sur chaque tranche (on ne dit pas que A_i est connexe).

<u>Preuve</u> : On procède par induction sur d, le degré de $P(\underline{x},y)$ en y. Si d = 0, P ne dépend pas de y, et il suffit de découper $\mathbb{R}^n$ en trois morceaux : celui où P est nul, celui où il est strictement positif, et celui où il est strictement négatif. Supposons maintenant le théorème montré pour tout entier strictement plus petit que d. On peut l'utiliser en particulier pour $\frac{\partial P}{\partial y}(\underline{x},y)$: ceci nous donne une partition de $\mathbb{R}^n$ en semi-algébriques $A_1,\ldots,A_m$, et pour chaque i tel que $\frac{\partial P}{\partial y}$ ne soit

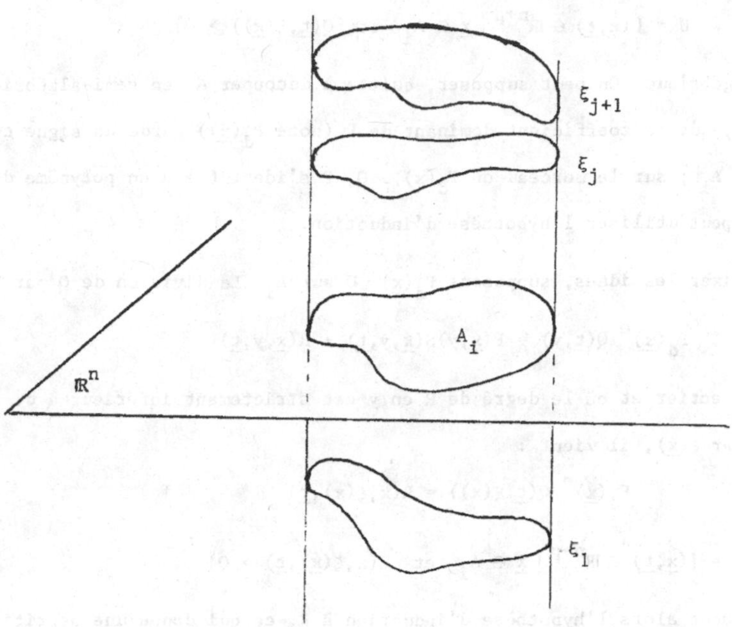

pas identiquement nul sur A_i, des fonctions semi-algébriques continues $\zeta_1 < \ldots < \zeta_{r_i}$ qui décrivent les racines de $\frac{\partial P}{\partial y}$ sur A_i. Le cas où $\frac{\partial P}{\partial y}$ garde un signe constant sur A_i se traite sans difficulté. Quitte à découper A_i en morceaux semi-algébriques plus petits, on peut supposer que pour tout j, $P(\underline{x},\zeta_j(\underline{x}))$ garde un signe constant (>0, <0, ou $=0$) sur A_i ; on utilise ici la semi-algébricité des fonctions ζ. Les zéros de P au-dessus de A_i sont alors donnés par des fonctions $\xi_1 < \ldots < \xi_{\ell_i}$, chaque fonction ξ étant soit égale à une fonction ζ, soit comprise entre deux ζ consécutives, soit plus petite (resp. plus grande) que la plus petite (resp. plus grande) fonction ζ. Par ailleurs, le signe de P est bien constant sur chacune des tranches délimitées par les fonctions ξ : on se convaint facilement de ce qui précède en remarquant qu'au-dessus de A_i on connaît le nombre et la position des racines de $\frac{\partial P}{\partial y}$, le signe de $\frac{\partial P}{\partial y}$ sur les intervalles entre ces racines (y compris les intervalles $]-\infty,\xi_1[$ et $]\zeta_{r_i}, +\infty[$) et le signe de P sur chacune de ces racines. Le point important à montrer est que les fonctions ξ sont semi-algébriques. Il suffit pour cela de voir, qu'étant donné un polynôme $Q(\underline{t},y)$, l'ensemble

$$U = \{(\underline{x},\underline{t}) \in \mathbb{R}^{n+p} \mid \underline{x} \in A_i \quad \text{et} \quad Q(\underline{t},\xi(\underline{x})) > 0\}$$

est semi-algébrique. On peut supposer, quitte à découper A_i en semi-algébriques plus petits, que le coefficient dominant de P (noté $P_d(\underline{x})$) garde un signe constant non nul sur A_i ; sur le morceau où $P_d(x) \equiv 0$, P s'identifie à un polynôme de degré d-1, et on peut utiliser l'hypothèse d'induction.

Pour fixer les idées, supposons $P_d(\underline{x}) > 0$ sur A_i. La division de Q par P donne :

$$P_d(\underline{x})^e \; Q(\underline{t},y) = P(\underline{x},y)S(\underline{x},y,\underline{t}) + R(\underline{x},y,\underline{t})$$

où e est un entier et où le degré de R en y est strictement inférieur à d. En remplaçant y par $\xi(\underline{x})$, il vient :

$$P_d(\underline{x})^e \; Q(\underline{t},\xi(x)) = R(\underline{x},\xi(x),\underline{t})$$

et donc

$$U = \{(\underline{x},\underline{t}) \in \mathbb{R}^{n+p} \mid \underline{x} \in A_i \quad \text{et} \quad R(x,\xi(\underline{x}),\underline{t}) > 0\} \; .$$

Appliquons alors l'hypothèse d'induction à R, ce qui donne une partition de $A \times \mathbb{R}^p$ en un nombre fini de semi-algébriques B_k, les zéros de R au-dessus de chaque B_k où $R \not\equiv 0$ étant décrits pas des fonctions semi-algébriques $\eta_1(x,t) < \ldots < \eta_{s_k}(x,t)$. Quitte à les découper en semi-algébriques C_k plus petits, on peut supposer que sur chaque C_k les fonctions $\zeta_i - \eta_j$ et $P(x,\eta_j)$, $i=1,\ldots,r$ et $j=1,\ldots,s_k$ gardent un signe (>0, <0 ou $=0$) constant. Mais alors le signe des $R(\underline{x},\xi(\underline{x}),\underline{t})$ est aussi constant sur chaque C_k et U, qui est réunion d'un nombre fini de C_k, est bien semi-algébrique.

Il reste à voir que les fonctions ξ sont continues. Si ξ coïncide avec une fonction ζ, il n'y a pas de problème. Sinon, on a par exemple $\zeta_j < \xi < \zeta_{j+1}$ avec ξ seule racine de P entre ζ_j et ζ_{j+1}. Fixons $\underline{x}_o \in A_i$ et choisissons des réels a et b avec $\zeta_i(\underline{x}_o) < a < \xi(\underline{x}_o) < b < \zeta_{i+1}(\underline{x}_o)$. Il y a un voisinage V de $\underline{x}_o$ dans A_i sur lequel on a $\zeta_i(\underline{x}) < a$, $b < \zeta_{i+1}(\underline{x})$ et sur lequel $P(\underline{x},a)$ et $P(\underline{x},b)$ gardent un signe constant. Sur V, on a encore $a < \xi(\underline{x}) < b$. ∎

On peut améliorer un peu le résultat obtenu :

COROLLAIRE 2.4 : *Soit* $P_1(\underline{x},y),\ldots,P_t(\underline{x},y)$ *une famille finie de polynômes. Il existe une partition de* $\mathbb{R}^n$ *en semi-algébriques* $A_1,\ldots,A_m$ *tels que, pour chaque i, les zéros de* $P_1,\ldots,P_t$ *sur* A_i *sont donnés par des fonctions semi-algébriques continues* $\xi_1 < \ldots < \xi_{\ell_i}$*, et que le signe de* $P_j(\underline{x},y)$, $j = 1,\ldots,t$, *sur* A_i *ne dépend que des signes de* $y - \xi_k(x)$, $k = 1,\ldots,\ell_i$.

Preuve : Par induction sur t en utilisant 2.3 : si on ajoute $P_{t+1}(\underline{x},y)$ à la liste, on sait que l'on a une partition de $\mathbb{R}^n$ en $B_1,\ldots,B_p$, et sur chaque B_j des fonctions semi-algébriques $\zeta_1,\ldots,\zeta_{q_j}$ qui décrivent les zéros de P_{t+1}. Il suffit alors de diviser chaque $A_i \cap B_j$ en semi-algébriques, où toutes les fonctions $\xi - \zeta$ gardent un signe constant. ∎

Venons-en maintenant au "principe de Tarski-Seidenberg" ([24], [23]) pour $\mathbb{R}$. Voici d'abord la version géométrique.

PROPOSITION 2.5 : *L'image par la projection* $\mathbb{R}^{n+1} \longrightarrow \mathbb{R}^n$ *d'un semi-algébrique de* $\mathbb{R}^{n+1}$ *est un semi-algébrique de* $\mathbb{R}^n$.

Preuve : Soit S un semi-algébrique de $\mathbb{R}^{n+1}$ donné par une combinaison booléenne de conditions de signe sur $P_1(\underline{x},y),\ldots,P_t(\underline{x},y)$. Il est clair que la projection de S sur $\mathbb{R}^n$ est réunion de certains des semi-algébriques A_i du corollaire 2.4. ∎

Voici maintenant la version logique.

PROPOSITION 2.6 : $\mathbb{R}$ *admet l'élimination des quantificateurs dans le langage des corps ordonnés.*

Ceci nécessite quelques explications : une formule du premier ordre du langage des corps ordonnés est une formule obtenue à partir de conditions de signe sur des polynômes par conjonction, disjonction, négation, quantification existentielle ou universelle sur les variables. Une formule sans quantificateur est une combinaison booléenne de conditions de signe sur des polynômes. La proposition 2.6 dit que, dans $\mathbb{R}$, toute formule du premier ordre du langage des corps ordonnés est équivalente à une formule sans quantificateur, c'est-à-dire encore que l'extension d'une formule du premier ordre du langage des corps ordonnés est toujours un semi-algébrique.

Il faut bien faire attention à ce que les quantifications ne portent <u>que sur des variables réelles</u> : on a vu que

$$\{(x,y) \in \mathbb{R}^2 \mid \exists\, n \in \mathbb{N},\ y = nx\}$$

n'est pas un semi-algébrique.

<u>Preuve de 2.6</u> : On raisonne par induction sur la construction des formules à partir des conditions de signe sur les polynômes. Si $\Phi(\underline{x})$, $\Psi(\underline{x})$ sont deux formules telles que $\{\underline{x} \in \mathbb{R}^n \mid \Phi(x)\}$ et $\{\underline{x} \in \mathbb{R}^n \mid \Psi(\underline{x})\}$ sont semi-algébriques, il est clair que $\{x \in \mathbb{R}^n \mid \Phi(\underline{x})$ et $\Psi(\underline{x})\}$, $\{\underline{x} \in \mathbb{R}^n \mid \Phi(\underline{x})$ ou $\Psi(\underline{x})\}$ et $\{\underline{x} \in \mathbb{R}^n \mid$ non $\Phi(\underline{x})\}$ sont semi-algébriques. Par ailleurs, si $\{(\underline{x},y) \in \mathbb{R}^{n+1} \mid \Theta(\underline{x},y)\}$ est semi-algébrique, son image par la projection sur $\mathbb{R}^n$ $\{\underline{x} \in \mathbb{R}^n \mid \exists y\ \Theta(x,y)\}$ est semi-algébrique par 2.5, ainsi que $\{\underline{x} \in \mathbb{R}^n \mid \forall y\ \Theta(\underline{x},y)\}$ qui est le complémentaire de la projection du complémentaire. ∎

Voici un premier exemple qui montre l'utilité du principe de Tarski-Seidenberg :

PROPOSITION 2.7 : *L'adhérence (et donc l'intérieur) d'un semi-algébrique est un semi-algébrique.*

<u>Preuve</u> : Soit S un semi-algébrique donné par une combinaison booléenne de conditions de signes sur des polynômes que nous noterons $\Phi(\underline{x})$. L'adhérence de S est donnée par

$$\mathrm{adh}(S) = \{\underline{x} \in \mathbb{R}^n \mid \forall r\ [r > 0 \Longrightarrow \exists y_1 \ldots \exists y_n\ (\sum_{i=1}^{n} (x_i - y_i)^2 < r \text{ et } \Phi(y))]\}\ .$$

D'après 2.6, adh(S) est bien semi-algébrique. ∎

Revenons aux fonctions semi-algébriques pour montrer que celles que nous avons considérées sont bien les fonctions semi-algébriques de tout le monde :

PROPOSITION 2.8 : *Une fonction est semi-algébrique si et seulement si son graphe est semi-algébrique.*

Preuve : Il est clair que si $f : S \longrightarrow \mathbb{R}$ est semi-algébrique

$$\text{Graphe}(f) = \{(\underline{x},y) \in \mathbb{R}^{n+1} \mid f(\underline{x}) = y\}$$

est semi-algébrique. Réciproquement, supposons Graphe(f) semi-algébrique, donné par la combinaison booléenne de conditions de signe $\Phi(\underline{x},y)$, et soit T un semi-algébrique de $\mathbb{R}^{p+1}$ donné par $\Psi(\underline{t},y)$. Alors :

$$\{(\underline{x},\underline{t}) \in \mathbb{R}^{n+p} \mid \underline{x} \in S \text{ et } (\underline{t},f(\underline{x})) \in T\} = \{(\underline{x},\underline{t}) \in \mathbb{R}^{n+p} \mid \exists\, y(\Phi(\underline{x},y) \text{ et } \Psi(\underline{t},y))\}$$

est d'après 2.6, un semi-algébrique. ∎

DEFINITION 2.9 : *Soient* $A \subset \mathbb{R}^m$ *et* $B \subset \mathbb{R}^n$ *deux semi-algébriques. Un morphisme* $f : A \longrightarrow B$ *est dit semi-algébrique quand les fonctions coordonnées de* f *sont semi-algébriques ou de façon équivalente, si pour toute fonction semi-algébrique* $g : B \longrightarrow \mathbb{R}$, *la composée* $g \circ f$ *est semi-algébrique, ou encore si son graphe est semi-algébrique dans* $\mathbb{R}^{m+n}$.

Le principe de Tarski-Seidenberg pour $\mathbb{R}$ nous donne immédiatement le résultat suivant :

PROPOSITION 2.10 : *L'image d'un semi-algébrique par un morphisme semi-algébrique est semi-algébrique.*

<u>Remarque</u> : Dans [17], Knebusch et Delfs réservent le nom de morphisme semi-algébriques aux morphismes semi-algébriques <u>continus</u>.

Les fonctions semi-algébriques générales sont peu utilisées. On considère plutôt des classes de fonctions plus restreintes :

a) Les fonctions semi-algébriques continues, déjà rencontrées lors du saucissonnage. Voici un autre exemple de fonction semi-algébrique continue : la fonction $\underline{x} \longmapsto$ distance de $\underline{x}$ à S, où S est un semi-algébrique non vide donné par $\Phi(\underline{t})$. Elle est bien sûr continue, et son graphe est

$$\{(\underline{x},y) \in \mathbb{R}^{n+1} \mid y \geq 0 \text{ et } \forall t_1 \ldots \forall t_n \ (\Phi(\underline{t}) \Longrightarrow \sum_{i=1}^{n} (x_i-t_i)^2 \geq y^2)$$

$$\text{et } \forall\, \varepsilon > 0 \ \exists t_1 \ldots \exists t_n \ (\Phi(\underline{t}) \text{ et } \sum_{i=1}^{n} (x_i-t_i)^2 < y^2 + \varepsilon\} .$$

b) Les fonctions semi-algébriques analytiques (ou fonctions de Nash) sur un ouvert semi-algébrique. Voir pour plus de détails, l'article de J. Bochnak et G. Efroymson [2] et le "survey" [2b] dans ce volume.

c) Pour mémoire, les fonctions rationnelles régulières qui sont bien sûr semi-algébriques.

Une propriété importante des fonctions semi-algébriques continues est l'"inégalité de Lojasiewicz" (cf. [14] et [19]) :

PROPOSITION 2.11 : *Soient* K *un compact semi-algébrique de* $\mathbb{R}^n$, f *et* g *deux fonctions semi-algébriques continues sur* K, *telles que tout zéro de* f *dans* K *est aussi zéro de* g. *Il existe des constantes strictement positives* c *et* r *telles que* $|f| \geq c|g|^r$ *sur* K.

<u>Preuve</u> : Soit $H = \{(u,v) \in \mathbb{R}^2 \mid \exists \underline{x} \in K \;\; u = |g(x)| \;\text{ et }\; v = |f(x)|\}$, H est semi-algébrique d'après 2.6, et donc peut être donné par une combinaison booléenne de conditions de signe sur un nombre fini de polynômes $P_1(u,v),\ldots,P_t(u,v)$. On peut trouver un $\varepsilon > 0$ tel qu'au-dessus de $]0,\varepsilon[$ les racines distinctes des P_i sont données par des séries de Puiseux $v_1(u) < \ldots < v_p(u)$ en une puissance fractionnaire de u (voir par exemple [26], chap. IV). Si $H \cap (]0,\varepsilon[\times\mathbb{R})$ n'est pas vide, il est borné inférieurement par le graphe d'une de ces fonctions $v_j(u)$:

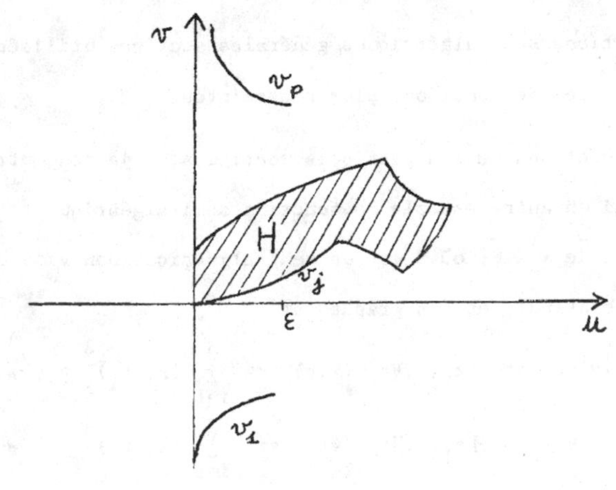

v_j est strictement positive sur $]0,\varepsilon[$ car pour $\delta \in\,]0,\varepsilon[$, $\{\underline{x} \in K \mid |g(x)| = \delta\}$ est un compact sur lequel $|f|$ ne s'annule pas et a donc un minimum non nul. L'inégalité $|f| \geq v_j(|g|) > 0$ pour $0 < |g| < \varepsilon$ montre que l'on peut bien trouver des constantes strictement positives c et r, telles que l'on ait $|f| \geq c\,|g|^r$ pour $|g| \leq \frac{\varepsilon}{2}$. Comme $\{\underline{x} \in K \mid |g(\underline{x})| \geq \frac{\varepsilon}{2}\}$ est un compact sur lequel $|f|$ ne s'annule pas, l'inégalité est vérifiée sur K tout entier, quitte à reviser la constante c en baisse. ∎

<u>Remarques</u> : a) La preuve montre que l'on peut affaiblir l'hypothèse "K compact". Il suffit de demander que pour tout $\varepsilon > 0$, $\{\underline{x} \in K \mid |g(\underline{x})| \geq \varepsilon\}$ soit un compact. Ceci se produit par exemple si K est n'importe quel fermé semi-algébrique et g la fonction $\underline{x} \to \frac{1}{1+\|\underline{x}\|^2}$.

b) On ne peut cependant pas oter toute hypothèse de compacité : prendre $K = \mathbb{R}$, $f = \frac{1}{1+x^2}$, $g = 1+x^2$. De façon générale, si K est localement fermé, on a des majorations sur tout compact de K uniformes pour l'exposant r, mais pas pour la constante c.

c) Cette preuve, qui utilise les séries de Puiseux, ne paraît pas élémentaire (elle l'est en fait, mais l'explication de ceci nous entraînerait trop loin).
H. Delfs a donné une preuve élémentaire valable pour tout corps réel clos ([8], lemma 3.2).

III - <u>LE LEMME DE THOM</u>

Le "Lemme de Thom" à une variable est le résultat suivant :

PROPOSITION 3.1 : *Soit* $P_1,\ldots,P_m$ *une famille finie de polynômes de* $\mathbb{R}[X]$, *stable par dérivation. Soit* A *un semi-algébrique de* $\mathbb{R}$, *donné par une condition de signe sur chaque* P_i, $i = 1,\ldots,m$. A *est de la forme*

$$A = \bigcap_{i=1}^{m} \{x \in \mathbb{R} \mid P_i(x)\ ?_i\ 0\}$$

où $?_i$ *est* $>$, $<$ *ou* $=$. *Alors*

a) A *est soit vide, soit connexe (et donc forcément un point si une condition de signe sur un polynôme non constant est "=0", ou un intervalle ouvert sinon).*

b) *Si* A *est non vide, on obtient son adhérence en relachant les inégalités strictes :*

$$\mathrm{adh}(A) = \bigcap_{i=1}^{m} \{x \in \mathbb{R} \mid P_i(x) \ ?_i' \ 0\}$$

où $?_i'$ *est* $\geq$ *(resp.* $\leq$*) si* $?_i$ *est* > *(resp.* <*) et* = *si* $?_i$ *est* = .

Le lemme de Thom figure chez Lojasiewicz [19] page 69, sans la précision apportée par le b).

<u>Preuve</u> : Par induction sur m. Il n'y a rien à montrer pour m = 0. Supposons le lemme de Thom montré pour m, et P_{m+1} de degré maximal dans la famille $P_1,\ldots,P_{m+1}$. La famille $P_1,\ldots,P_m$ est encore stable par dérivation. Soit

$$A' = \bigcap_{i=1}^{m} \{x \in \mathbb{R} \mid P_i(x) \ ?_i \ 0\}, \quad A = A' \cap \{x \in \mathbb{R} \mid P_{m+1}(x) \ ?_{m+1} 0\}.$$

Supposons A' non vide. Si A' est un point, il n'y a pas de problème. Si A' est un intervalle ouvert, la dérivée de P_{m+1} y garde un signe constant. P_{m+1} est strictement monotone (ou constant) sur adh(A'), ce qui permet de montrer a et b pour A. ∎

Voici maintenant ce qui se passe quand on a plusieurs variables :

THEOREME 3.2 : *On dira qu'une famille finie* $P_1,\ldots,P_m$ *de polynômes de* $\mathbb{R}[X_1,\ldots,X_n]$ *est séparante si pour tout semi-algébrique* A *de* $\mathbb{R}^n$ *donné par une condition de signe sur chaque* P_i :

$$A = \bigcap_{i=1}^{m} \{\underline{x} \in \mathbb{R}^n \mid P_i(\underline{x}) \ ?_i 0\} \quad (\text{où } ?_i \text{ est } >,\ < \text{ ou } =)$$

a) A *est soit vide, soit connexe ;*

b) *et si* A *n'est pas vide, son adhérence s'obtient en relachant les inégalités strites.*

b) *peut se reformuler ainsi :*

b') *Si* A *n'est pas vide et si* B *est aussi donné par une condition de signe sur chaque* P_i, B *est inclus dans l'adhérence de* A *ssi toute condition de signe stricte* (> 0 *ou* < 0) *sur les* P_i *vraie dans* B *est vraie dans* A.

Soit maintenant $P_1,\dots,P_r$ *une famille finie quelconque de polynômes de* $\mathbb{R}[X_1,\dots,X_n]$. *On peut trouver des polynômes* $P_{r+1},\dots,P_{r+s}$ *tels que la famille* $P_1,\dots,P_{r+s}$ *soit séparante.*

Ce résultat est dû à Efroymson [11]. Il est redémontré dans [5]. La démonstration suivante est empruntée à Houdebine [15].

Preuve : Par induction sur n. L'idée (dans toutes les preuves) est d'utiliser l'hypothèse d'induction sur $\mathbb{R}^n$ et le lemme de Thom à une variable sur les fibres de la projection $\mathbb{R}^{n+1} \longrightarrow \mathbb{R}^n$, en faisant appel à un "saucissonnage" convenable. Pour $n = 1$, le lemme de Thom à une variable dit : ajoutez toutes les dérivées non constantes à tous les ordres et vous aurez une famille séparante. Passons maintenant de n à $n+1$:

On peut toujours supposer que les polynômes $P_1,\dots,P_r$ sont unitaires en X_{n+1}, quitte à faire un changement de variables bien choisi du genre $X_i = X'_i + a_i X'_{n+1}$ pour $i = 1,\dots,n$ et $X_{n+1} = X'_{n+1}$. Ceci fait, on ajoute toutes les dérivées non constantes à tous les ordres des polynômes $P_1,\dots,P_r$. Ceci nous donne une liste $P_1,\dots,P_{r+t}$ de polynômes tous unitaires en X_{n+1} (à un facteur constant près). Le corollaire 2.4 nous donne une partition de $\mathbb{R}^n$ en semi-algébriques $B_1,\dots,B_m$ tels que, pour chaque $i = 1,\dots,m$, les zéros de $P_1,\dots,P_{r+t}$ au-dessus de B_i soient donnés par des fonctions semi-algébriques continues $\xi_1 < \dots < \xi_{\ell_i}$. Les B_i sont donnés par des combinaisons booléennes de conditions de signe sur un nombre fini de polynômes en $X_1,\dots,X_n$. On peut leur appliquer l'hypothèse d'induction, et compléter cette liste en une liste séparante $P_{r+t+1},\dots,P_{r+t+u}$. Quitte à découper les B_i, on peut supposer que ceux-ci sont donnés par la conjonction de conditions de signe sur chaque P_j, $j = r+t+1,\dots,r+t+u$:

$$B_i = \bigcap_{j=r+t+u}^{r+t+u} \{\underline{x} \in \mathbb{R}^n \mid P_j(\underline{x}) \ ?_{i,j} 0\}.$$

Nous allons montrer que la liste complète, $P_1,\dots,P_{r+t+u}$ qui a été construite, est une liste séparante dans $\mathbb{R}^{n+1}$. Considérons donc un semi-algébrique A de $\mathbb{R}^{n+1}$ donné par la conjonction de conditions de signe sur chaque P_j, $j = 1,\dots,r+t+u$:

$$A = \bigcap_{j=1}^{r+t+u} \{(\underline{x},y) \in \mathbb{R}^{n+1} \mid P_j(\underline{x},y) \ ?_j 0\} .$$

Soit $B = \bigcap_{j=r+t+1}^{r+t+u} \{\underline{x} \in \mathbb{R}^n \mid P_j(\underline{x}) \ ?_j 0\}$ et soient $\xi_1 < \ldots < \xi_\ell$ les racines des $P_1,\ldots,P_{r+t}$ au-dessus de B.

D'après le lemme de Thom à une variable, la fibre $A \cap \pi^{-1}(\underline{x}_o)$ en un point donné $\underline{x}_o \in B$ est soit vide, soit une racine $(\underline{x}_o,\xi_k(\underline{x}_o))$, soit un intervalle $\{(\underline{x}_o,y) \mid \xi_k(\underline{x}) < y < \xi_{k+1}(\underline{x})\}$ (les bornes pouvant éventuellement être $-\infty$ et $+\infty$). Comme les signes des polynômes $P_1,\ldots,P_{r+t}$ sont constants sur les tranches, A est respectivement soit vide, soit $\{(\underline{x},\xi_k(\underline{x})) \mid \underline{x} \in B\}$, soit $\{(\underline{x},y) \mid \underline{x} \in B$ et $\xi_k(x) < y < \xi_{k+1}(x)\}$. Dans tous les cas, A est connexe (ou vide). Supposons A non vide. Il reste à voir que l'adhérence de A est

$$A' = \bigcap_{j=1}^{r+t+u} \{(\underline{x},y) \in \mathbb{R}^{n+1} \mid P_j(\underline{x},y) \ ?'_j 0\}$$

où $?'_j$ est $\geq$ (resp. $\leq$) si $?_j$ est $>$ (resp. $<$) et $=$ si $?_j$ est $=$. On a bien sûr adh(A) $\subset$ A'. L'hypothèse d'induction nous dit que

$$B' = \bigcap_{j=r+t+1}^{r+t+u} \{\underline{x} \in \mathbb{R}^n \mid P_j(\underline{x}) \ ?'_j 0\}$$

est l'adhérence de B. Soit $\underline{x}_o$ un point de B'. Comme les polynômes P_j, $j = 1,\ldots,r+t$, sont unitaires en X_{n+1}, on peut trouver un voisinage V de $\underline{x}_o$ dans $\mathbb{R}_n$ et une constante positive M tels que les racines $\xi_1,\ldots,\xi_\ell$ soient bornées en valeur absolue par M sur $V \cap B$. Ceci entraîne que pour tout $\underline{x}$ de $V \cap B$, on a $\{\underline{x}\} \times [-M,+M] \cap \text{adh } A \neq \emptyset$, et donc $\{\underline{x}_o\} \times [-M,+M] \cap \text{adh } A \neq \emptyset$; la fibre de adh(A) au-dessus de $\underline{x}_o$ est donc non vide (ceci n'est pas vrai si on ne suppose pas $P_1,\ldots,P_{r+t}$ unitaires en X_{n+1}). Le lemme de Thom à une variable nous montre qu'il y a deux possibilités pour la fibre de A' au-dessus de $\underline{x}_o$:

a) Cette fibre est réduite à un point, et elle est donc forcément égale à la fibre de adh A au-dessus de x_o.

b) Cette fibre est un intervalle fermé d'intérieur non vide. Si $(\underline{x}_o,y)$ est un point de cet intérieur, on a surement $P_j(\underline{x}_o,y) \ ?_j 0$ pour $j = 1,\ldots,r+t$, et $?_j$ est une inégalité stricte (en effet, puisque P_j est unitaire en X_{n+1}, $P_j(\underline{x}_o,X_{n+1})$ n'est pas identiquement nul). Ceci montre que $(\underline{x}_o,y)$ est dans l'adhérence de A, et donc l'intervalle fermé tout entier est dans l'adhérence de A.

Au total on a bien $A' \subset adh(A)$, ce qui achève la preuve. ∎

Voici maintenant quelques conséquences immédiates du lemme de Thom à plusieurs variables :

COROLLAIRE 3.3 : *Un ensemble semi-algébrique de $\mathbb{R}^n$ est localement connexe, et a un nombre fini de composantes connexes qui sont semi-algébriques.*

Preuve : Soit S un semi-algébrique donné par une combinaison booléenne de conditions de signe sur les polynômes $(P_i)_{i=1,...,r}$. Complétons cette liste en une liste séparante $P_1,...,P_{r+s}$. S est réunion disjointe d'un nombre fini de semi-algébriques connexes qui sont donnés par une condition de signe sur chaque P_i, $i=1,...,r+s$. La connexité locale vient alors de ce que tout point de S admet une base de voisinages semi-algébriques, possédant eux-mêmes des composantes connexes en nombre fini (et donc ouvertes). ∎

Le résultat ci-dessus est dans Lojasiewicz [19], page 110. Il précise le "théorème de Whitney" (th. 4 de [28]). Voir aussi [9].

Remarque : On pourrait, en réexaminant la preuve du "saucissonnage" et celle du lemme de Thom à plusieurs variables, apporter la précision suivante : étant donné le nombre n de variables, le nombre r de polynômes et le degré maximal de ceux-ci, on peut borner le nombre des polynômes à ajouter pour obtenir une liste séparante ainsi que leur degré maximal. (On peut voir aussi la preuve du lemme de Thom à plusieurs variables dans [7]). Ceci donne aussi une borne sur le nombre de composantes connexes d'un semi-algébrique S de $\mathbb{R}^n$ en fonction de n, du nombre de polynômes entrant dans la définition de S et de leur degré maximal. L'existence d'une telle borne pourrait aussi se montrer à partir du théorème de trivialité locale des morphismes semi-algébriques que nous verrons plus loin (théorème 5.1).

COROLLAIRE 3.4 : *Soient* S *un semi-algébrique de* $\mathbb{R}^n$, U *un semi-algébrique ouvert dans* S. U *est réunion finie de semi-algébriques du genre*

$$\{\underline{x} \in S \mid P_1(\underline{x}) > 0 \ \textit{et} \dots \textit{et} \ P_k(\underline{x}) > 0\}$$

qui sont l'intersection de S *avec un semi-algébrique ouvert donné par une conjonction finie de conditions de signe strictes.*

Remarque : Il est clair que tout semi-algébrique de la forme ci-dessus est bien ouvert dans S. La réciproque n'est pas évidente comme le montre l'exemple $= \mathbb{R}^2$, $U = \{x^2+y^2<1 \text{ et } y \neq 0\} \cup \{y=0 \text{ et } x^2 < \frac{1}{2}\}$. On peut, pour cet exemple, grossir le $y = 0$ en $y^2 < \frac{1}{3}$ sans sortir de U. Bochnak et Efroymson suivent cette idée dans le cas général, en utilisant l'inégalité de Lojasiewicz (voir [2]). La preuve qui suit est celle de [5]. Le résultat a été énoncé par Brumfiel dans [3] et montré aussi par Recio [22] et Delzell [9].

Preuve : Complétons la liste de polynômes entrant dans la définition de S et U en une liste séparante $P_1,\dots,P_m$. U s'écrit alors comme une union finie

$$U = \bigcup_{i=1}^{p} T_i \quad \text{où} \quad T_i = \bigcap_{j=1}^{m} \{\underline{x} \in \mathbb{R}^n \mid P_j(\underline{x}) \ ?_{i,j} 0\} ,$$

avec T_i non vide. Soit V_i l'ouvert semi-algébrique donné par la conjonction des conditions de signe strictes (> 0 ou < 0) sur les P_j vérifiées sur T_i. On a bien sûr $T_i \subset V_i$, et donc $U \subset (\bigcup_{i=1}^{p} V_i) \cap S$. Il reste à voir que $V_i \cap S$ est contenu dans U. $V_i \cap S$ lui-meme est réunion de semi-algébriques donnés par une condition de signe sur chaque P_j, $j = 1,\dots,m$. Soit A un de ces semi-algébriques. D'après la définition de V_i, toute condition de signe stricte vérifiée sur T_i l'est aussi sur A, et donc $T_i \subset \mathrm{adh}(A)$ d'après le b') du théorème 3.2. Mais alors on a $U \cap A \neq \emptyset$, et donc nécessairement $A \subset U$. ∎

PROPOSITION 3.5 (Stratification, [19] [28]) : *Un semi-algébrique* S *de* $\mathbb{R}^n$ *est réunion disjointe d'un nombre fini de variétés analytiques connexes* $A_1,\dots,A_p$ *qui vérifient :*

$$A_i \cap \mathrm{adh}(A_j) \neq \emptyset \Longrightarrow A_i \subset \mathrm{adh}(A_j) \ \textit{et} \ \dim A_i < \dim A_j .$$

Preuve : Ce résultat est une conséquence de la preuve du lemme de Thom à plusieurs variables. Il suffit de voir que les semi-algébriques, donnés par une condition de signe sur chaque polynôme de la liste séparante construite dans cette preuve, sont en fait des variétés analytiques. En reprenant les notations de la preuve, on voit qu'une racine ξ_k d'un polynôme P_j, $j = 1,\dots,r+t$ au-dessus de B est toujours racine simple d'un de ces polynômes, puisque la famille est stable par dérivation par rapport à X_{n+1}. Si B est une variété analytique (hypothèse d'induction), les ξ_k sont analytiques, et A est une variété analytique. L'assertion sur les dimensions des strates se vérifie aisément sur la preuve du lemme de Thom à plusieurs variables.

Remarque 1 : Les ξ_k sont en fait des fonctions de Nash, et donc on peut ajouter que les A_i de la stratification sont des "variétés de Nash" (voir Artin-Mazur [1]).

Remarque 2 : La stratification permet de définir la dimension d'un semi-algébrique S (il y a d'autres moyens, de nature plus algébrique [6] [17]). Si $S = \bigcup_{i=1}^{p} A_i$ est une stratification, on pose $\dim S = \sup \dim A_i$. La dimension ainsi définie est indépendante de la stratification choisie : on peut vérifier que $\dim S = d$, si et seulement si S contient un ouvert homéomorphe à la boule unité ouverte de $\mathbb{R}^d$, et pas d'ouvert homéomorphe à la boule unité ouverte de $\mathbb{R}^{d'}$ pour $d' > d$.

IV - LA TRIANGULATION DES SEMI-ALGEBRIQUES.

Nous allons maintenant montrer qu'un semi-algébrique compact est homéomorphe à un complexe fini ([18], [13]). Dans sa thèse [8], Delfs a donné la première preuve élémentaire de la triangulation. Fixons d'abord la terminologie. Un simplexe ouvert S, de dimension k dans $\mathbb{R}^n (n \geq k)$, est donné par k+1 points $\underline{a}_o,\dots,a_k$ qui ne sont pas contenus dans un sous-espace affine de dimension k-1. On a :

$$S = \{\underline{x} \in \mathbb{R}^n \mid t_o,\dots,t_k > 0 \quad \sum_{i=o}^{k} t_i = 1 \quad \text{et} \quad \underline{x} = \sum t_i \underline{a}_i\}.$$

On pourra désigner S par l'ensemble $(a_o,\dots,a_k)$ de ses sommets. Une face itérée de S est un simplexe ouvert, dont l'ensemble des sommets est une partie de $(a_o,\dots,a_k)$. Un complexe fini K de $\mathbb{R}^n$ est une réunion finie disjointe de simplexes ouverts, $K = \cup S_i$, telle que toute face itérée d'un S_i soit un $S_{i'}$. Un simplexe ouvert ou un complexe fini sont des ensembles semi-algébriques.

THEOREME 4.1 : *Soient* A *un semi-algébrique compact de* $\mathbb{R}^n$, *et* $A = \cup A_k$ *une partition finie de* A *en semi-algébriques. Il existe un homéomorphisme semi-algébrique* Φ *de* A *sur un complexe fini* $K = \cup S_i$ *tel que chaque* A_k *soit réunion de* $\Phi^{-1}(S_i)$.

Remarque : Soit A un semi-algébrique quelconque de $\mathbb{R}^n$. A est semi-algébriquement homéomorphe (au moyen de $\underline{x} \longmapsto \dfrac{x}{1+\|x\|}$) à un semi-algébrique borné. Supposons donc A borné. Le théorème 4.1, appliqué à la partition $adh(A) = A \cup (adh(A)-A)$ donne une triangulation $\Phi : adh(A) \longrightarrow K$ telle que, $\Phi(A)$ soit une réunion de simplexes ouverts de K.

Le théorème 4.1 est une conséquence immédiate du

LEMME 4.2 : *Soient* A *un semi-algébrique compact de* $\mathbb{R}^n$, $A = \cup A_k$ *une partition finie de* A *en semi-algébriques, et supposons que* A *et les* A_k *soient donnés par des combinaisons de conditions de signe sur des polynômes tous unitaires en* X_n. *Soit* $\pi : \mathbb{R}^n \longrightarrow \mathbb{R}^{n-1}$ *la projection qui oublie le dernier facteur,* $B = \pi(A)$. *Il existe des triangulations de* A *et* B *données par des homéomorphismes semi-algébriques :*

$$\Phi : A \longrightarrow K = \cup S_i \subset \mathbb{R}^n$$

$$\Psi : B \longrightarrow L = \cup T_j \subset \mathbb{R}^{n-1}$$

telles que le diagramme

$$\begin{array}{ccc} A & \xrightarrow{\ \Phi\ } & K \\ \pi\downarrow & & \downarrow\pi \\ B & \xrightarrow{\ \Psi\ } & L \end{array} \qquad \textit{commute,}$$

que pour tout i, $\pi(S_i)$ *est un simplexe ouvert* T_j, *et que chaque* A_k *est réunion de* $\Phi^{-1}(S_i)$.

Preuve : Pour n fixé, le théorème 4.1 est bien une version affaiblie du lemme 4.2 puisqu'on peut toujours, par un changement de variables du genre de celui effectué au début de la preuve de 3.2, rendre un nombre fini de polynômes en $X_1,\ldots,X_n$ tous unitaires en X_n.

Supposant maintenant le théorème 4.1 acquis pour un n donné, nous allons montrer le lemme 4.2 pour n+1. Soient $P_1,\ldots,P_\ell$ les polynômes unitaires en X_{n+1} qui interviennent dans les définitions de A et des A_k, auxquels on a ajouté toutes les dérivées non constantes à tous les ordres par rapport à X_{n+1}. L'application du "saucissonnage" nous donne une partition finie de B en semi-algébriques B_j, telle qu'au-dessus de chaque B_j, les racines de $P_1,\ldots,P_\ell$ sont données par des fonctions semi-algébriques continues $\xi_1 < \ldots < \xi_{p_j}$. Le théorème 4.1 pour n montre que l'on peut supposer, quitte à diviser chaque B_j en un nombre fini de semi-algébriques plus petits, que l'on a une triangulation $\Psi : B \longrightarrow L = \cup T_j$ avec $B_j = \Psi^{-1}(T_j)$. Dans la suite, on identifiera sans vergogne B_j avec le simplexe ouvert T_j. Il est bon de remarquer les deux faits suivants :

a) Si ξ est une racine au-dessus de B_j, et $B_{j'}$ une face itérée de B_j, ξ a une limite ξ' au-dessus de $B_{j'}$ (ceci parce que les $P_1,\ldots,P_\ell$ sont unitaires en X_{n+1}). On peut de plus supposer - quitte à subdiviser la triangulation - que deux racines distinctes sur B_j ont des limites distinctes en au moins un des sommets de B_j.

b) Si ζ' est une racine au-dessus d'une face itérée $B_{j'}$ de B_j, ζ' est limite d'au moins une racine ζ sur B_j.

Le fait (a) est une conséquence de la démonstration du lemme de Thom à plusieurs variables. Le graphe de ξ est donné au-dessus de B_j par une conjonction de conditions de signes sur chacun des $P_1,\ldots,P_\ell$. L'adhérence de ce graphe au-dessus de $B_{j'}$ s'obtient en relachant les inégalités strictes. Comme ce ne peut être le "segment" fermé compris entre deux racines au-dessus de $B_{j'}$, cette adhérence est forcément le graphe d'une racine ξ' au-dessus de $B_{j'}$. Le fait (b) vient du théorème des fonctions implicites (ζ' est racine simple d'un des $P_1,\ldots,P_\ell$) et du fait (a).

Venons-en maintenant à la construction de Φ. Φ se construit au-dessus de chaque

B_j en commençant par les sommets de B, et en grimpant dans la dimension des simplexes. On suppose d'abord que l'on a ordonné l'ensemble de tous les sommets de B. On se fixe un j, et on suppose que Φ a été construit au-dessus de toutes les faces itérées de B_j. Soient $(\underline{a}_o,\ldots,\underline{a}_k)$ les sommets de B_j énumérés dans l'ordre. Soient ξ et ζ deux racines consécutives au-dessus de B_j, dont les limites au-dessus des sommets sont respectivement $\beta_o,\ldots,\beta_k$ et $\gamma_o,\ldots,\gamma_k$. On note $\underline{b}_i$ (resp. $\underline{c}_i$) le point $(\underline{a}_i,\beta_i)$ (resp. $(\underline{a}_i,\gamma_i)$) de $\mathbb{R}^{n+1}$,

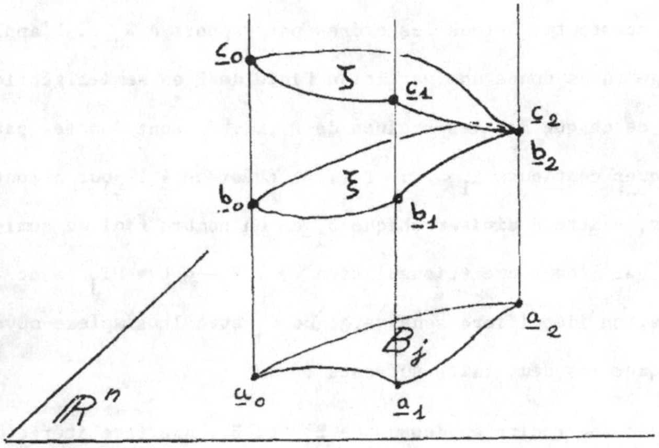

Soit alors $\underline{x} \in B_j$, et $t_o,\ldots,t_k$ les coordonnées barycentriques de $\Psi(\underline{x})$. Si $\xi(x) \leq y \leq \zeta(x)$, on pose

$$\Phi(\underline{x},y) = (\Psi(\underline{x}),z)$$

où

$$z = \left(\frac{y-\xi(\underline{x})}{\zeta(x)-\xi(\underline{x})}\right) \sum_{i=o}^{k} t_i\,\gamma_i + \left(\frac{\zeta(\underline{x})-y}{\zeta(x)-\xi(x)}\right) \sum_{i=o}^{k} t_i\,\beta_i$$

Φ envoie donc $\xi(B_j)$ sur le simplexe $(\underline{b}_o,\ldots,\underline{b}_k)$ et $\zeta(B_j)$ sur le simplexe $(\underline{c}_o,\ldots,\underline{c}_k)$, et pour chaque $\underline{x}$ de B_j, Φ envoie l'intervalle $[(\underline{x},\xi(\underline{x}),(\underline{x},\zeta(\underline{x})]$ de façon affine sur l'intervalle de la fibre de π au-dessus de $\Psi(\underline{x})$ qui est compris entre les deux simplexes ci-dessus. Il faut trianguler la tranche entre les deux simplexes $(b_o,\ldots,\underline{b}_k)$ et $(\underline{c}_o,\ldots,c_k)$. Ceci se fait en considérant les simplexes

$$(\underline{b}_o,\ldots,\underline{b}_m,\underline{c}_m,\ldots,\underline{c}_k)$$

où m est tel que $\underline{b}_m \neq \underline{c}_m$ (il en existe au moins un) et leurs faces situées au-dessus de $(\underline{a}_o,\ldots,\underline{a}_k)$.

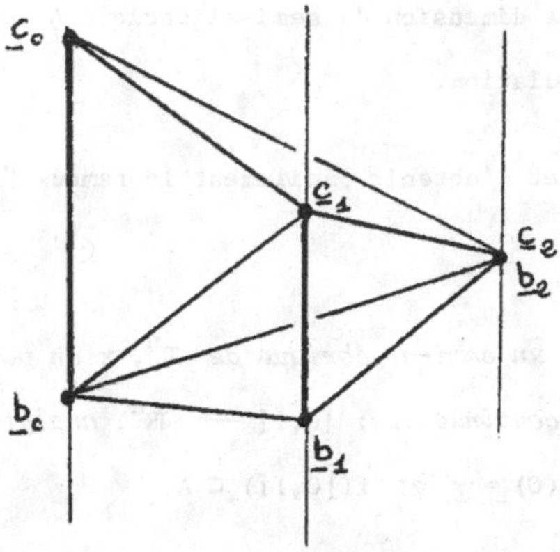

Puisque les limites de ξ et ζ au-dessus d'une face itérée de B_j sont deux racines consécutives ou confondues, et puisque l'on a à chaque étape respecté l'ordre des sommets, les autres faces itérées de ces simplexes ont déjà été construites au-dessus des faces itérées de B_j.

Comme A est compact, un point de $A \cap \pi^{-1}(B_j)$ est bien compris entre deux racines consécutives, et on a bien défini Φ sur $A \cap \pi^{-1}(B_j)$ tout entier.

Au total, on a construit un complexe $K = \cup S_i$ de $\mathbb{R}^{n+1}$ et une bijection semi-algébrique $\Phi : A \longrightarrow K$ telle que le diagramme

$$\begin{array}{ccc} A & \xrightarrow{\Phi} & K \\ \pi \downarrow & & \downarrow \pi \\ B & \xrightarrow{\Psi} & L \end{array}$$

commute, que chaque A_k soit réunion de $\Phi^{-1}(S_i)$, et que chaque $\pi(S_i)$ soit un T_j. Par ailleurs, Φ restreint à chaque $A \cap \pi^{-1}(B_j)$ est clairement un homéomorphisme. Il ne reste plus à voir que Φ "passe bien aux faces", et pour cela il suffit dans la définition ci-dessus de Φ, de faire tendre les coordonnées barycentriques t_i convenables vers 0. ∎

<u>Remarque 1</u> : On peut combiner triangulation et stratification, de telle sorte que $A = \cup \Phi^{-1}(S_i)$ soit une stratification (cf. proposition 3.5), et que Φ restreint à $\Phi^{-1}(S_i)$ soit un isomorphisme analytique (et même de Nash). La triangulation construite plus haut a bien ces propriétés.

Remarque 2 : On retrouve la dimension du semi-algébrique A comme sup de la dimension des simplexes d'une triangulation.

La triangulation permet d'obtenir facilement le fameux "lemme de sélection des courbes" ([20] [19]) :

PROPOSITION 4.3 : *Soient* A *un semi-algébrique de* $\mathbb{R}^n$, $\underline{x}$ *un point de l'adhérence de* A. *Il existe une fonction continue* $f : [0,1[\longrightarrow \mathbb{R}^n$, *analytique sur* $]0,1[$ *(et même de Nash), telle que* $f(0) = \underline{x}$ *et* $f(]0,1[) \subset A$.

Preuve : On peut supposer que A est borné, et trianguler adh A de telle façon que $\Phi(\underline{x})$ soit un sommet et $\Phi(A)$ une réunion de simplexes ouverts. $\Phi(\underline{x})$ est donc le sommet d'un de ces simplexes, et on peut bien trouver un segment $[\Phi(\underline{x}),y[$, tel que $]\Phi(\underline{x}),\underline{y}[$ soit à l'intérieur du simplexe. L'image réciproque de ce segment par Φ donne la courbe cherchée (modulo la remarque 1 ci-dessus). ∎

V - TRIVIALITE LOCALE DES MORPHISMES SEMI-ALGEBRIQUES.

La trivialité locale est montrée par Hardt dans [12], et on trouve des énoncés moins complets chez Varchenko [25] et Wallace [27] (voir aussi [21]). Le résultat énoncé ci-dessous a été montré par H. Delfs et M. Knebusch pour un corps réel clos quelconque dans [17b]. Ils l'utilisent de très jolie manière pour obtenir, entre autres, la généralisation de la majoration de Milnor de la somme des nombres de Betti d'une variété algébrique réelle au cas d'un corps réel clos quelconque. Je remercie M. Knebusch pour m'avoir indiqué le résultat de Hardt.

THEOREME 5.1 : *Soient* $f : A \longrightarrow B$ *un morphisme semi-algébrique continue entre deux semi-algébriques* A *et* B, $A = \cup A_k$ *une partition finie de* A *en semi-algébriques. Il existe une partition finie de* B *en semi-algébriques* $B = \cup B_\ell$, *et pour chaque* ℓ *un homéomorphisme semi-algébrique*

$$\begin{array}{ccc} \Theta_\ell : f^{-1}(B_\ell) & \xrightarrow{\sim} & B_\ell \times F_\ell \\ & {\scriptstyle f}\searrow \quad \swarrow & \\ & B_\ell & \end{array}$$

au-dessus de B_ℓ*, où* F_ℓ *est un semi-algébrique qui admet une partition* $F_\ell = \cup F_{\ell,k}$ *en semi-algébriques (éventuellement vides), telle que* $\theta_\ell(f^{-1}(B_\ell) \cap A_k) = B_\ell \times F_{\ell,k}$.

On peut, dans le théorème, raffiner la partition $B = \cup B_\ell$ en une stratification. Si la dimension de B est n, la réunion des strates de dimension n est un ouvert semi-algébrique U de B. Les composantes connexes de U sont précisément les strates de dimension n, et la dimension de B-U est strictement inférieure à n. On a donc :

COROLLAIRE 5.2 : *Soit* $f : A \longrightarrow B$ *un morphisme semi-algébrique continu entre deux semi-algébriques* A *et* B. *Il existe un semi-algébrique* B' *fermé dans* B, *de dimension strictement inférieure à celle de* B, *tel que sur chaque composante connexe* U_ℓ *de* $B - B'$, f *soit semi-algébriquement trivial : il existe un semi-algébrique* F_ℓ *et un homéomorphisme semi-algébrique*

$$\begin{array}{ccc} \theta_\ell : f^{-1}(U_\ell) & \xrightarrow{\ \sim\ } & U_\ell \times F_\ell \\ & {}_{f}\searrow \quad \swarrow & \\ & U_\ell & \end{array}$$

au-dessus de U_ℓ.

Venons-en maintenant à la preuve du théorème. On peut supposer que f est surjectif puisque f(A) est semi-algébrique d'après 2.10. En utilisant la triangulation et la remarque qui suit le théorème 4.1, on se ramène au cas où B est un simplexe ouvert de dimension n supposé plongé dans $\mathbb{R}^n$. On peut supposer que A est borné dans un $\mathbb{R}^m$, puis remplacer A par le graphe de f qui est borné dans $\mathbb{R}^{m+n}$. On en est à la situation suivante : A borné dans $\mathbb{R}^{m+n}$, $\pi_m : \mathbb{R}^{m+n} \longrightarrow \mathbb{R}^n$ la projection qui oublie les m premiers facteurs, $B = \pi_m(A)$ de dimension n. Si on remplace A par adh(A) avec la partition $\mathrm{adh}(A) = (\mathrm{adh}(A)-A) \cup \bigcup_k A_k$, et B par adh(B) on est ramené à montrer le lemme suivant :

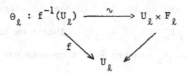

LEMME 5.3 : *Soient* A *un semi-algébrique compact de* $\mathbb{R}^{m+n}$, $A = \bigcup_k A_k$ *une partition*

finie de A *en semi-algébriques,* $\pi_m : \mathbb{R}^{m+n} \longrightarrow \mathbb{R}^n$ *la projection qui oublie les* m *premiers facteurs. On suppose que* $B = \pi_m(A)$ *est de dimension* n. *Alors la conclusion du théorème* 5.1 *est vérifiée (avec* $f = \pi_m$*).*

Preuve : Par induction sur l'ordre lexicographique des couples (n,m) : A et les A_k sont donnés par des combinaisons booléennes de conditions de signe sur un nombre fini de polynômes $P_i(X_1,\dots,X_m,Y_1,\dots,Y_n)$. Quitte à faire un changement de variables linéaire sur les $\underline{X}$, on peut supposer que chaque P_i s'écrit :

$$X_m^{d_i} Q_{o,i}(\underline{Y}) + X_m^{d_i-1} Q_{1,i}(X_1,\dots,X_{m-1},\underline{Y}) + \dots$$

Le semi-algébrique $\{\underline{y} \in B \mid \prod_i Q_{o,i}(y) = 0\}$ est de dimension strictement plus petite que n, on peut, après une réduction analogue à celle effectuée avant l'énoncé du lemme, utiliser l'hypothèse d'induction au-dessus de ce semi-algébrique. Sur une composante connexe du complémentaire dans B, les $Q_{o,i}$ gardent un signe constant non nul. On peut ainsi se ramener au cas où les $Q_{o,i}$ gardent un signe constant non nul sur B. Le changement de variable $X'_m = X_m \prod_i Q_{o,i}(\underline{Y})$ induit un homéomorphisme semi-algébrique sur A, et les polynômes P_i peuvent être remplacés par des polynômes unitaires en X'_m. On se ramène ainsi au cas où tous les P_i sont unitaires en X_m. On n'a pas fait attention dans ce qui précède à préserver la compacité de A. A reste bien sûr borné. Son adhérence sera compacte, et elle est aussi donnée par une combinaison de conditions de signe sur des polynomes unitaires en X_m (les polynômes que l'on pourrait avoir besoin d'ajouter aux P_i pour obtenir une liste séparante étant aussi unitaires en X_m , cf. la démonstration du lemme de Thom à plusieurs variables). Quitte à remplacer A par adh(A) avec la partition donnée par adh(A)-A et les A_k , on a toujours A compact, et donné (ainsi que la partition de A) par des combinaisons de conditions de signe sur des polynômes unitaires en X_m. On peut alors appliquer le lemme 4.2 à la projection $\pi : \mathbb{R}^{m+n} \longrightarrow \mathbb{R}^{m-1+n}$ qui oublie le m^{ème} facteur. On a ainsi des triangulations :

$$\Phi : A \xrightarrow{\sim} K = \cup S_i \subset \mathbb{R}^{m+n}$$

$$\Psi : C = \pi(A) \xrightarrow{\sim} L = \cup T_i \subset \mathbb{R}^{m-1+n}$$

telles que $\pi\circ\Phi = \Psi\circ\pi$, que chaque $\pi(S_i)$ est un T_j, et que chaque A_k est réunion de $\Phi^{-1}(S_i)$. On peut maintenant appliquer l'hypothèse d'induction à $C = \cup\Psi^{-1}(T_j)$ et à la projection $\pi_{m-1} : \mathbb{R}^{m-1+n} \longrightarrow \mathbb{R}^n$. On obtient une partition finie de $B = \pi_{m-1}(C)$ en semi-algébriques $B = \cup B_\ell$ et des homéomorphismes semi-algébriques :

$$\begin{array}{ccc} \rho_\ell : \pi_{m-1}^{-1}(B_\ell) & \xrightarrow{\sim} & B_\ell \times G_\ell \\ & \searrow \quad \swarrow & \\ & B_\ell & \end{array}$$

avec $G_\ell = \cup G_{\ell_j}$ et $\rho_\ell(\pi_{m-1}^{-1}(B_\ell) \cap \Psi^{-1}(T_j)) = B_\ell \times G_{\ell_j}$. Fixons ℓ, et soit $\underline{y}_o$ un point de B. Posons $F_\ell = \pi_m^{-1}(\underline{y}_o)$ et $F_{\ell,k} = \pi_m^{-1}(\underline{y}_o) \cap A_k$. On peut bien supposer que $G_\ell = \pi_{m-1}^{-1}(\underline{y}_o)$ et que si $\pi_{m-1}(\underline{z}) = \underline{y}_o$, on a $\rho_\ell(\underline{z}) = (\underline{y}_o,\underline{z})$.

Il nous faut maintenant construire $\Theta_\ell : \pi_m^{-1}(B_\ell) \longrightarrow B_\ell \times F_\ell$.
Soient $\underline{x} \in \pi_m^{-1}(B_\ell)$, $\underline{y} = \pi_m(x) \in B_\ell$, $\underline{z} = \pi(\underline{x})$. $\Phi(\underline{x})$ est dans un certain S_{i_o}, et $\Psi(\underline{z})$ dans $T_{j_o} = \pi(S_{i_o})$. Soient $\underline{a}_o,\ldots,\underline{a}_t$ les sommets de T_{j_o}, $\underline{b}_{o,o},\ldots,\underline{b}_{o,s(o)},\underline{b}_{1,o},\ldots,\underline{b}_{t,s(t)}$ ceux de S_{i_o} avec $\pi(\underline{b}_{i,j}) = \underline{a}_i$. Notons $\lambda_{o,o},\ldots,\lambda_{t,s(t)}$ les coordonnées barycentriques de $\Phi(\underline{x})$ dans S_{i_o} ; celles de $\Psi(\underline{z})$ dans T_{j_o} sont donc $\lambda_{o,o}+\ldots+\lambda_{o,s(o)},\ldots,\lambda_{t,o}+\ldots+\lambda_{t,s(t)}$. On sait que $\rho_\ell(\underline{z}) = (\underline{y},\underline{u})$ avec $\underline{u} \in \pi_{m-1}^{-1}(y_o) \cap \Psi^{-1}(T_j)$. Soient $\mu_1,\ldots,\mu_t$ les coordonnées barycentriques de $\Psi(\underline{u})$ dans T_j. Soit enfin $\underline{v}$ le point de $\Phi^{-1}(S_{i_o})$, tel que les coordonnées barycentriques de $\Phi(\underline{v})$ dans S_{i_o} sont $\eta_{i,j} = \lambda_{i,j} \dfrac{\mu_i}{\sum_j \lambda_{i,j}}$. On a bien sûr $\pi(\underline{v}) = \underline{u}$.

La construction paraitra peut être plus explicite sur le dessin suivant, où l'on a identifié S_{i_o} et T_{j_o} à leur images réciproques par Φ et Ψ

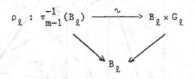

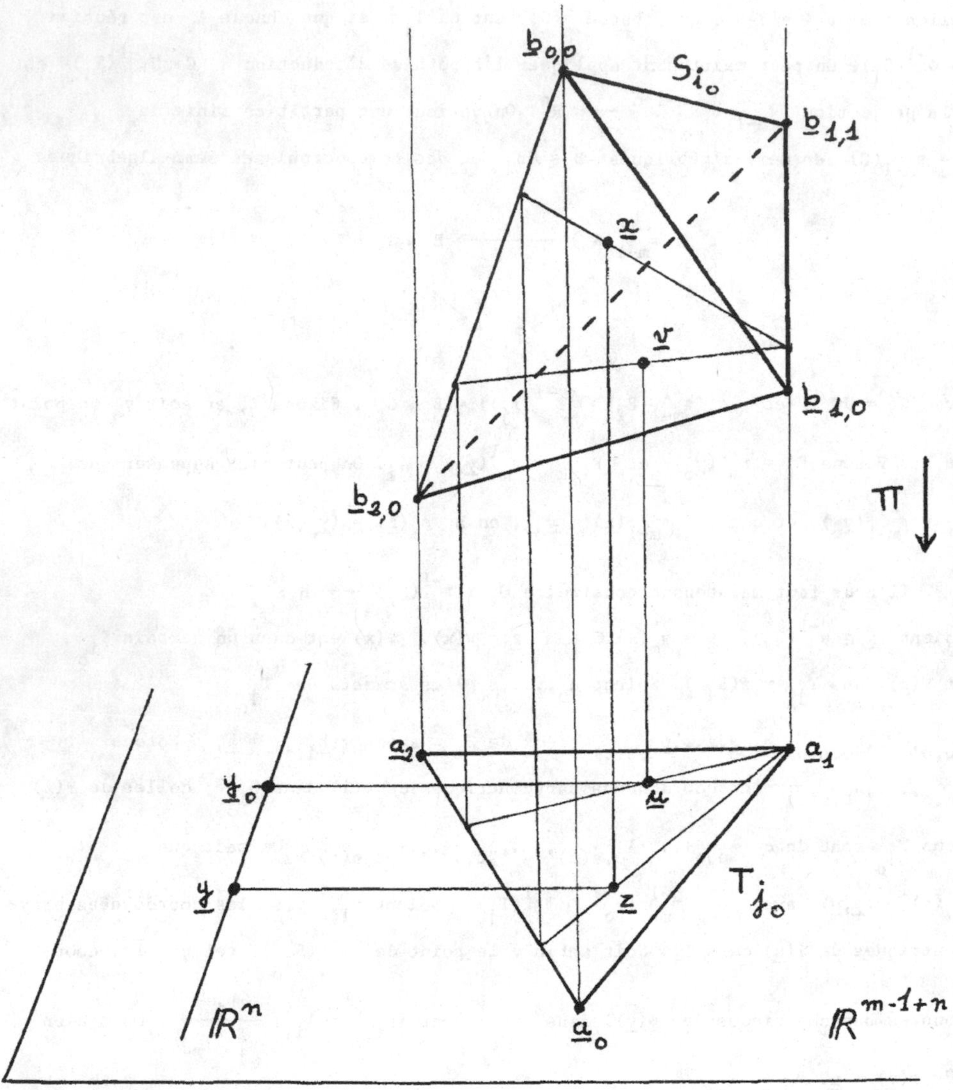

On pose alors $\Theta_\ell(\underline{x}) = (\underline{y},\underline{v})$. Il est clair que Θ_ℓ est un homéomorphisme semi-algébrique de $S_{i_0} \cap \pi_m^{-1}(B_\ell)$ sur $B_\ell \times (F_\ell \cap S_{i_0})$. La seule chose qui reste à voir est que Θ_ℓ "passe bien aux faces" (en restant bien sûr dans $\pi_m^{-1}(B_\ell)$).

Par exemple, si $\Phi'(\underline{x}')$ est sur une face itérée de S_{i_0} avec $\underline{x}' \in \pi_m^{-1}(B_\ell)$, et si $\underline{x}$ tend vers $\underline{x}'$ dans $\Phi^{-1}(S_{i_0}) \cap \pi_m^{-1}(B_\ell)$, alors :

- les coordonnées barycentriques (λ_{i_j}) tendent vers (λ'_{i_j}) (certaines de ces dernières sont nulles) ;

- $\underline{z}$ tend vers $\underline{z}'$, donc $\underline{u}$ tend vers $\underline{u}'$, et les coordonnées barycentriques (μ_i) tendent vers (μ'_i) (on a $\mu'_i = 0$ ssi $\sum_j \lambda'_{ij} = 0$) ;

- les coordonnées barycentriques (η_{ij}) tendent donc vers (η'_{ij}), où $\eta'_{ij} = 0$ si $\sum_j \lambda'_{i,j} = \mu'_i = 0$, ce qui montre que $\Theta_\ell(\underline{x})$ tend vers $\Theta_\ell(\underline{x}')$.

Un raisonnement du même genre s'applique à Θ_ℓ^{-1}. La démonstration est terminée. ∎

Remarque : Le théorème 5.1 et son corollaire 5.2 ne sont plus valables si l'on supprime l'hypothèse de continuité du morphisme f. Soit par exemple $f : \mathbb{R}^2 \longrightarrow \mathbb{R}$ le morphisme semi-algébrique donné par $f(x,y) = \frac{1}{y}$ si $y \neq 0$ et $f(x,y) = x$ si $y = 0$. Le corollaire 5.2 nous donnerait un $M > 0$ tel que $f^{-1}([M,+\infty[)$ soit homéomorphe à $f^{-1}(M) \times [M,+\infty[$. Or $f^{-1}([M,+\infty[)$ est connexe tandis que $f^{-1}(M)$ est la somme disjointe d'une droie et d'un point.

Le seul endroit où l'on a utilisé la continuité de f dans la démonstration est le remplacement de A par le graphe de f ; ce dernier n'est pas homéomorphe à A si f n'est pas continu. Il reste néanmoins, que pour tout $\underline{y} \in B$, la fibre $f^{-1}(\underline{y})$ est homéomorphe à la fibre de la projection $\pi_m^{-1}(\underline{y}) \cap$ graphe(f) . On obtient donc, dans le cas où f n'est pas continu, le résultat plus faible suivant : on peut trouver une partition finie de B en semi-algébrique B_ℓ , telle que les fibres de f en deux points quelconques d'un même B_ℓ soient homéomorphes semi-algébriquement.

La trivialité locale des morphismes semi-algébriques, permet d'obtenir la structure conique locale des semi-algébriques comme dans [21].

PROPOSITION 5.4 : *Soient* A *un semi-algébrique de* $\mathbb{R}^n$, $\underline{x}$ *un point non isolé de* A. *Il existe un* $\varepsilon > 0$ *tel que l'intersection de la boule fermée* $\overline{B}(\underline{x},\varepsilon)$ *de centre* $\underline{x}$ *et de rayon* ε *avec* A *soit semi-algébriquement homéomorphe au cone de sommet* $\underline{x}$ *et de base l'intersection de* A *avec la sphère* $S(\underline{x},\varepsilon)$ *de centre* $\underline{x}$ *et de rayon* ε

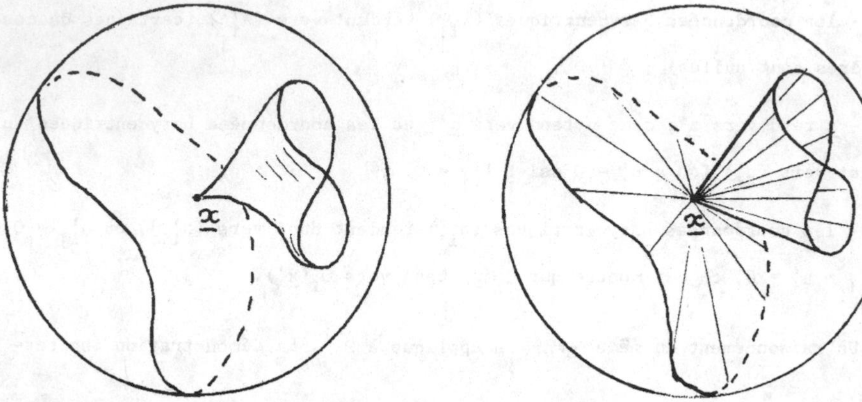

$A \cap \bar{B}(x,\varepsilon)$

Preuve : Appliquons le théorème 5.1 à l'application

$$f : y \longrightarrow \|y - x\|$$

de A dans $\mathbb{R}$. On peut trouver $\varepsilon > 0$, tel que $f^{-1}(]0,\varepsilon])$ soit semi-algébriquement homéomorphe à $]0,\varepsilon] \times f^{-1}(\varepsilon)$, ce qui donne bien le résultat voulu. ∎

La structure conique locale est utile dans l'étude de la topologie des singularités de variétés algébriques réelles ou complexes (cf. Milnor [20]).

REFERENCES

[1] M. ARTIN, B. MAZUR : On periodic points, Annals of Math. 81 (1965), 82-99.

[2] J. BOCHNAK, G. EFROYMSON : Real algebraic geometry and the 17th Hilbert problem, Math. Annalen 251 (1980), 213-242.

[2b] J. BOCHNAK, G. EFROYMSON : An Introduction to Nash Functions, ce volume.

[3] G.W. BRUMFIEL : Partially ordered rings and semi-algebraic geometry, Cambridge University Press, 1979.

[4] P.J. COHEN : Decision procedures for real and p-adic fields, Commun. Pure & Applied Math. 22 (1969), 131-151.

[5] M. COSTE, M.-F. COSTE-ROY : Topologies for real algebraic geometry, dans Topos theoretic methods in geometry, Aarhus Univ. Various Publication Séries 30, A. Kock éd., 1979.

[6] M. COSTE, M.-F. ROY : La topologie du spectre réel, à paraître dans Contemporary Mathematics.

[7] M. COSTE : Ensembles semi-algébriques et fonctions de Nash, Prépublications mathématiques 18, Université Paris-Nord, 1981.

[8] H. DELFS : Kohomologie affiner semi-algebraischer Räume, Thèse, Univ. Regensburg, 1980.

[9] C. DELZELL : A constructive, continous solution to Hilbert's 17th problem, and other result in semi-algebraic geometry, These, Stanford Univ., 1980.

[10] G. EFROYMSON : A Nullstellensatz for Nash rings, Pacific J. Math. 54 (1974), 101-112.

[11] G. EFROYMSON : Substitution in Nash functions, Pacific J. Math. 63 (1976), 137-145.

[12] R. HARDT : Semi-algebraic local triviality in semi-algebraic mappings, American American Journal of Mathematics 102 (1980), 291-302.

[13] H. HIRONAKA : Triangulation of semi-algebraic sets, Proc. Symp. in Pure Math. 29 (A.M.S. 1975), 165-185.

[14] L. HORMANDER : Linear partial differential operators, Springer-Verlag.

[15] J. HOUDEBINE : Lemme de séparation, multigraphié, Univ. Rennes (1980).

[16] H. DELFS, M. KNEBUSCH : Semi-algebraic topology over a real closed field I : Paths and components in the set of rational points of an algebraic variety, Math. Z. 177 (1981), p. 107-129.

[17] H. DELFS, M. KNEBUSCH : Semi-algebraic topology over a real closed filed II : Basic theory of semi-algebraic spaces, Math. Z.

[17b] H. DELFS, M. KNEBUSCH : On the homology of algebraic varieties over real closed fields, preprint.

[18] S. LOJASIEWICZ : Triangulation of semi-analytic sets, Ann. Scuola Norm. Sup. Pisa 18 (1964), 449-474.

[19] S. LOJASIEWICZ : Ensembles semi-analytiques, multigraphié I.H.E.S., 1965.

[20] J. MILNOR : Singular points of complex hypersurfaces, Annals of Math. Studies 61, Princeton University Press 1968.

[21] S. PNEUMATIKOS : Introduction à la géométrie algébrique réelle, d'après un séminaire de J. Bochnak, Univ. Dijon, 1977.

[22] T. RECIO : Actas de la IV reunion de Mathematicos de Expresion Latina, Mallorca, 1977.

[23] A. SEIDENBERG : A new decision method for elementary algebra, Ann. of Math. 60 (1954), 365-374.

[24] A. TARSKI : A decision method for elementary algebra and geometry. Univ. of Calif. Press, 1951.

[25] VARCHENKO : Equisingularités topologiques, Izvestia 36 (1972), 957-1019.

[26] R. WALKER : Algebraic curves, Dover.

[27] WALLACE : Linear sections of algebraic varieties, Indiana Univ. Math. J. 20 (1971), 1153-1162.

[28] H. WHITNEY : Elementary structures of real algebraic varieties, Annals of Math. 66 (1957), 545-556.

B.P. 11573
Niamey
NIGER

THE REAL HOLOMORPHY RING AND SUMS OF 2n-TH POWERS

Eberhard Becker

This paper is concerned with the role the real holomorphy ring of a formally real field K plays in the study of the sums of $2n$-th powers in K. By definition, the holomorphy ring H of K is the intersection of all valuation rings of K with a formally real residue field. At a first look there seems to exist little or no connection with the sums of $2n$-th powers in K. But, in fact, the ideal structure of H and the structure of its group $\mathbb{E}$ of units have great influence.

Set $\sum_1^k K^{2n} = \{\sum_1^k x_i^{2n} \mid x_1,\ldots,x_k \in K\}$, $\Sigma K^{2n} = \bigcup \sum_1^k K^{2n}$, $\mathbb{E}^+ = \mathbb{E} \cap \Sigma K^2$.

In the first section we describe the group $\mathbb{E}^+$ and derive the two basic results:

$$\mathbb{E}^+ \subset \bigcap_n \Sigma K^{2n} \quad , \quad \Sigma K^{2n} = \mathbb{E}^+ \cdot (\Sigma K^2)^n \; .$$

It is further shown that $\mathbb{E}^+ = \bigcap_n \Sigma K^{2n}$ holds if K is, e.g., a function field over $\mathbb{Q}$ or $\mathbb{R}$. The proof of this latter statement depends on the existence of real prime divisors in certain function fields.

The second section is devoted to quantitative investigations. More precisely, we shall study the $2n$-th <u>Pythagoras number</u> of K defined as

$$P_{2n}(K) = \min\{k \in \mathbb{N} \cup \{\infty\} \mid \sum_1^k K^{2n} = \Sigma K^{2n}\} \; .$$

Hereby, certain generalized Hilbert's identities play a decisive role. Apart from their existence very little is known about them.

But a deeper insight into their construction is absolutely necessary to get better estimates for these Pythagoras numbers. The main result in this section states that all the higher Pythagoras numbers $P_{2n}(K)$ are finite provided the ordinary Pythagoras number $P_2(K)$ is finite. In particular $P_{2n}(K) < \infty$ for any function field $K|\mathbb{R}$ or $K = \mathbb{Q}(X)$ and all $n \in \mathbb{N}$.

As stated above, we have $(\Sigma K^2)^n \subset \Sigma K^{2n}$. In the third section we are dealing with the question of equality of these two sets. Set $S(K) = \{n \mid (\Sigma K^2)^n = \Sigma K^{2n}\}$. It turns out that $S(K)$ is a multiplicative semigroup generated by a certain set of primes. Moreover, this set of primes can be prescribed arbitrarily. The semigroup $S(K)$ is of particular interest for a pythagorean field K; then $n \in S(K)$ amounts to saying $\Sigma K^{2n} = K^{2n}$. A field with this latter property is called n-pythagorean. The well known result of Diller and Dress on finite pythagorean extensions likewise holds in the present situation: if L is a finite n-pythagorean extension of K, then K is also n-pythagorean. Finally, strictly n-pythagorean fields are introduced. Setting $\tilde{S}(K) = \{n \mid K$ is strictly n-pythagorean$\}$, the main results can be stated as follows: $\tilde{S}(K)$ is either empty or a semigroup generated by a set of primes; every set of primes occur; and if $L|K$ is a finite extension, then $\tilde{S}(L) \subset \tilde{S}(K)$.

1. The real holomorphy ring

Throughout this work we adopt the following notations:

K a formally real field,

H(K) the real holomorphy ring of K ,

M(K) the compact Hausdorff space of all places

$\lambda : K \to \mathbb{R} \cup \infty$.

When no confusion is to be suspected, we shall drop the reference to K ; e.g., we shall write H instead of H(K) .

By definition, H is the intersection of all valuation rings V with a formally real residue field. (For short, these valuation rings are called real.) We freely make use of the results of [B4 , Sections 2,3], [Sch 1], [Sch 2] concerning the holomorphy ring. In addition to the above conventions we introduce the notations

$\mathbb{E}(K)$ = group of units of H(K) ,

$\mathbb{E}^+(K) = \mathbb{E}(K) \cap \Sigma K^2$.

The elements of $\mathbb{E}(K)$ are called the formally real units of K - for short, real units. Accordingly, $\mathbb{E}^+(K)$ consists of the totally positive real units of K .

(1.1) Remark. By [B-H-R, (2.10)] we have for $a \in K^\times$:

$a \in \mathbb{E}^+ \iff \chi(a) = 1$ for all $\chi \in \mathrm{Sgn}(K)$.

Given a topological space X , denote by B(X) the attached group of the clopen subsets of X under the symmetric difference Δ . Hereby, $U \Delta V = (U \setminus V) \cup (V \setminus U)$ and "clopen" means: closed and open at the same time. Set

$$\mathbb{R}_+ := \{r \in \mathbb{R} \mid r \geq 0\} \ , \ \mathbb{R}_- = \{r \in \mathbb{R} \mid r \leq 0\} \ , \ \mathbb{R}_+^\times = \mathbb{R}_+ \setminus \{0\} \ ,$$

equally we deal with $\mathbb{Q}_+$, $\mathbb{Q}_+^\times$.

(1.2) Proposition.

i) The assignment $\varepsilon \mapsto \varepsilon^{-1}(\mathbb{R}_-)$ induces an isomorphism $\mathbb{E}/\mathbb{E}^+ \simeq B(M)$,

ii) $\mathbb{E}^+ = \{r \frac{s+q}{t+q} \mid r,s,t \in \mathbb{Q}_+^\times \ , \ q \in \Sigma K^2\}$.

Proof. We make use of the results on the natural representation $H \to C(M,\mathbb{R})$, $a \mapsto \hat{a}$, $\hat{a}(\lambda) = \lambda(a)$, proved in [B 4, (2.20)] or [Sch 2, (1.6)] . Furthermore, we need the following statements on elements of H :

$$(1.3) \quad \begin{array}{ll} \text{i)} & a \in \mathbb{E} \iff \hat{a}(\lambda) \neq 0 \ \text{ for all } \lambda \in M \ , \\ \text{ii)} & a \in \mathbb{E}^+ \iff \hat{a}(\lambda) > 0 \ \text{ for all } \lambda \in M \ . \end{array}$$

Statement ii) implies the first one: consider a^2 instead of a . Given $a \in \mathbb{E}^+$, we apply the fact $\lambda(\Sigma K^2) \subset \mathbb{R}_+ \cup \{\infty\}$ for $\lambda \in M$ to see $\lambda(a) = \hat{a}(\lambda) > 0$. Conversely, choose, by the compactness of M , $r \in \mathbb{Q}_+^\times$ such that $\hat{a} - r > 0$ everywhere on M . By [B 4, (2.20),ii)] , we get $a = \frac{1}{2}r + q$, $q \in \Sigma K^2$. Since H is generated by the elements $\frac{1}{1+q'}$, $q' \in \Sigma K^2$, we finally see $a \in \mathbb{E}^+$.

To prove the first statement, note that the mapping $\varepsilon \mapsto \varepsilon^{-1}(\mathbb{R}_-)$ is a homomorphism with kernel $\mathbb{E}^+$, according to (1.3) . Given a clopen subset U of M , let χ_U denote the continuous characteristic function of U and choose $a \in H$ such that $\|2\chi_U - 1 + \hat{a}\| < \frac{1}{2}$ holds. Hereby, $\| \ \|$ denotes the maximum norm on $C(M,\mathbb{R})$. The approximation is possible by [B 4, (2.20), i)] . From (1.3) we obtain $a \in \mathbb{E}$ and we have $U = \hat{a}^{-1}(\mathbb{R}_-)$. Thus, $\mathbb{E}/\mathbb{E}^+ \simeq B(M)$ is proved. To prove that $r\,\frac{s+q}{t+q} \in \mathbb{E}^+$ holds, we make use of the symmetry in s and t . It is, therefore, enough to show that $r\,\frac{s+q}{t+q} \in V$ for any real valuation ring V of K . But this is easily checked. Conversely, pick $\varepsilon \in \mathbb{E}^+$. Then $\hat{\varepsilon} > 0$ on M everywhere. By (1.3) and the compactness of M we find $n \in \mathbb{N}$ with $\varepsilon - \frac{1}{n} = \eta \in \mathbb{E}^+$. Again by applying (1.3) , we find $m \in \mathbb{N}$ with $\eta^{-1} = \frac{1}{m} + q'$, $q' \in \Sigma K^2$. Putting it all together, we obtain $\varepsilon = \frac{1}{n} + \frac{m}{1+mq'} = r\,\frac{s+q}{t+q}$, as required.

We add the following obvious corollary to (1.2):

<u>(1.4) Corollary.</u> $\mathbb{E}^+$ has a finite index in $\mathbb{E}$ iff the number of connected components of M is finite. If there are s components, then $[\mathbb{E} : \mathbb{E}^+] = 2^s$.

<u>(1.5) Remark.</u> In the geometric situation of a function field $K/\mathbb{R}$, the number of connected components of M is finite; furthermore, it equals the number of the connected components of $V(\mathbb{R})$, V being any smooth projective model of K . (See, e.g., Schülting's contribution in these Proceedings.) In [Sch 2, 4.11] it is proved that for any base field k , the space $M(k(T))$ has the same number of components as $M(k)$; consequently, we have $[\mathbb{E}(k(T)) : \mathbb{E}^+(k(T))] = [\mathbb{E}(k) : \mathbb{E}^+(k)]$.

We are now going to investigate the role of $\mathbb{E}^+$ for the study of the sums of 2n-th powers in K . Hereby, we shall apply the general representation theorem of Kadison-Dubois as stated in [B 4 , Section 3] . By reformulating Theorem (3.3) [B 4 , loc. cit.] , we first state

<u>(1.6) Theorem.</u> $\mathbb{E}^+ \subset \bigcap_{n\in\mathbb{N}} \Sigma K^{2n}$.

In the following section a second proof will be given. Instead of the Kadison-Dubois Theorem we then make use of Hilbert's identities. Theorem (1.6) provides many examples of elements which are sums of 2n-th powers for every n .

<u>(1.7) Examples.</u>

i) Choose $k,\ell \in \mathbb{N}$ and $d \in K$, $d \neq -1$. Then

$$\frac{d^{2k}+1}{d^{2\ell}+1} \circ \frac{d^{2\ell-1}+1}{d^{2k-1}+1} \in \bigcap_n \Sigma K^{2n} .$$

Denote this element by c . To see $c \in \mathbb{E}^+$, we show $\lambda(c) > 0$, $\lambda(c) \neq \infty$ for every $\lambda \in M$. Then $c \in H$ and, in view of (1.3) , $c \in \mathbb{E}^+$ follows. Given $\lambda \in M$, then $\lambda(c) = 1$ in the cases of $\lambda(d) = 0$ or $\lambda(d) = \infty$. If $\lambda(d) \neq \infty$, but $\lambda(d) > 0$, then clearly $\lambda(c) > 0$; if $\lambda(d) < 0$, but $\lambda(d) \neq -1$, then $\lambda(c) =$

$$= \frac{\lambda(d)^{2k}+1}{\lambda(d)^{2\ell}+1} \cdot \frac{|\lambda(d)|^{2\ell-1}-1}{|\lambda(d)|^{2k-1}-1} > 0 .$$

So, finally, consider $\lambda(d) = -1$. Then

$$\lambda\left(\frac{d^{2\ell-1}+1}{d^{2k-1}+1}\right) = \frac{2\ell-1}{2k-1}$$

which implies $\lambda(c) > 0$.

ii) In the rational function field $\mathbb{Q}(X)$ we derive from (1.2), ii) the result:

$$\frac{1+X^2}{2+X^2} \in \bigcap_n \Sigma\, \mathbb{Q}(X)^{2n} .$$

For none of these examples are explicit representations as sums of 2nth-powers known, apart from very special values for n.

The next lemma slightly improves a result of Schülting [Sch 2, 3.3].

(1.8) Lemma. Let $n = 2^r m$, $2 \nmid m$ and $a \in K^\times$. Then the following statements are equivalent:

i) $v(a) \in \Gamma^n$ for all Krull valuation v of K with value group Γ and a real valuation ring;

ii) $a \in \mathbb{E} \cdot K^n$ if $r = 0$ or

$a \in \mathbb{E} \cdot (\Sigma K^2)^{\frac{n}{2}}$ if $r \geq 1$.

Proof. i) => ii). Let $\mathfrak{m}_v$ denote the maximal ideal of the valuation ring V of v. By assumption, we have $a = \varepsilon_v \cdot x_v^n$, ε_v a unit for v, $x_v \in K^\times$. Since H is a Prüfer ring and $H \subset V$, we see $V = H_{\mathfrak{m}_v \cap H}$. Hence, $\varepsilon_v = \frac{r_v}{s_v}$ with $r_v, s_v \in H \setminus \mathfrak{m}_v$. By writing $a = (r_v s_v^{n-1}) \cdot (x_v s_v^{-1})^n$, we may assume $a = \varepsilon_v x_v^n$ with $\varepsilon_v \in H$!
Now consider the ideal $\mathfrak{a}$ of H generated by all the elements $a x_v^{-n} = \varepsilon_v$, v ranging over all valuations with a real valuation ring. By [B 4, (2.16), iv)] every maximal ideal of H occurs among the ideals $\mathfrak{m}_v \cap H$. Because of $\varepsilon_v \notin \mathfrak{m}_v \cap H$, we get $\mathfrak{a} = H$. But then $\mathfrak{a} = H$ is already generated by finitely many elements, say $\varepsilon_1, \dots, \varepsilon_s$: $H = (a) \cdot (\frac{1}{x_1^n}, \dots, \frac{1}{x_s^n})$. In Prüfer rings we have the relation $(\mathfrak{a}_1 + \dots + \mathfrak{a}_k)^\ell = \mathfrak{a}_1^\ell + \dots + \mathfrak{a}_k^\ell$ for arbitrary ideals. [G, § 24, Exerc.]. This applies to H. Additionally, in H the relation $(a_1, \dots, a_k)^2 = (a_1^2 + \dots a_k^2)$ holds for any fractional ideal; this easily follows from the

facts $(a_1,\dots a_k)^2 = (a_1^2,\dots a_k^2)$ and $a_i^2/(a_1^2+\dots+a_k^2) \in H$. (Also see [Sch 2, 3.1]). Consider first the case of n being odd, then $(\frac{1}{x_1},\dots,\frac{1}{x_s})^n$ is a principal ideal. As just seen the class group of H has exponent 2, hence $(\frac{1}{x_1},\dots,\frac{1}{x_s}) = (\frac{1}{x})$ for some $x \in K^\times$. This means $a = \varepsilon x^n$, $\varepsilon \in \mathbb{E}$. Now assume $n = 2m'$. Then

$$(1) = (a)(\frac{1}{x_1^2},\dots,\frac{1}{x_s^2})^{m'} = (a)((\sum_1^s x_i^{-2})^{m'}),$$

which implies $a \in \mathbb{E}\cdot(\Sigma K^2)^{m'}$. ii) $\Rightarrow$ i). This follows easily from the fact $v(\sum_1^k x_i^{2\ell}) \in \Gamma^{2\ell}$, where v is any valuation with a real valuation ring.

<u>(1.9) Theorem.</u> The following three sets coincide:

i) ΣK^{2n},

ii) $\mathbb{E}^+\cdot(\Sigma K^2)^n$,

iii) $\{a \in \Sigma K^2 \mid a = 0$ or $v(a) \in \Gamma^{2n}$ for all valuations v with a real valuation ring$\}$.

<u>Proof.</u> As just remarked, $v(a) \in \Gamma^{2n}$ for $a \in \Sigma K^{2n}$, $a \neq 0$ and one of the valuations in question. Therefore, ΣK^{2n} is contained in the third set. By (1.8), this latter one is contained in $\mathbb{E}^+\cdot(\Sigma K^2)^n$. Finally, it remains to be shown that $\mathbb{E}^+\cdot(\Sigma K^2)^n \subset \Sigma K^{2n}$. That $\mathbb{E}^+ \subset \Sigma K^{2n}$ holds is stated in (1.6).

Given $x_1,\dots,x_r \in K$, not all zero, we have

$$(\sum_1^r x_i^2)^n = \frac{(\sum_1^r x_i^2)^n}{\sum_1^r x_i^{2n}} \cdot \sum_1^r x_i^{2n} .$$

The first factor is readily checked to lie in $\mathbb{E}^+$ which completes the proof.

The characterization of the sums of 2nth powers as given in (1.9) iii) has already been obtained in [B2, Satz 2.14] where applications can also be found.

Recall that a formally real field K is called pythagorean if $\Sigma K^2 = K^2$ holds. In such a field we have $\mathbb{E}^+ = \mathbb{E}^2$ because of $\mathbb{E}^+ = \mathbb{E} \cap K^2$ and the fact that H, as a Prüfer ring, is integrally closed. Putting this together with (1.9) we obtain

(1.10) Corollary. If K is pythagorean, then

$$\Sigma K^{2n} = \mathbb{E}^2 K^{2n} .$$

We next return to the statement $\mathbb{E}^+ \subset \bigcap_n \Sigma K^{2n}$ for a closer look. The main objective is to study the equality of these sets. In general they do not coincide. To see this, consider a real closed field K with a non-Archimedean order P. Then H equals the valuation ring $A(P)$ attached to P, and we have $K^2 = \cap \Sigma K^{2n}$. But $A(P) = H \neq K$ and hence $\mathbb{E}^+ \underset{\neq}{\subset} \bigcap_n \Sigma K^{2n}$.

The elements of $\bigcap_n \Sigma K^{2n}$ allow a characterization by means of valuations which resembles the one given in (1.9),iii). Given an abelian group A, denote its maximal divisible subgroup by A_{div}.

(1.11) Proposition. Given $a \in K^\times$, then the following statements are equivalent:

i) $a \in \bigcap_n \Sigma K^{2n}$,

ii) $a \in \Sigma K^2$ and for all valuations v with value group Γ and a real valuation ring we have $v(a) \in \Gamma_{div}$.

Proof. i) => ii). We have $v(a) \in \bigcap_n \Gamma^{2n}$. But, given any torsion-free group Γ, its maximal divisible subgroup is just $\bigcap_n \Gamma^{2n}$. ii) => i). Apply (1.9).

This last result will be applied in the following way. Suppose a family $\{V_\alpha\}_{\alpha \in I}$ of real valuation rings is given with the two properties

i) $H = \bigcap_{\alpha \in I} V_\alpha$

ii) for all α $(\Gamma_\alpha)_{div} = 1$, Γ_α the value group of V_α.

Then $\mathbb{E}^+ = \bigcap \Sigma K^{2n}$ must follow because an element in the set on the right-hand side has to be a unit in all rings V_α. Therefore, we have to look for fields where this hypothesis can be checked. We shall see that this method works for certain function fields.

Suppose k is a subring of the field K. Then set

$$H(K|k) = \bigcap V, \ V \text{ ranging over all real valuation rings containing } k.$$

$H(K|k)$ is called the real holomorphy ring of K over k. Note $H(K) = H(K|\mathbb{Q})$.

(1.12) Lemma.

$$H(K|k) = k \cdot H(K) = \{a \in K \mid r \pm a \in \Sigma K^2 \text{ for some } r \in k\}.$$

Proof. Denote the occurring sets, from left to right, by

A_1, A_2, A_3. Given $r \pm a \in \Sigma K^2$, then $r \in \Sigma K^2$ and hence $1 \pm r^{-1} a \in \Sigma K^2$, i.e., $r^{-1} a \in H(K)$ and $a \in A_2$. Clearly, $k \cdot H(K) \subset H(K|k)$. Thus, $A_3 \subset A_2 \subset A_1$. Using the identity $(x+y)(u+v) + (x-y)(u-v) = 2[xy+uv]$, one checks that A_3 is a ring. Because of $H(K) \subset A_3$, A_3 turns out to be a Prüfer ring. Hence, A_3 is the intersection of the valuation rings $V \supset A_3$. Such a V contains H and is therefore real. Moreover, $k \subset A_3$ as we have $(1+a^2) \pm a \in \Sigma K^2$ for every $a \in k$. Thus, $V \supset A_3$ implies $V \supset k$ and that V is real. Consequently, $A_3 = \bigcap_{A_3 \subset V} V$ is an overring of $H(K|k)$, which completes the proof.

In particular, we shall deal with the case that K is a function field over the base field k. Hereby, adopting the terminology of Zariski-Samuel [Z-S, Ch. VI, §14], we call a <u>real prime divisor</u> of $K|k$ any real valuation ring V of $K|k$ which is discrete of rank 1. The following theorem was first proved by Schülting using profound results from Hironaka. (See his contribution in these Proceedings.) The proof to be given here does not need the techniques of Hironaka. Rather, it appeals to the more common methods of the theory of formally real fields; particularly, Knebusch's trace formula for real places

$$(1.13) \qquad \lambda_*(\mathrm{tr}^* \varphi) = \sum_{\mu|\lambda} \mu_*(\varphi)$$

is the essential device. This formula is concerned with a place $\lambda : K \to R \cup \infty$, R real closed, and all its extensions $\mu : L \to R \cup \infty$, where $L|K$ is a finite extension, and a quadratic form φ over L. Proofs can be found in [Kn, §3] and [Sch 1, Sect. 4].

<u>(1.14) Theorem.</u> $H(K|k) = \bigcap V$, where V ranges over all real prime divisors of $K|k$.

<u>Proof.</u> In fact, we are going to prove the stronger result:

(1.15) if $a \in K$ is contained in the maximal ideal $\mathfrak{m}_V$ of a real valuation ring V of $K|k$, then it is even contained in the maximal ideal of a real prime divisor.

The result (1.15) implies the theorem. Namely, assume a to lie in the intersection of the real prime divisors but not in $H(K|k)$. Then, by the definition of $H(K|k)$, a^{-1} meets the hypothesis of (1.15) , the conclusion of which then leads to a contradiction. In order to prove (1.15) we first reduce the problem to the situation where the transcendence degree $tr(K|k)$ equals 2 and V is discrete of rank 2 . Obviously, (1.15) holds in case $tr(K|k) = 1$ or $a = 0$. So, assume $a \neq 0$ and proceed by induction on $n = tr(K|k)$. Since $a \in \mathfrak{m}_V$, we find an order P of K with $a \in P$ such that a is infinitely small as compared to elements of k . The element a is transcendental over k . Denote by P_1 the restriction of P to $k(a)$. By the theorem of Artin-Lang (see, e.g., [B4, Section 1]) we find a discrete rank $(n-1)$ valuation ring U of $K|k(a)$ with a residue field contained in the real closure $(R, \tilde{P}_1)$ of $(k(a), P_1)$. Restrict the valuation ring

$$A(\tilde{P}_1, k) := \{x \in R \mid s \pm x \in \tilde{P}_1 \quad \text{for some} \quad s \in k\}$$

to the residue field of U which is finite over $k(a)$. This restriction $\bar{V}$ is discrete of rank 1 since it extends the valuation ring of $k(a)$ which belongs to the prime polynomial a of $k(a)$. Pulling back $\bar{V}$ by means of U , we end up having a discrete, rank n, real valuation ring V with $a \in \mathfrak{m}_V$, $\mathfrak{m}_V$ the maximal ideal of V. Let W denote the rank 1 valuation ring over V. If $a \in \mathfrak{m}_W$, we are through. Otherwise, pass to the residue field L of W which is a function field of transcendence degree $n-1$ over k. The residue class $\bar{a}$ belongs to the maximal ideal of the image of V in L. Hence, the induction hypothesis applies and we find a real prime divisor $\bar{W}_2$ in L with $\bar{a}$ in its maximal ideal. Since $\bar{W}_2$ is real, there is a chain of discrete real valuation rings:

$$\bar{W}_2 \underset{\neq}{\supset} \bar{W}_3 \underset{\neq}{\supset} \dots \underset{\neq}{\supset} \bar{W}_n .$$

By means of W we have an induced chain of discrete real valuation rings in K:

$$W = W_1 \underset{\neq}{\supset} W_2 \underset{\neq}{\supset} \dots \underset{\neq}{\supset} W_n .$$

In the corresponding series of maximal ideals

$$\mathfrak{m}_1 \underset{\neq}{\subset} \mathfrak{m}_2 \underset{\neq}{\subset} \dots \underset{\neq}{\subset} \mathfrak{m}_n$$

we have $a \in \mathfrak{m}_2 \setminus \mathfrak{m}_1$ and we choose $a_i \in \mathfrak{m}_{i+1} \setminus \mathfrak{m}_i$ for $i = 2,\dots,n-1$. Then the rational function field $k(a_2,\dots,a_{n-1})$ is contained in W_2. Hence, replacing k by $k(a_2,\dots,a_{n-1})$, we are now concerned with the following situation:

$$\operatorname{tr}(K|k) = 2 , \mathfrak{m}_W \subset \mathfrak{m}_V \subset V \subset W , a \in \mathfrak{m}_V \setminus \mathfrak{m}_W .$$

Choose b with $\mathfrak{m}_W = Wb$. Then, on $k(a,b)$, the valuation ring V induces a valuation v with value group $\mathbb{Z} \times \mathbb{Z}$. Hereby, $v(b) = (1,0)$, $v(a) = (0,1)$ and $v(b)$ is infinitely larger than $v(a)$. We shall apply the formula (1.13) to the finite extension $K|k(a,b)$. Given $c_1,\dots,c_k \in K^\times$, the symbol $\langle c_1,\dots,c_k \rangle$ denotes the quadratic form $\sum_1^k c_i X_i^2$. We choose a real place $\lambda : K \to R \cup \infty$ with valuation ring V. We have

(1.16) $\quad tr^*\langle 1\rangle = \langle \ldots, f_i(a,b), \ldots \rangle$ with $f_i \in k[a,b] \setminus \{0\}$.

Each $f(a,b) \in k[a,b] \setminus \{0\}$ has a unique "normal form".

(1.17) $\quad f(a,b) = b^k[a^\ell g(a) + bh(a,b)]$, $g(a) \in k[a]$, $g(0) \neq 0$,
$h(a,b) \in k[a,b]$, $k,\ell \geq 0$.

Given f in normal form, we get $v(f) = (k,\ell)$. This implies for $\lambda' := \lambda|_{k(a,b)}$ the formula

(1.18) $\quad \lambda'_* tr^*\langle 1\rangle = \sum_{k_i,\ell_i \text{ even}} \operatorname{sgn}_{R^2}(g_i(0))$.

Hereby, k_i, ℓ_i, g_i are the constituents of the normal form of f_i . By (1.13) we know that $\lambda'_* tr^*\langle 1\rangle$ equals the number of the extensions of λ' to K , which is obviously ≥ 1 .
Now choose $t \in \mathbb{N}$ and consider the place over $k(ba^{-t})$

$$\tau : k(a,b) = k(ba^{-t},a) \to k(ba^{-t}) \cup \infty , \; a \mapsto 0 .$$

Let $\tilde{v}$ be the corresponding valuation. Then $\tilde{v}(a) = 1$, $\tilde{v}(b) = t$.
If t is sufficiently large, then

(1.19) $\quad \tilde{v}(f_i) = tk_i + \ell_i$

holds for all f_i occurring in $tr^*\langle 1\rangle$. Choose t in this way and as an even number. Set $P_0 = R^2 \cap k$ and let P_1 be an order of $k(ba^{-t})$ which extends P_0 and contains ba^{-t} . The place $\tau : k(a,b) \to k(ba^{-t}) \cup \infty$ will be considered as a real place into the real closure of $(k(ba^{-t}),P_1)$. We want to show that τ extends to a real place to K , the valuation ring of which is then the real prime divisor we are looking for. To show that τ extends we compute $\tau_* tr^*\langle 1\rangle = \Sigma\, \tau_*\langle f_i\rangle$. If $tk_i + \ell_i$ is odd, then $\tau_*\langle f_i\rangle = 0$. If k_i and ℓ_i are even, then $\tau_*\langle f_i\rangle = \mathrm{sgn}_{P_0} g_i(0) = \mathrm{sgn}_{R^2} g_i(0) = \lambda'_*\langle f_i\rangle$. Finally, assume that k_i is odd, but ℓ_i even. Then

$$f_i = (b^{k_i-1} a^{\ell_i+t}) \cdot ba^{-t} g_i(0) \cdot \eta$$

where the first factor is a square and η a 1-unit. Hence, $\tau_*\langle f_i\rangle = \mathrm{sgn}_{R^2} g_i(0) = \lambda'_*\langle bf_i\rangle$. Putting this and (1.18) together we have

$$\text{(1.20)} \qquad \tau_* \mathrm{tr}^*\langle 1\rangle = \lambda'_* \mathrm{tr}^*\langle 1\rangle + \sum_{\substack{k_i \text{ odd} \\ \ell_i \text{ even}}} \lambda'_*\langle bf_i\rangle .$$

Because of $b \in k(a,b)$ we get $\mathrm{tr}^*\langle b\rangle = \sum_i \langle bf_i\rangle$ which implies $\lambda'_*\mathrm{tr}^*\langle b\rangle = \sum_{\substack{k_i \text{ odd} \\ \ell_i \text{ even}}} \lambda'_*\langle bf_i\rangle$.

But, on the other hand, by (1.13), $\lambda'_*\mathrm{tr}^*\langle b\rangle = \sum_{\sigma|\lambda'} \sigma_*\langle b\rangle$.

Among those σ's there is the place λ, and for λ we have $\lambda_*\langle b\rangle = 0$. This follows from the fact that the value of b in the value group of V cannot be a square; otherwise, b would not be a prime element for W. In consequence, the absolute value of the second term of the right-hand side in (1.20) is strictly less than the number of extensions of λ' which is just $\lambda'_*\mathrm{tr}^*\langle 1\rangle$. Thus, $\tau_*\mathrm{tr}^*\langle 1\rangle > 0$ and τ is extendible as to be shown.

The theorem (1.14) finds many applications. Consider first, as a base field, a field k which has no real valuation rings $\neq k$, equivalently which has only Archimedean orders. Such a field is called <u>totally Archimedean</u>. Number fields and $\mathbb{R}$ are examples. Given a function field K over a totally Archimedean field, then $H(K|k) = H(K)$. Hence, the method as stated before (1.12) works and we get

<u>(1.21) Theorem.</u> If K is a function field over a totally Archimedean field k, then

$$\mathbb{E}^+ = \bigcap_n \Sigma K^{2n} .$$

We conclude this section by briefly mentioning further consequences of (1.14). Hereby, the base field k need not be totally Archimedean and K is assumed to be a function field over k. If $tr(K|k) = d$, then $H(K|k)$ has the Krull dimension d: it is $\leqslant d$ by [Z-S, VI, § 10], it is $\geqslant d$ since there are real prime divisors, e.g.

<u>(1.22) Proposition.</u>

i) Given $f_1,\dots,f_n \in H(K|k)$ with $(f_1,\dots,f_n) \neq H(K|k)$, then there is a prime ideal $\mathfrak{y}$ of $H(k|k)$ of height 1 and dimension $tr(K|k)-1$ with $f_1,\dots,f_n \in \mathfrak{y}$,

ii) $H(K|k)$ is completely integrally closed,

iii) if $tr(K|k) \geq 2$, then no prime ideal $\neq \{0\}$ is finitely generated.

<u>Proof.</u>

i) Apply the proof of (1.14) to find a real prime divisor with $\sum_1^n f_i^2$ in its maximal ideal $\mathfrak{m}$. Then $\mathfrak{y} = \mathfrak{m} \cap H(K|k)$ is the required prime ideal.

ii) The condition of [G, Th. (22.8), p. 315] is satisfied in view of i) .

iii) By [G, Th. (19.3) a), p. 269] a finitely generated prime ideal $\mathfrak{p} \neq \{0\}$ has to be maximal. But i) then shows $\dim(\mathfrak{p}) \geq 1$, which is a contradiction.

In [Sch 3] Schülting has shown that in $H(\mathbb{R}(X,Y))$ the fractional ideal $(1,X,Y)$ cannot be generated by two elements. It is highly conceivable that in $H(\mathbb{R}(X_1,\ldots X_n))$ the fractional ideal $(1,X_1,\ldots,X_n)$ cannot be generated by n elements but no proof has been found so far. Equivalently, one has to prove that there is no expression $1+X_1^2+\ldots+X_n^2 = \varepsilon\cdot(f_1^2+\ldots+f_n^2)$ with $\varepsilon \in \mathbb{E}^+, f_1,\ldots,f_n \in \mathbb{R}(X_1,\ldots,X_n)$. From quadratic form theory it is already known that there is no identity with $\varepsilon = 1$.

2. The higher Pythagoras numbers

Let K be a field of characteristic zero. We shall be concerned with certain identities over K .

$$\underline{I(k,n):} \quad \left(\sum_1^k X_i^2\right)^n = \sum_{j=1}^{\ell(k,n)} \alpha_j f_j(X_1,\ldots,X_k)^{2n}$$

with $\alpha_j \in \mathbb{Q}_+^\times$, $f_j \in K(X_1,\ldots,X_k)$,

$\underline{I^*(k,n,\alpha):}$

$$\left(\sum_1^k X_i^2\right)^n + \alpha X_1^2\left(\sum_1^k X_i^2\right)^{n-1} = \sum_{j=1}^{\ell^*(k,n,a)} \beta_j g_j(X_1,\ldots,X_k)^{2n}$$

with $\alpha,\beta_j \in \mathbb{Q}_+^\times$, $g_j \in K(X_1,\ldots,X_k)$.

(2.1) Theorem. For every choice of $k,n \in \mathbb{N}$, $\alpha \in \mathbb{Q}_+^\times$ there are identities $I(k,n)$ and $I^*(k,n,\alpha)$.

Proof. Obviously, it is enough to consider the case $K = \mathbb{Q}$. Denoting the rational function field $\mathbb{Q}(X_1,\dots,X_k)$ by F, we have to show that the elements of the left-hand sides of the identities are contained in ΣF^{2n}. We appeal to Theorem (1.9). In consequence, the first identity is established. As to the second, write the left-hand side as

$$\left(\sum_1^k X_i^2\right)^n \left[1+\alpha\cdot\left(1+\sum_2^k (X_i X_1^{-1})^2\right)^{-1}\right].$$

The second factor lies in $\mathbb{E}^+$, which in view of (1.9) proves the second identity to exist.

(2.2) Remark. We have used (1.9) to derive the identities. But one can proceed in the reverse order. First show, in some way, the existence of the identities $I(k,n)$ and $I^*(k,n,\alpha)$. Then, let K be any formally real field. In order to make use of the identities one wants to insert elements of K. In fact, this works but not in the direct way. One has to apply the following result:

(2.3) Lemma. Given $a_1,\dots,a_k \in K$ there is a place

$$\lambda : \mathbb{Q}(X_1,\dots,X_k) \to K \cup \infty \quad \text{with} \quad \lambda(X_i) = a_i,\ i=1,\dots,k.$$

Proof. As well known, there is a place $\lambda' : K(X_1,\dots,X_k) \to K \cup \infty$ with $\lambda'(X_i) = a_i$. Then $\lambda := \lambda' \circ i$ is the required place where $i : \mathbb{Q}(X_1,\dots,X_k) \to K(X_1,\dots,X_k)$ is the natural map.

Now given $a_1,\dots,a_k \in K$ and having chosen λ as in (2.3), the polynomials on the left-hand sides in the identities I and I^* are contained in the valuation ring V of λ. V has to be real as its residue field is a subfield of the formally real field K. But a sum Σy_i^{2n} lies in a real valuation ring if and only if all y_i's are equally contained in it. Returning to our situation, this means $\lambda(f_j)$, $\lambda(g_j) \neq \infty$ for all f_j, g_j in question. Consequently, we have:

$$(2.4) \qquad (\sum_1^k a_i^2)^n = \sum_1^\ell \alpha_j b_j^{2n} \in \Sigma K^{2n} \text{ and}$$

$$(2.5) \qquad (\sum_1^k a_i^2)^n + \alpha a_1^2 (\sum_1^k a_i^2)^{n-1} = \sum_1^{\ell^*} \beta_j c_j^{2n} \in K^{2n}.$$

From (2.4) one deduces $(\Sigma K^2)^n \subset \Sigma K^{2n}$. To derive $\mathbb{E}^+ \subset \Sigma K^{2n}$ we extend an idea of Siegel which he used in [S, Satz 2]. (See also [K].) Let $\varepsilon \in \mathbb{E}^+$ be given, then $\hat{\varepsilon} > \frac{1}{s}$ for some $s \in \mathbb{N}$ (compare with the proof of (1.2)). This implies $\varepsilon = \frac{1}{s} + \sum_1^u \omega_i^2$ with ω_i necessarily in H. Choose $\alpha \in \mathbb{Q}_+^\times$ with $\hat{\omega}_i^2 < \frac{\alpha}{su}$ on M(K). Again, this means $\frac{\alpha}{su} - \omega_i^2 = \sum_{j=1}^{k-1} \omega_{ij}^2$ for some $\omega_{ij} \in H$.

Now apply (2.5) with $(\omega_i, \omega_{i1}, \ldots, \omega_{ik-1})$ as $(a_1, \ldots, a_k)$ to get

$$(\frac{\alpha}{su})^n + \alpha\omega_i^2 (\frac{\alpha}{su})^{n-1} \in \Sigma K^{2n}.$$

In view of $\mathbb{Q}_+^\times \subset \Sigma K^{2n}$, one obtains

$$\frac{1}{su} + \omega_i^2 \in \Sigma K^{2n}$$

and, because of $\varepsilon = \sum_{i=1}^u (\frac{1}{su} + \omega_i^2)$, the desired result $\varepsilon \in \Sigma K^{2n}$.

These arguments have shown that the statement $\Sigma K^{2n} = \mathbb{E}^+ (\Sigma K^2)^n$

is equivalent to the existence of the identities I and I^*. It is therefore of methodological importance to find different proofs for the existence of these identities. Such proofs are known and they date back to Hilbert's famous solution of the Waring problem [H] . In this context the identities are called Hilbert's identities. A comprehensive account of the variety of known proofs can be found in [E]; for a detailed proof the reader is also referred to [K-P, Section 5] .

<u>(2.6) Theorem (Hilbert).</u> There are identities $I(k,n)$ and $I^*(k,n,2n)$ with the additional properties:

i) $\ell(k,n) = \binom{2n+k-1}{k-1}$, $\ell^*(k,n,2n) = \binom{2n+k+1}{k-1}$

ii) all rational functions f_j, g_j are linear homogeneous polynomial in $X_1,\dots,X_k$.

Usually, only the identities $I(k,n)$ of (2.6) are called Hilbert's identities. The identities I^*, also already used by Hilbert, are obtained from $I(n+1,k)$ by taking twice the derivative with respect to X_1. The advantage of Hilbert's identities is to be seen in the fact that we know values for ℓ and ℓ^*. But if we allow rational identities, there may be smaller upper bounds for the sums of the right-hand sides in I and I^*. Since these upper bounds play a crucial role in the applications to follow, it is of great interest to derive shorter identities than Hilbert's. A look at the proof that the existence of I and I^* implies the result $\mathbb{E}^+(\Sigma K^2)^n = \Sigma K^{2n}$ might lead to the idea that Hilbert's identities are not general enough to give this result. But, in fact, they can be used likewise. Choose a multiple $2\ell n$ for α and show, by applying $I(\ell n,k,2\ell n)$, that $\frac{1}{su} + \omega_i^2 \in \Sigma K^{2\ell n} \subset \Sigma K^{2n}$ holds.

In the sequel we shall make use of the identity

$$(2.7)\qquad (X^2+Y^2+Z^2)^2 + 4X^2(X^2+Y^2+Z^2) = \frac{1}{12}(X+Y+Z)^4 + \frac{1}{12}(X+Y-Z)^4$$
$$+ \frac{1}{12}(X-Y+Z)^4 + \frac{1}{12}(X-Y-Z)^4 + \frac{1}{3}(X+Y)^4 + \frac{1}{3}(X-Y)^4$$
$$+ \frac{1}{3}(X+Z)^4 + \frac{1}{3}(X-Z)^4 + \frac{10}{3}X^4 .$$

Its "length" is 9 , whereas $\ell^* = 28$ according to (2.6) .

We are now going to define and investigate the higher Pythagoras numbers. For the sake of simplicity, we only deal with fields K of characteristic zero. Define

$$P_n(K) = \infty \quad \text{or} \quad \min \{k \mid \Sigma K^n = \sum_1^k K^n\}$$

as the <u>n-th Pythagoras number of K</u>. Hereby, we set $\sum_1^k K^n = \{\sum_1^k x_i^n \mid x_1,\dots,x_k \in K\}$. We shall need the <u>n-th level of K</u> to be defined as

$$s_n(K) = \infty \quad \text{or} \quad \min \{k \mid -1 \in \sum_1^k R^n\} .$$

In the terminology of [Jo] our $P_n(K)$ is just the invariant $w(n;K)$ and $s_n(K)$ equals $u(n;K)$. Many results concerning P_n and s_n , even for rings, can be found there. We have $P_1 = 1$ and $s_n = 1$ if n is odd. We denote by $G(n)$ the Waring constant for the representation of sufficiently large numbers as sums of n-th powers of natural numbers (as to this notation we follow [H-W, Chap. XX]).

<u>(2.8) Proposition.</u> If $s_n < \infty$, then

$$P_n \le (n+1)G(n)s_n .$$

Proof. We make use of the identity

$$n!X = \sum_{h=0}^{n-1} (-1)^{n-1-h} \binom{n-1}{h} [(X+h)^n - h^n] .$$

Divide both sides by $n!$ and write every occurring positive rational number as a fraction of a large numerator and a n-th power as the denominator. So, these positive rational numbers require $G(n)$ n-th powers. For every factor -1 insert its representation as a sum of s_n n-th powers. The remaining sum $\frac{1}{n!} \sum_h (-1)^{n-h} \binom{n-1}{h} h^n$ is treated equally. Putting it all together, we get the inequality $P_n \le (n+1)G(n)s_n$.

In particular, if n is odd, then $P_n \le (n+1)G(n)$. This is not a sharp bound, e.g., we have $P_n(\mathbb{Q}) \le G(n)$ and $P_n(\mathbb{R}) = 1$ for every $n \in \mathbb{N}$. We next consider the case of an even exponent $2n$. Using the Hilbert's identities $I(k,n)$ of (2.6) , Joly in [Jo, (6.16)] was the first to prove the following result.

(2.9) Theorem.

a) The following statements are equivalent:

   i) $s_2(K) < \infty$,

   ii) $s_{2n}(K) < \infty$ for some n ,

   iii) $s_{2n}(K) < \infty$ for all n ,

b) there is a function $f : \mathbb{N} \times \mathbb{N} \to \mathbb{N}$ such that

$$s_{2n}(K) \le f(s_2(K), n)$$

holds if $s_2(K)$ is finite.

As stated in [B4, Section 3] there is a proof of Part a) of (2.9) using the Kadison-Dubois Theorem instead of the Hilbert's identities.

As a consequence of (2.8) and (2.9), the finiteness of P_n is proved for all odd n and for all non-formally real fields. Therefore, we now turn to a formally real field K and its numbers $P_{2n}(K)$. In this situation, another invariant comes in. Define

$$\mu(K) = \infty \text{ or } \min\{k \mid \text{every finitely generated ideal of } H(K) \text{ can be generated by } k \text{ elements}\}.$$

Since $H(K)$ is a Prüfer ring, one can apply the theorem of Heitmann [He] to get:

(2.10) Proposition. If $\dim H(K) = d$, then $\mu(K) \leq d + 1$.

The importance of μ can already be seen from the next result.

(2.11) Proposition.

i) $\mu \leq P_{2n}$ for all n,

ii) if $\mu < \infty$, then

$$\Sigma K^{2n} = \mathbb{E}^{+}\left(\sum_{1}^{\mu} K^{2n}\right) = \mathbb{E}^{+}\left(\sum_{1}^{\mu} K^{2}\right)^{n}.$$

Proof.

i) Let $\mathfrak{a} = (a_1,\dots,a_k)$ be an ideal of H and assume $\ell := P_{2n} < \infty$. Then, as in the proof of (1.8), we get $\mathfrak{a}^{2n} = \left(\sum_1^k a_i^{2n}\right) = \left(\sum_1^{\ell} b_i^{2n}\right) = (b_1,\dots,b_\ell)^{2n}$. But in a Prüfer ring the equality $\mathfrak{a}^{2n} = \mathfrak{b}^{2n}$ implies $\mathfrak{a} = \mathfrak{b}$ where $\mathfrak{a}$ and $\mathfrak{b}$ may be any ideal [G, § 22, Exerc.]. Hence $\mu \leq P_{2n}$.

ii) Given $a_1,\dots,a_k$, then $(a_1,\dots,a_k) = (b_1,\dots,b_\mu)$ which implies $(a_1,\dots,a_k)^{2n} = (\sum_1^k a_i^{2n}) = (\sum_1^\mu b_i^{2n})$. This shows $\Sigma K^{2n} \subset \mathbb{E}^+(\sum_1^\mu K^{2n})$. Noting

$$(b_1,\dots,b_\mu)^{2n} = (b_1^2,\dots,b_\mu^2)^n = (\sum_1^\mu b_i^2)^n = ((\sum b_i^2)^n),$$

we get $\Sigma K^{2n} \subset \mathbb{E}^+(\sum_1^\mu K^2)^n$ which completes the proof.

In [Br 2] Bröcker has shown that there is a field K with $P_2 = \infty$ but only one ordering P . Then $H(K)$ is the valuation ring $A(P)$ attached to P . Hence, $\mu(K) = 1$ but $P_{2n} = \infty$ for all n in view of the following theorem.

It is now all prepared to prove the main result of this section. We shall apply the identities $I^*(k,n,\alpha)$ and set $\ell^*(k,n) = \min\{\ell^*(n,k,\alpha) \mid \alpha \in \mathbb{Q}_+^\times\}$.

By (2.6) we know $\ell^*(k,n) \le \binom{2n+k+1}{k-1}$.

<u>(2.12) Theorem.</u> Let K be a formally real field. Then the following statements are equivalent:

i) $P_2(K) < \infty$,

ii) $P_{2n}(K) < \infty$ for some n ,

iii) $P_{2n}(K) < \infty$ for all n .

If $P_2(K) < \infty$, then

$$P_{2n}(K) \le P_2(K)\,\ell^*(P_2(K)+1,n)\,G(2n)\,\mu(K) .$$

<u>Proof.</u> We show ii) => i) , then i) => iii) and that we have this upper bound for $P_{2n}(K)$. ii) => i) . More precisely, we prove $P_{2m} \le \mu P_{2n}$ provided $m|n$. Given $a \in \Sigma K^{2m}$, then by

(2.11), ii) $a = \varepsilon \sum_1^{\mu} x_i^{2m}$, $\varepsilon \in \mathbb{E}^+$. Since $\mathbb{E}^+ \subset \Sigma K^{2n}$, we have by assumption

$$\varepsilon = \sum_1^k y_i^{2n} = \sum_1^k (y_i^t)^{2m} ,$$

where $k = P_{2n}$, $mt = n$. Hence $P_{2m} \le \mu P_{2n}$ is proved. i => iii). Again, by applying (2.11), ii), one obtains the factor μ in the upper bound. So we concentrate on a unit ε. Pick $\alpha \in \mathbb{Q}_+^\times$ with $\ell^*(k+1,n) = \ell^*(k+1,n,\alpha)$ where $k := P_2$. Then consider the strictly positive continuous functions $k\hat{\varepsilon}^{-1}$ and $(k+\alpha)\hat{\varepsilon}^{-1}$ on $M(K)$. There is a positive function f whose 2n-th power lies strictly between these two functions. f can be approximated as closely as possible by a function $\hat{\gamma}$, $\gamma \in H$. Choose $\gamma \in H$ with $k < \hat{\varepsilon}\hat{\gamma}^{2n} < (k+\alpha)$. Then, setting

$$\eta = (\varepsilon\gamma^{2n} - k)/\alpha ,$$

we have found $\eta \in \mathbb{E}^+$ with the additional property $\hat{\eta} < 1$. By assumption, we have $\eta = \sum_1^k \omega_i^2$. These ω_i's are necessarily elements of H and $\hat{\omega}_i^2 < 1$ holds. This implies $1 - \omega_i^2 = \sum_1^k \omega_{ij}^2$. Now apply the identity $I^*(k+1,n,\alpha)$ to obtain from (2.5)

$$1 + \alpha\omega_i^2 = \sum_1^{\ell^*} \beta_j c_{ij}^{2n} .$$

Summing up the equalities for $i = 1,\dots,k$, we get

$$\varepsilon\gamma^{2n} = k + \alpha\eta = \sum_{i=1}^{k} \sum_{j=1}^{\ell^*} \beta_j c_{ij}^{2n} .$$

This means $\varepsilon \in \sum_1^s K^{2n}$ with $s = k\ell^* G(2n)$, treating the β_j's as was done in the proof of (2.8). This was to be shown.

For example, let $K/\mathbb{R}$ be a real algebraic function field such that $\operatorname{tr}(K/\mathbb{R}) = t$. By Pfister [L 1, p. 302] we know $P_2(K) \le 2^t$. The factor $G(2n)$ does not occur in our situation because the β_j's are 2n-th powers of real numbers, hence, 2n-th powers in K. By (2.10) we know $\mu(K) \le t + 1$. To see this, note that a real valuation ring of K is trivial on $\mathbb{R}$ and is therefore of Krull dimension $\leq t$. Putting this all together, we obtain

$$P_{2n}(K) \leq 2^t \binom{2n+2^t+1}{2n+2} (t+1) .$$

In general, this bound seems too large. For example, in case $K = \mathbb{R}(T)$, we have by (2.7) $\ell^*(3,2) \leq 9$, which gives the better estimate $P_4(\mathbb{R}(T)) \leq 36$.
The precise value ins unknown. According to Landau $P_2(\mathbb{Q}(T)) < \infty$, hence $P_{2n}(\mathbb{Q}(T)) < \infty$ for all $n \in \mathbb{N}$, see [P] for a proof of $P_2(Q(T)) = 5$

3. The semigroups $S(K)$ and $\tilde{S}(K)$

Given a formally real field K, we have seen in (1.9) that $(\Sigma K^2)^n \subset \Sigma K^{2n}$ holds. In this section we are concerned with the equality of these sets. Let us define

$$S(K) := \{n \in \mathbb{N} \mid \Sigma K^{2n} = (\Sigma K^2)^n\}.$$

Hereby, we assume K to be a formally real field. Obviously, $1 \in S(K)$. The definition makes sense for non-formally real fields K too. But, if, further, char $K = 0$ is assumed, then $n \in S(K)$ iff $K = K^n$ because of $K = \Sigma K^2$ for every $t \in \mathbb{N}$. (Compare with [B 2].) Hence, $S(K)$ is a much more interesting set for formally real fields.

Recall our convention that the reference to K is dropped when no confusion is to be expected.

<u>(3.1) Proposition.</u> $n \in S$ if and only if $\mathbb{E}^+ = (\mathbb{E}^+)^n$.

Proof. In view of (1.9) we have $n \in S$ iff $\mathbb{E}^+ \subset (\Sigma K^2)^n$. As a Prüfer ring H is integrally closed. Hence, $\mathbb{E}^+ \subset (\Sigma K^2)^n$ implies $\mathbb{E}^+ \subset (\mathbb{E}^+)^n$.

The statement " $2 \in S(K)$ " has a surprising interpretation.

(3.2) Proposition. Consider the statements

i) $2 \in S(K)$,

ii) $M(K)$ is a connected topological space.

Then i) implies ii) and they are equivalent if, additionally, K is pythagorean.

Proof. i) => ii). If $2 \in S$, then $\mathbb{E}^+ = (\mathbb{E}^+)^2$. This implies $\mathbb{E}^+ = \mathbb{E}^2$ and $\mathbb{E}^2 = \mathbb{E}^4$. From this: $\mathbb{E} = \mathbb{E}^2 \cup -\mathbb{E}^2$, and, in view of (1.4), the assertion follows. Now, let K be pythagorean and $M(K)$ connected. In a pythagorean field we have $\mathbb{E}^+ = \mathbb{E}^2$. Again, by (1.4), we see $\mathbb{E} = \mathbb{E}^+ \cup -\mathbb{E}^+$ implying $\mathbb{E}^+ = \mathbb{E}^2 = (\mathbb{E}^+)^2$ and hence $2 \in S(K)$.

The structure of the set $S(K)$ is completely described in the next theorem.

(3.3) Theorem.

i) $S(K)$ is a multiplicative semigroup generated by a set of prime numbers $\mathbb{P}(K)$;

ii) every set of prime numbers occurs as a set $\mathbb{P}(K)$, hereby K may be chosen as a pythagorean number field of infinite degree over $\mathbb{Q}$.

Proof.

i) Given any abelian group A, written multiplicatively,

the set $\{n \mid A = A^n\}$ is closed under multiplication. Furthermore, if n is in this set, then all divisors of n are equally contained in it. Hence, this set is a semigroup generated by a set of prime numbers. Apply this to $A := \mathbb{E}^+$ to obtain the result.

ii) Let k be the intersection of all real closures of $\mathbb{Q}$ in $\mathbb{C}$. We make use of the following facts concerning k.

(3.4) i) $k|\mathbb{Q}$ is a Galois extension,

ii) k is pythagorean, $S(k) = \{1\}$,

iii) $k(\sqrt{-1})$ contains all roots of unity.

Proof of (3.4). $k|\mathbb{Q}$ is a Galois extension as k is the intersection of all real closures of $\mathbb{Q}$. We next show $p \notin S(K)$ for every prime number p. Obviously, k is pythagorean as an intersection of pythagorean fields. Hence, $p \in S(k)$ is equivalently expressed by $\Sigma k^{2p} = k^{2p}$. We have $2 \in \Sigma k^{2p}$ but $2 \notin k^{2p}$. Otherwise, the irreducible polynomial $X^{2p} - 2$ would completely split in the Galois extension k, thus showing that the formally real fields k contain roots of unities $\neq 1, -1$, which is absurd. To prove iii) first note that, given any root of unity ζ, the element $\zeta + \zeta^{-1}$ lies in every real closure of $\mathbb{Q}$ and hence $\zeta + \zeta^{-1} \in k$. Now, ζ is a root of the quadratic polynomial $X^2 - (\zeta+\zeta^{-1})X + 1$ over k. As a pythagorean field k has a unique non-formally real quadratic extension, namely $k(\sqrt{-1})$. In consequence, $\zeta \in k(\sqrt{-1})$ for every root of unity ζ. We now return to the proof of (3.3). Let $\mathbb{P}$ be the given set of prime numbers. Choose a real closure R of $\mathbb{Q}$, kept fixed in the sequel. Given an element $a \in R$, denote by $\sqrt[2p]{1+a^{2p}}$ the unique positive 2p-th root of $1 + a^{2p}$ in R. Then define for $n \geq 0$:

$$k_0 = k \ , \ k_{2n+1} = k_{2n}(\{ \sqrt[2p]{1+a^{2p}} \mid a \in k_{2n} \ , \ p \in \mathbb{P} \})$$

$$k_{2n+2} = \text{pythagorean closure of } k_{2n+1} \ .$$

(See [L 1, p. 235] for the notion of the pythagorean closure.)
Then set

$$K = \bigcup_{n \geq 0} k_n \ .$$

We first need some information on the degree $[K:k]$, considered as a so-called "supernatural" number. Since k_{2n} is a pythagorean field, we have $1 + a^{2p} = b^2$, $b \in k_{2n}$. By using the fact that $k_{2n}(\sqrt{-1})$ contains all roots of unities, we conclude that $[k_{2n+1}:k_{2n}]$ contains only $p \in \mathbb{P}$ as prime divisors. Hence, the prime divisors of $[K:k]$ are all contained in $\mathbb{P} \cup \{2\}$. We now turn to the proof that $S(K)$ is generated by $\mathbb{P}$. This amounts to proving $\mathbb{P} \subset S(K)$ and that $S(K)$ contains no other primes than those in $\mathbb{P}$. Given $p \in \mathbb{P}$, we have, by the construction of K , the result $K^{2p} + K^{2p} = K^{2p}$ which implies

$$(\Sigma K^2)^p \subset \Sigma K^{2p} = K^{2p} \ ,$$

showing first $\Sigma K^{2p} = (\Sigma K^2)^p$, i.e., $p \in S(K)$, and secondly, $\Sigma K^2 = K^2$, i.e., that K is pythagorean. Next, let p be a prime number in $S(K) \setminus \mathbb{P}$. First consider the case $p \neq 2$. Pick $a \in k$ and then $1 + a^{2p} = c^{2p}$ with $c \in K$. Because of $p \nmid [K:k]$, we see $c^2 = d \in k$ and $d^p = 1 + a^{2p}$. As p is odd, this implies $d \in \Sigma k^2 = k^2$, i.e., $d = e^2$, $e \in k$. Thus, we have shown that $p \in S(k) = \{1\}$. Finally, we assume $2 \in S(K) \setminus \mathbb{P}$. Then, $k_{2n+1} \mid k_{2n}$ is an extension of odd degree. Since k_{2n+2} is the pythagorean closure of k_{2n+1} , we see that every order is extendible from k to K . Consequently, the natural continuous restriction map $M(K) \to M(k)$ is surjective. By (3.2), $M(K)$ is connected and, hence, $M(k)$ as well, which, again by (3.2), implies $2 \in S(k)$. But we have $S(k) = \{1\}$.

(3.5) Examples. (All fields are assumed to be formally real.)

i) If R is real closed, then $S(R) = \mathbb{N}$.

ii) If A is a real henselian valuation ring of K with residue field k, then $S(K) = S(k)$. This follows from the fact that the group of 1-units is divisible. For example, $S(\mathbb{R}((X_1,\ldots,X_k))) = \mathbb{N}$.

iii) In the next theorem (3.8) we show $S(L) \subset S(K)$ for every finite extension $L|K$. This implies $S(K) = \{1\}$ for

$K|\mathbb{Q}$ finite or

$K|k$ finitely generated, $\mathrm{tr}(K|k) \geq 1$.

By the above-mentioned result one only has to show $S(K) = \{1\}$ for $K = \mathbb{Q}$ and $K = k(X_1,\ldots,X_t)$, $t \geq 1$. But this is readily checked:
$2 \notin (\Sigma\mathbb{Q}^2)^p$, $1 + X_1^{2p} \notin (\Sigma K^2)^p$.

(3.6) Theorem. If L is a finite formally real extension of K, then

$$S(L) \subset S(K).$$

Remark. The author's original proof was considerably simplified by J.-L. Colliot-Thélène. With his kind permission, his proof will be presented.

<u>Proof.</u> We may assume that $L|K$ has no proper intermediate extension and that there exists a prime $p \in S(L)$, not in $S(K)$. First note that $p^i \in S(L)$ holds as well, by (3.3), i). From $H(K) \subset H(L)$ we derive $\mathbb{E}^+(K) \subset \mathbb{E}^+(L)$. Assume $p \neq 2$. By assumption and (3.1), we find $\alpha \in \mathbb{E}^+(K) \setminus [\mathbb{E}^+(K)]^p$ such that $\alpha = \beta^p$ holds for a suitable $\beta \in L^\times$. If $\beta \in K^\times$, then $\beta \in \mathbb{E}(K)$ since $\alpha \in \mathbb{E}(K)$ and $\beta \in \Sigma K^2$ since p is odd and $\alpha \in \Sigma K^2$. Hence, $\beta \in \mathbb{E}^+(K)$: a contradiction. Therefore, $\beta \notin K$ and $L = K(\beta)$, $[L:K] = p$. Now using $p^2 \in S(L)$, we find $\gamma \in L^\times$ with $\alpha = \gamma^{p^2}$. Taking norms for $L|K$, we get $\beta^{p^2} = \alpha^p = N(\gamma)^{p^2}$ which yields $\beta = N(\gamma) \in K$ since p is odd. But as just seen, this is impossible.

Now consider the case $p = 2$. This time we find $\alpha \in \mathbb{E}^+(K) \setminus [\mathbb{E}^+(K)]^2$ and $\beta \in L^\times$ with $\alpha = \beta^4$. If $\beta \in K$, then $\alpha = (\beta^2)^2$ with β^2 necessarily in $\mathbb{E}^+(K)$: a contradiction. Hence, $\beta \notin K$ and $L = K(\beta)$. The extension $L|K$ must be of degree 2, not 4; otherwise we would have the chain $L = K(\beta) \underset{\neq}{\supset} K(\beta^2) \underset{\neq}{\supset} K$, contradicting our assumption on $L|K$. Again by using $p^i \in S(L)$ for every $i \in \mathbb{N}$, we choose $\gamma \in L^\times$ with $\alpha = \gamma^8$. Taking norms for $L|K$, we obtain $\beta^8 = \alpha^2 = N(\gamma)^8$; hence, $\beta = \pm N(\gamma)$ since L is formally real. But this means $\beta \in K$, already seen to be impossible.

In [B1, Chapter I] the author has introduced the notions of n-pythagorean and strictly n-pythagorean fields. One should also consider the paper [J2] where many results concerning these fields can be found. In the spirit of the definitions as given in [B1], we first extend them to all even exponents. We begin by redefining a field K to be <u>n-pythagorean</u> if it is formally real and satisfies

$$\Sigma K^{2n} = K^{2n} .$$

Obviously, one has only to check $1 + a^{2n} \in K^{2n}$ for every $a \in K$ in order to know $\Sigma K^{2n} = K^{2n}$. Note: 1-pythagorean = = pythagorean.

<u>(3.7) Proposition.</u> K is n-pythagorean if and only if K is pythagorean and $n \in S(K)$.

<u>Proof.</u> Make use of $(\Sigma K^2)^n \subset \Sigma K^{2n}$.

We restate the results (3.3) and (3.6) for n-pythagorean fields. Note that the field in (3.3), ii) was already pythagorean.

<u>(3.8) Theorem.</u> Let K be a formally real field. Then the following statements hold:

i) the set $\{n \in \mathbb{N} \mid K^{2n}+K^{2n} = K^{2n}\}$ is either empty or a multiplicative semigroup generated by a set of primes; every set of primes occur

ii) if $L|K$ is a finite extension and if L is n-pythagorean, then K is also n-pythagorean.

<u>Proof.</u> The only argument not given so far is the statement that K has to be pythagorean as its finite extension L is pythagorean. This result is due to Diller-Dress [D-D] .

An astonishing consequence of (3.8) ought to be mentioned. Given $n \in \mathbb{N}$ with $n = p_1^{\alpha_1} \dots p_r^{\alpha_r}$ as its decomposition in powers of the distinct primes $p_1, \dots, p_r$. If K is then n-pythagorean, i.e., satisfies

$$K^{2n} + K^{2n} = K^{2n} ,$$

then
$$K^{2m} + K^{2m} = K^{2m}$$

follows for every $m = p_1^{\beta_1} \dots p_r^{\beta_r}$, $\beta_i \geq 0$. In the case of $n = 2$ this was proved by Jacob [J, Section 4] for fields with a finite number of square classes and for arbitrary fields by J. Harman in his Ph.D. thesis, Berkeley, 1980.

In the theory of formally real fields the class of <u>strictly pythagorean</u> or <u>superpythagorean</u> fields has been studied by various authors. (See [L2, §11] for a survey.) We shall introduce and study here the strictly n-pythagorean fields. For $n = 2^t$ these fields have already been studied in [B1] . We first need some definitions. Following [B 3] , a subset $T \subset K$ is called a <u>preordering</u> of K if it satisfies

(3.9) $T + T \subset T$, $TT \subset T$, $0,1 \in T$, $-1 \notin T$, $T^\times := T \setminus \{0\}$ is a subgroup of $K^\times$.

A preordering T is called a <u>fan</u> if every subgroup U of $K^\times$ with $T^\times \subset U$, $-1 \notin U$ is additively closed. We denote by X_T the set of all signatures, (see [B-H-R]), $\chi : K^\times \to \mu$ which are trivial on $T^\times$: $\chi(T^\times) = 1$. By definition, a signature χ is any character on $K^\times$ with values in $\mu := \{\zeta \in \mathbb{C} \mid \zeta^n = 1 \text{ for some } n \in \mathbb{N}\}$ whose kernel is additively closed.

(3.10) Proposition. Let T be a preordering of K. Then the following statements are equivalent:

i) T is a fan;

ii) every character $\chi : K^\times \to \mu$ with $\chi(T^\times) = 1$, $\chi(-1) = -1$ is a signature.

If T is a fan, then

$$T^\times = \bigcap_{\chi \in X_T} \operatorname{Ker} \chi .$$

Proof. The following result is basic for the proof. Let A be a multiplicative abelian group, $\varepsilon \in A$, $\varepsilon \neq 1$. Then

$\{1\} = \bigcap \ker \chi$, where χ ranges over

all characters $\chi : A \to \mu$ with $\chi(\varepsilon) \neq 1$. This follows from the fact that μ is divisible and hence an injective $\mathbb{Z}$-module. Apply this result to $A = K^\times/U$ and $\varepsilon = -U \in A$.

A fan T is called trivial if it is the intersection of at most two (ordinary) orders. The next theorem generalizes the result [B3, (4.3)] which in turn extends Bröcker's result [Br 1] that every quadratic fan is a "pullback" of a trivial one. We set $A(\chi) := A(\ker \chi \cup \{0\})$ as in the notation of [B3].

(3.11) Theorem. Let T be a fan of K. Denote the compositum of all valuation rings $A(\chi)$, $\chi \in X_T$ by A_T. Then A_T is a valuation ring with a formally real residue field. Let I be its maximal ideal and set $\overline{T} := \{a+I \mid a \in T \cap A_T\}$. Then $\overline{T}$ is a trivial fan of A_T/I.

Proof. Check that in the proof of [B3, (4.3)] the hypothesis "T is a torsion fan" was only needed to show $T^\times = \cap \ker \chi$, $\chi \in X_T$. But in our context this is provided by (3.10).

This theorem has the following important consequence.

(3.12) Theorem. A preordering T of K is a fan if and only if for every $a \notin -T$

$$T + Ta = T \cup Ta .$$

Proof. The difficult part of the proof is concerned with the statement that a fan T meets the condition as given above. First note $1 + I \subset T$ because of $A(\chi) \subset A_T$, $1 + I \subset \ker \chi$ and $T^\times = \cap \ker \chi$, $\chi \in X_T$. This implies $1 + a \in T \cup Ta$ if a is not a unit of A_T. Then assume $a \in A_T^\times$. By assumption and by using $1 + I \subset T$, we see $\overline{a} \notin -\overline{T}$, $\overline{a} := a + I$. Since $\overline{T}$ is a trivial fan, we get $1 + \overline{a} \in \overline{T} \cup \overline{T}\overline{a}$ and, from this, $1 + a \in T \cup Ta$, again by using $1 + I \subset T$.

In the spirit of the definition of a strictly pythagorean field we call a field K strictly n-pythagorean if

$$K^{2n} \text{ is a fan,}$$

or, equivalently expressed, if

every character $\chi : K^\times \to \mu$ with $\chi(K^{\times 2n}) = 1$

and $\chi(-1) = -1$ is in fact a signature.

Note: $K^{\times 2n} = \cap \ker \chi$, $\chi : K^{\times} \to \mu$ with $\chi(K^{\times 2n}) = 1$, $\chi(-1) = -1$.

Now set

$$\tilde{S}(K) = \{n \mid K^{2n} \text{ is a fan}\} .$$

Obviously, $1 \in \tilde{S}$ iff K is strictly pythagorean. If $n \in \tilde{S}(K)$, then $K^{2n} + K^{2n} = K^{2n}$ and hence $n \in S(K)$ and K is pythagorean. Thus we obtain:

<u>(3.13) Proposition.</u> If $\tilde{S}(K) \neq \emptyset$, then K is pythagorean and

$$\tilde{S}(K) \subset S(K) .$$

In general $\tilde{S}(K) \neq S(K)$ for pythagorean fields K as we shall see in a moment. First, we derive the main result on $\tilde{S}(K)$.

<u>(3.14) Theorem.</u> Let K be a formally real field. Then the following statements hold:

i) $\tilde{S}(K) = \emptyset$ or $\tilde{S}(K)$ is a multiplicative semigroup generated, as a semigroup with a unit, by a set of primes $\tilde{\mathbb{P}}(K)$; every set of primes occurs as a set $\tilde{\mathbb{P}}(K)$ where K can be chosen as a pythagorean number field.

ii) if $L|K$ is a finite extension and L formally real, then

$$\tilde{S}(L) \subset \tilde{S}(K) .$$

<u>Proof.</u> i) Let $\tilde{V}$ be the valuation ring of K which is generated by the valuation rings V_λ of all the real

places $\lambda \in M(K)$. Let $\tilde{I}$ denote its maximal ideal and $\tilde{k}$ its residue field. Obviously, we have

$$(3.15) \qquad 1 + \tilde{I} \subset \mathbb{E}^+ .$$

Given a <u>pythagorean field K</u> , we next prove

$$(3.16) \qquad n \in \tilde{S} \text{ iff } 1 + \tilde{I} \text{ n-divisible, } \tilde{k}^2 = \tilde{k}^{2n} \text{ , } \tilde{k}^2 \text{ a trivial fan.}$$

To prove this, first note that $1 + \tilde{I}$ is 2-divisible since $\mathbb{E}^+ \subset K^2$ as K is pythagorean, i.e. $\tilde{V}$ is 2-henselian.

(In view of this, (3.16) extends the Brown-Bröcker characterization of strictly pythagorean fields. (See [L 2 , Theorem 11.12] .) Assume $n \in \tilde{S}$. By definition, $A_{K^{2n}}$ is generated by the valuation rings $A(\chi)$, $\chi \in X_{K^{2n}}$. But by [B3, (3.4)] , these rings are just the valuation rings V_λ , λ ranging over $M(K)$. Hence $A_{K^{2n}} = \tilde{V}$. From $\mathbb{E}^+ \subset \Sigma K^{2n} = K^{2n}$ we get that $1 + \tilde{I}$ is n-divisible. Furthermore, by (3.11) , $\overline{K^{2n}} = \tilde{k}^{2n}$ is a trivial fan. This shows $\tilde{k}^2 = \tilde{k}^{2n}$. To prove the converse implication in (3.16) extend, for example, the proof of [B-K , Lemma 7] to our situation.

Now, (3.16) being proved, argue as in the proof of (3.3),i) to see that $\tilde{S}(K)$ is a semigroup. Since $n \in \tilde{S}$, $m|n$ implies $m \in \tilde{S}$, we get that $\tilde{S}$ is generated by a set of primes (as a semigroup with a unit element). In order to realize a given set $\tilde{\mathbb{P}}$ of primes we first apply [Ge , Theorem 4.3] to exhibit a field $K = R_1 \cap R_2$, where R_1, R_2 are real closures of $\mathbb{Q}$, with an absolute Galois group

$$G(\overline{K} \mid K(\sqrt{-1})) \simeq \prod_{p \notin \tilde{\mathbb{P}}} \mathbb{Z}_p \quad ,$$

$\mathbb{Z}_p$ the group of p-adic integers. K is a hereditarily pythagorean field with at most two orders. We have

$$K^2 = K^{2p} \text{ for } p \in \tilde{\mathbb{P}},\ [K^{\times 2} : K^{\times 2p}] = p \text{ for } p \notin \tilde{\mathbb{P}} .$$

To see this appeal, e.g., to [B1, Theorem 16, p. 120] . Since K is a number field, we have $\tilde{V} = K$. In view of this, (3.16) shows $\tilde{\mathbb{P}}(K) = \tilde{\mathbb{P}}$.

We now turn to the proof of ii) . We have to show that

$$p \in \tilde{S}(L) \text{ implies } \tilde{p} \in S(K)$$

for $p = 1$ or p a prime number. In [B1, p. 136, Corollary] this was proved for $p = 1$ and $p = 2$. Hence, $p \neq 1,2$ from now on. We may assume that $L|K$ has no proper intermediate extension. Now assume that $p \in \tilde{S}(L) \setminus \tilde{S}(K)$ holds. By (3.12) we find $a \in K$ with $a \notin -K^{2p}$, $1 + a \in L^{2p}$ but $1 + a \notin K^{2p} \cup K^{2p}a$. Set $1 + a = x^{2p}$, $x \in L$. We have $x \notin K$. First assume $x^2 \in K$. Then $L = K(x)$ and we shall apply the norm $N: L \to K$. We have $N(x) = -x^2$ and $N(x^p) = -x^{2p} = -1 -a$. If $x^p \in -L^{2p}$ applies, then $N(x)^p \in K^{2p}$ follows showing $-1 -a \in K^{2p}$, $-a \in 1 + K^{2p} \subset K^{2p}$ as, by (3.8), ii) K is p-pythagorean. This is a contradiction and we have $1 + x^p \in L^{2p} \cup L^{2p}x^p$. Taking norms, we get $-a = N(1+x^p) \in K^{2p} \cup K^{2p}(-1-a)$, contrary to our assumption on a . Hence, $x^2 \notin K$ and we get $L = K(x^2)$, $[L:K] = p$. From $x^{2p} = 1 + a$ we derive $N(x^2) = 1 + a$. Using $1 + a = (x^2)^p = \sum_{i=0}^{p} (x^2-1)^i(-1)^{p-i}$ we further have $N(x^2-1) = a$. Now, $x^2 \in L^{2p}$ would imply $1 + a = N(x^2) \in K^{2p}$. Hence, $x^2 \notin L^{2p}$ and $x^2 - 1 \in -L^{2p} \cup L^{2p}x^2$. Taking norms, one gets the contradictions $a \in -K^{2p}$ or $a \in K^{2p}(1+a)$. Thus, the proof is complete.

As an application we prove

(3.17) Proposition. For a formally real field K the following statements are valid:

i) if $\tilde{S}(K) \neq \emptyset$, then $\# M(K) \leq 2$,

ii) if $\# M(K) < \infty$ and if K is 2-pythagorean, then K is even strictly 2-pythagorean.

Proof.

i) Every real place of K obviously factors over the residue field $\tilde{k}$ of $\tilde{V}$. By (3.16) $\tilde{k}$ has at most two orders which implies $\# M(K) = \# M(\tilde{k}) \leq 2$.

ii) If K is 2-pythagorean, then, by (3.2), $M(K)$ is connected. Therefore, $\# M(K) = 1$. From this one gets $H(\tilde{k}) = \tilde{k}$, $\mathbb{E}^+(\tilde{k}) = \Sigma\tilde{k}^2$. Now, since K is 2-pythagorean, $\Sigma\tilde{k}^2 = \tilde{k}^2$ and $\Sigma\tilde{k}^4 = \tilde{k}^4$. Because of (1.6) we get $\tilde{k}^2 = \tilde{k}^4$, $\tilde{k} = \tilde{k}^2 \cup -\tilde{k}^2$. Hence, $\tilde{k}^2$ is the unique order of $\tilde{k}$. In order to apply (3.16) one further needs that $1 + \tilde{I}$ is 2-divisible. But this holds as K is pythagorean. Note, e.g., (3.15).

The condition $\# M(K) < \infty$ is essential in ii). To see this consider the field K of the real meromorphic functions on $\mathbb{R}$, where $\mathbb{R}$ is considered as a real analytic curve. One checks that K is n-pythagorean for every n. Hence $S(K) = \mathbb{N}$. But, using [Sch 2, 1.11], one sees that every $x \in \mathbb{R}$ induces, by evaluation in x, a homomorphism $H(K) \to \mathbb{R}$. Thus $\# M(K) = \infty$ and, by (3.17),i), $\tilde{S}(K) = \emptyset$.

(3.18) Remark. After this paper was nearly completed,

A. Wadsworth suggested that the open mapping theorem of [E-L-W] could also be used in the proof of (1.14), instead of the trace formula (1.13). One would show that the place $\tau : k(a,b) \to k(ba^{-t}) \cup \infty$, for t sufficiently large and even, is compatible with an order P' of $k(a,b)$ which is extendible to K. Then this τ also extends to a real place of K.

References

[B1] Becker, E.: Hereditarily pythagorean fields and orderings of higher level. IMPA Lecture Notes, No. 29 (1978), Rio de Janeiro.

[B2] Becker, E.: Summen n-ter Potenzen in Körpern, J. reine angew. Mathematik 307/308 (1979), 8-30.

[B3] Becker, E.: Partial orders on a field and valuation rings, Comm. Alg. 7 (1979), 1933-1976.

[B4] Becker, E.: Valuations and real places in the theory of formally real fields, these Proceedings.

[B-H-R] Becker, E. and Harman, J. and Rosenberg, A.: Signatures of fields and extension theory, J. reine angew. Mathematik, to appear.

[B-K] Becker, E. and Köpping, E.: Reduzierte quadratische Formen und Semiordnungen reeller Körper, Abh. Math. Sem. Univ. Hamburg 46 (1977), 143-177.

[Br 1] Bröcker, L.: Characterization of fans and hereditarily pythagorean fields, Math. Z. 151 (1976), 149-163.

[Br 2] Bröcker, L.: Über die Pythagoraszahl eines Körpers, Arch. Math. 31 (1978), 133-136.

[D-D] Diller, J. and Dress, A.: Zur Galoistheorie pythagoreischer Körper, Arch. Math. 16 (1965), 148-152.

[E] Ellison, W.J.: Waring's problem, Amer. Math. Monthly 78 (1971), 10-36.

[E-L-W] Elman, R. and Lam, T.Y. and Wadsworth, A.: Orderings under field extensions, J. reine angew. Math. 306 (1979), 7-27.

[G] Gilmer, R.: Multiplicative ideal theory, Kingston 1968.

[Ge] Geyer, W.-D.: Galois groups of intersections of local fields, Israel J. Math. 30 (1978), 382-396.

[H-W] Hardy, G.H. and Wright, E.M.: An introduction to the theory of numbers, Oxford 1960.

[H] Hilbert, D.: Beweis für die Darstellbarkeit der ganzen Zahlen durch eine feste Anzahl n-ter Potenzen (Waringsches Problem), Math. Ann. 67 (1909), 281-300.

[He] Heitmann, R.C.: Generating ideals in Prüfer domains, Pac. J. Math. 62 (1976), 117-126.

[Jo] Joly, J.R.: Sommes des puissances d-ièmes dans un anneau commutatif, Acta Arithm. 17 (1970), 37-114.

[J] Jacob, B.: On the structure of pythagorean fields, J. Algebra 68 (1981), 247-267.

[K] Kamke, E.: Zum Waringschen Problem für rationale Zahlen und Polynome, Math. Ann. 87 (1922), 238-245.

[K-P] Koch, H. and Pieper, H.: Zahlentheorie. Ausgewählte Methoden und Ergebnisse, Berlin 1976.

[Kn] Knebusch, M.: On the extension of real places, Comment. Math. Helv. 48 (1973), 354-369.

[L1] Lam, T.Y.: The algebraic theory of quadratic forms, Reading 1973.

[L2] Lam, T.Y.: The theory of ordered fields, in: Proceedings of the Algebra and Ring Theory Conference (ed. B. Mc Donald), Univ. of Oklahoma 1979, Lect. Not. in Pure and App. Math. Vol. 55.

[Sch 1] Schülting, H.-W.: Über reelle Stellen eines Körpers und ihren Holomorphiering, Ph.D. thesis, Dortmund 1979.

[Sch 2] Schülting, H.-W.: On real places of a field and their holomorphy ring, Comm. Alg., to appear.

[Sch 3] Schülting, H.-W.: Über die Erzeugendenzahl invertierbarer Ideale in Prüferringen, Comm. Alg. 7 (1979), 1331-1349.

[S] Siegel, C.L.: Darstellung total positiver Zahlen durch Quadrate, Math. Z. 11 (1921), 246-275.

[Z-S] Zariski, O. and Samuel, P.: Commutative algebra II, New York 1960.

Added in proof:

[J2] Jacob, B.: Fans, real valuations and hereditarily-Pythagorean fields, Pac. J. Math. 93 (1981), 95-105.

[P] Pourchet, Y.: Sur la représentation en somme de carrés des polynômes à une indéterminée sur un corps de nombres algébriques, Acta Arithm. 19 (1971), 89-104.

Mathematisches Institut
Universität Dortmund
Postfach 500500
4600 DORTMUND 50

ON CERTAIN TOPOLOGICAL SPACES ADMITTING STRONGLY COHERENT REAL ALGEBRAIC STRUCTURE

Riccardo Benedetti (PISA)(1)

Ist.Mat."L.Tonelli"- Università di Pisa

Introduction

By an algebraic variety we mean an affine real algebraic variety realized in some Euclidean space R^n . It is well known that every such a variety X can be "nicely" stratified and triangulated ; on the other hand , needless to say that there are particular "topological" conditions necessarily verified by any stratified , or polyhedral , space in order that it has the same topology type of an algebraic variety . For instance , D.Sullivan ([1]) first observed that :

Every algebraic variety X (actually every real analytic space) is locally homeomorphic to the cone over a polyhedron with even Euler-Poincaré characteristic .

Thus , X is a so-called Euler space (see [2] , [3]) ; moreover the algebraic resolution of singularities (Hironaka [4]) implies , of course , the existence of a certain kind of "good" topological resolution (so one has , on priciple,

---

(1)The author belongs to the G.N.S.A.G.A. of Italian C.N.R.

the vanishing of obstructive (co-) cycles in suitable (co-) bordism theories on X ; see [5] ,[6]) . A satisfactory general definition of "good" topological resolution is not available at present , but recent papers show that there are reasons to believe that just the existence of such a resolution characterizes the topology of algebraic varieties .
First of all , S.Akbulut and H.King (as announced in [7]) have obtained "polynomial equations" for certain stratified spaces (called A-spaces) admitting in fact a very good resolution by definition .This result is really remarkable because it implies a complete description , from the topology viewpoint , of algebraic varieties with isolated singularities and, mostly , because A-spaces are a so large class to contain at least all (compact) closed P.L. manifolds (see [8]) . So one has a fine extension to the general non-smoothable case of the Nash-Tognoli theorem (we recall that N.H.Kuiper solved the problem for P.L. manifolds of dimension 8 [9]) .
In [10] we have proved that every compact Euler space of dimension ≤ 2 can be resolved in a way , generally not so good as for A-spaces , but enough to achieve in these dimensions the converse of Sullivan's remark . However , it comes in handy to recall here that , for technical reasons (but not only for those) , we also distinguish a special class of(two three dimensional) Euler spaces (essentially equivalent to

the class of A-spaces) getting easier and more natural constructions , so that the "pathologies"which can occur in the general case become clear .

The purpose of this short and largely expositive note is to understand the algebraic-analytic meaning of the very good topological resolution of singularities , actually showing that those spaces admitting such resolutions (the A-spaces in fact) also admit a real algebraic structure with very strong coherence properties of the underlying analytic one . Moreover we prove also a sort of converse of this result .

Proofs are sketched or even omitted ; details (which may be tedious) will appear elsewhere . Anyway we believe that it is enough to clarify a little what was obtained up to now and how it is far from a complete and as simple as possible solution of the problem , and we hope that it can give some indication for further steps .

A preliminary remark For every algebraic variety X , its one-point compactification $\overline{X}$ can be regarded as an algebraic variety such that $X = \overline{X} - \{point\}$. So , even if we shall consider only compact spaces we don't lose in generality .

1. Spaces with very good resolutions .

We present some definitions of spaces with a natural very good topological resolution of singularities . All these examples can be regarded as possible (and fairly naive) topological analogue of the (complex) algebraic resolution .

(1) A-spaces

They are defined inductively ; an A-space is an A_k-space for some $k = 0,1,2,\dots,k,\dots$, where :

A_0- spaces are smooth compact manifolds ;

an A_k-space is a compact smooth stratified space X such that :

(i) the neighbourhood of each stratum X_i has a fixed <u>trivialization</u> $f_i : X_i \times \text{cone}(T_i) \longrightarrow X$ where T_i is an A_{k-1}-space and f_i is compatible with the trivializations of the neighbourhoods of the strata of T_i ;

(ii) Each T_i <u>bounds</u> a compact A_{k-1}-space with boundary V_i .

More-over one <u>proves</u> (see [7]) that every A_k-space T which bounds , in fact can be regarded as the boundary of an A_k-space W_i with a "<u>nice</u>" <u>spine</u> S (that is $W \setminus S$ is an open collar on T) consisting of codimension one closed (i.e without boundary) A_k-subspaces in general position .

Now it is clear how to construct the very good resolution of X : take a lowest dimensional stratum X_i and replace $f_i(X_i \times \text{cone}(T_i))$ by $X_i \times W_i$ .There is a natural map from the so obtained space Y_1 (which remains an A-space) to X which is the identity outside $X \setminus f_i(X_i \times \text{cone}(T_i))$ and collapses $X_i \times S_i$ to $X_i \times \{*\}$, $*$ being the vertex of the cone over T_i . After a finite number of such "blow ups" one obtains a smooth manifold $Y_h = \tilde{X}$ resolving X . The result of [7] asserts that :

1.1 THEOREM <u>Every closed A-space X is homeomorphic to an algebraic variety $\hat{X}$. Moreover the natural stratification of $\hat{X}$ (i.e. $\hat{X} \setminus \text{Sing}(\hat{X})$, $\text{Sing}(\hat{X}) \setminus \text{Sing}^2(\hat{X})$...)coincides with the stratification of X</u> .

2.1 REMARK The simplest examples of closed A_1-spaces are produced by the compact algebraic varieties with isolated singularities . In fact we know by the algebraic resolution that the neighbourhood of each singular point is of the form cone($\bigcup_{j=1}^{r} T_j$) where every T_j is a closed compact smooth manifold (of dimension d_j) which bounds . Thus theorem 1.1 contains , in particular , a complete topological characterization for these varieties .

(2) In [10] we considered , as special subclass of Euler spaces , the set $\mathfrak{F}$ of all compact smooth stratified spaces P of dimension 2 (which we may assume to be connected and "purely dimensional" i.e. without interior points of dimension one) having each point y contained in one of the following sets :

(i) $R(P) = \{$ smooth points of P $\}$;

(ii) $S_1(P) = \{$ points of P having a neighbourhood of the form : $[0,1] \times$ cone (even number of points) $\}$;

(ii) $S_0(P) = \{$points of P neither in R(P) nor in $S_1(P)$ and having a neighbourhood of the form : cone(G) where G is a graph with the following properties :

(a) Each singular point of G has as neighbourhood cone(even number of points) ;

(b) The number of singular points of G with associated the same neighbourhood's number as in (a) is even$\}$.

In [10] we proved the above theorem 1.1 by replacing in the statement "<u>closed A-space X</u>" by "<u>every X in $\mathfrak{F}$</u>" .

3.1 REMARK Every A-space of dimension 2 belongs to $\mathfrak{F}$.
On the other hand it is easy to see that every element of $\mathfrak{F}$ has structure of A-space up to refinements of the stratification . But in order to prove both points of the statement of

Thm. 1.1 for the elements of $\mathfrak{J}$ we had to make a different construction . In fact , the "<u>trivialized</u> <u>neighbourhoods</u>" assumption in the definition of A-spaces is fairly honest (for instance every algebraic variety can be stratified in this manner) ; doubtless , it allows technical advantages , expecially for the construction of "algebraic blow downs" (see [7]) . However it seems to us in some extent , not completely natural ; for example it is easy to produce a space in $\mathfrak{J}$ such that (by means of [10]) for the homeomorphic algebraic variety $\hat{X}$ one has $\mathrm{Sing}^2(\hat{X}) = \emptyset$ (we may assume , for instance , that $S_0(X) = \emptyset$ and $S_1(X)$ = "a single circle of singularities") , while , whenever one regards X as an A-space and applies theorem 1.1 , necessarily , $\mathrm{Sing}^2(\hat{X}) \neq \emptyset$.(See also [6] for a discussion about possible extra structures over stratified spaces) . This induces us to consider another acceptable definition of spaces with very good resolution , which contain $\mathfrak{J}$ as a particular case . We do it below .

(3) <u>$\tilde{A}$-spaces</u>

As for A-spaces they are defined by induction :

$\tilde{A}_0$-spaces are again compact smooth manifolds ;

an $\tilde{A}_k$-space is a compact smooth stratified space X given by:

$$X = X_0 \bigcup_h \coprod_{i=1}^{r} (B_i)$$

where : (i) X_0 is an $\tilde{A}_{k-1}$ -space with boundary ; each B_i is the mapping cylinder of a locally trivial fiber bundle $f_i : P_i \longrightarrow Y_i$, Y_i being a smooth manifold and the typical fiber an $\tilde{A}_{k-1}$ -space without boundary ; $h = \{h_i\}$ and each

$h_i : P_i \longrightarrow \partial X_0$ is a smooth embedding (preserving the strata and the links of the strata) ;

(ii) The couple (P_i, f_i) is a boundary for the relative $\tilde{A}_{k-1}$-bordism over $(Y_i, \partial Y_i)$;

(iii) (P_i, f_i) bounds (in the sense of (ii)) a couple (W_i, F_i) with a nice spine , that is a spine S_i of W_i consisting of $\tilde{A}_{k-1}$ -subspaces in general position and a smooth map $g_i : S_i \longrightarrow Y_i$ such that the space :

$$W_i \coprod Y_i \Big/ S_i \ni x \sim y \in Y_i \text{ iff } y = g_i(x)$$

is isomorphic to the mapping cylinder of f_i (in fact F_i follows from g_i).

We have not the general analogue of theorem 1.1 for closed $\tilde{A}$-spaces (in dimension 2 we worked by hands) . Furthermore we have probably to require more about the nice spine of (iii). In particular it seems harder to obtain the algebraic blow downs . In fact , in the case of A-spaces , roughly speaking, one has to "approximate" by algebraic varieties smooth stratified spaces ; here one needs approximations even for maps between such spaces . Anyway some natural questions come out which have , probably , adfirmative answers ; for example :

(a) Does every $\tilde{A}$-space admit a structure of A-space (by refining the stratification) ?

(b) Do (i) and (ii) in the definition of $\tilde{A}$-space imply (iii) ? (The analogue is true for A-spaces) .

2. Some "pathologies" for general Euler spaces of dimension 2 .

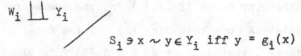

Consider compact Euler spaces of dimension 2 (we know that they are homeomorphic to algebraic varieties) which don't belong to the set $\mathfrak{F}$ defined above.

1.2 EXAMPLE The space X of fig.1 (with the natural minimal stratification) is an "irreducible" Euler space (the meaning is clear) ; also the germs of X at the point a and b are irreducible. Notice : (i) every neighbourhood of a or b contains points of lower dimension = 1 ; (ii) in every neighbourhood of a or b there are points of $S_1(X)$ (with the same sense as in the definition of $\mathfrak{F}$) with neighbourhoods $[0,1] \times$ cone(n points) and non constant n , even if the germ of $S_1(X)$ at a and b is irreducible ; (iii) the pathology appears in codimension two (at a and b) that is the germ of X at any other point coincides with the germ of an element of $\mathfrak{F}$.

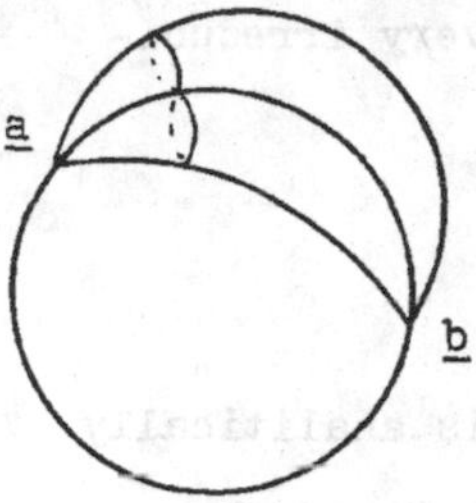

Fig. 1

2.2 EXAMPLE Let X be the suspension of the wedge of three circles as in fig.2 . Notice that the "singular set" of X , $S(X) = S_0(X) \cup S_1(X)$, is not an Euler space . On the other hand , in order to get a good resolution of X which enables us to find "polynomial equations" for X (see [10] for the details) , S(X) has to be completed as in fig.2 by the dashed arc L from a to b made by smooth points of X ; furthermore the resulting algebraic variety $\hat{X}$ homeomorphic to X has as singular set a subvariety corresponding to the "actual" singularities $\tilde{S}(X) = S(X) \cup L$ of X . Thus $\tilde{S}(X)$ has an "immersed component" (L) .

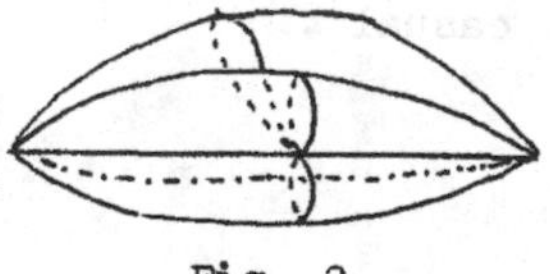

Fig. 2

We recall now some well known facts about the set of non coherence of a real analytic set .

3.2 (a) The set of non coherence of every real analytic set X defined in R^n by global equations is a semianalytic subset of codimension ≥ 2 . (See [11]).

(b) Let X be as in (a) , X_x its germ at the point x , $\tilde{X}_x$ the complexification of X_x . Assume that X_x is irreducible ; then X is coherent in x if and only if there exists a neighbourhood U of x in X such that : (i) If $y \in U$ then $\dim X_x = \dim X_y$; (ii) There exists a representative $\tilde{X}'$ of $\tilde{X}_x$ such that for each y in U the number of irreducible components of X_y equals the number of those of $\tilde{X}'_y$. In general X is coherent in x iff the above conditions hold for every irreducible component of X_x . (See [12]) .

(c) Remember the Cartan example ([13]) :

$$C = \left\{ z(x + y)(x^2 + y^2) = x^4 \right\}$$

Each point of this cone in R^3 near the origin is analitically regular ; but C is non coherent in 0 and the points of the form $(0,0,z)$ $z \neq 0$ are near the origin singular points for the complexification : they are immersed singular points for C . Consider also the following example

$$W = \left\{ y(x^2 - zy^2) = 0 \right\}$$

The points of the form $(0,0,z)$ $z < 0$ are immersed singular points for W .

We shall see soon that the analogies between "bad" Euler spaces and non coherent real analytic sets are not casual .

3. <u>Strong coherent real (algebraic) structures</u> .

We begin by recalling a definition .

1.3 DEFINITION Let X be an algebraic variety (in R^n) , $I(X)$ the ideal of polynomials in n-indeterminates vanishing on X , X_x the germ at x of X as analytic set and , finally , $\mathfrak{J}(X_x)$

the ideal of germs of analytic functions at x vanishing on X_x . We say that X is <u>quasi</u> <u>regular</u> iff for each x in X $\mathcal{J}(X_x) = I(X)\,\mathcal{O}^n_x$, $\mathcal{O}^n_x$ being the ring of germs of analytic functions at x . Notice that , in particular , a quasi regular variety is a <u>coherent</u> real analytic set . Furthermore , it is rather easy to see that X <u>is quasi regular iff the affine algebraic complexification</u> X_C <u>of</u> X <u>induces the analytic germ complexification</u> $\tilde{X}_x$ <u>of each point</u> x <u>of</u> X . (See [14] pag. 52) .

2.3 DEFINITION Let X be an algebraic variety ; we say that X is <u>strongly coherent</u> iff all the varieties X , Sing(X) ,..., $Sing^k(X)$ .... are quasi regular .

Now we can state the first main result of this note .

3.3 THEOREM <u>Let X be a closed A-space (an element of</u> $\mathcal{F}$) ; <u>Then there exists an algebraic variety</u> $\hat{X}$ <u>such that</u> : (i) $\hat{X}$ <u>satisfies the statement of theorem</u> 1.1 ; (ii) $\hat{X}$ <u>is strongly coherent</u> .

<u>Proof</u>.(Sketch) To explicate the kind of argument consider the simplest case of A_1-space arising from the compact algebraic varieties with isolated singulatities (see remark 2.1) . We may assume that X is such a space with only one singular point x_0 (that is $X\setminus\{x_0\}$ is a smooth manifold) and that x_0 has a neighbourhood U = cone(T) where T is a smooth manifold which bounds and is of dimension d-1 (thus X is of pure dimension d) . It comes from the constructions([7], [10]) that we can do so without losing of generality . There exists a very good resolution of X : $X_1 \xrightarrow{p} X_0 = X$, X_1 being a smooth manifold (of dimension d) given by $X_1 = \overline{X\setminus U} \bigcup_T W$ where W has T as boundary and contains a nice spine S (which we may assume to be connected) consisting of smooth hypersurfaces in general position ; p is the identity on $\overline{X\setminus U}$ and collapses S to x_0.

Recall now some known facts :

(a) the couple (X_1,S) is diffeomorphic to a couple $(\check{X},\check{S} = \bigcup_i \check{S}_i)$ where $\check{X}$ and each $\check{S}_i$ are regular algebraic varieties (see , for instance , pag. 133 of [15]) .

(b) let $V \subset W$, Z be compact algebraic varieties and $f: V \longrightarrow Z$ a regular algebraic surjective map . Then there exist an algebraic variety Q and a regular map $F: W \longrightarrow Q$ such that (i) $Z \subset Q$; (ii) $F|_V = f$; (iii) $F|_{W \setminus V}$ is an algebraic isomorphism onto $Q \setminus Z$ (see prop. 3.4 of [10]) .

It is clear that applying (b) to $W = \check{X}$, $V = \check{S}$, $Z = \{x_0\}$ and $f=\text{const.}: \check{S} \longrightarrow \{x_0\}$, one obtains $\hat{X} = Q$ which satisfies the statement of theorem 1.1 . We claim that the so obtained variety $\hat{X}$ is also strongly coherent . Actually we have to prove only that $\hat{X}$ is quasi regular . Notice that $\hat{X} \setminus \{x_0\}$ is quasi regular because it is made by algebraically regular points . Thus it remains to show that $\hat{X}$ is quasi regular at x_0 . Assume that it is not true . Then the analytic complexification $\tilde{\hat{X}}_{x_0}$ of $\hat{X}_{x_0}$ doesn't coincide with the germ at x_0 of the affine algebraic complexification $\hat{X}_C$ of $\hat{X}$. The regular morphism $F: \check{X} \longrightarrow \hat{X}$ (as in (b)) can be extended (in our hypotheses) to a regular <u>surjective</u> morphism $F_C: \check{X}_C \longrightarrow \hat{X}_C$ (see pag. 47 of [14]) . Thus if we consider the restriction of F_C to a suitable neighbourhood N of $\check{S}$ in $\check{X}_C$ onto a neighbourhood P of x_0 in $\hat{X}_C$ which contains a representative Y of $\tilde{\hat{X}}_{x_0}$ we obtain that $F_C^{-1}(Y)$ is a <u>proper</u> analytic subset of N containing $N \cap \check{X}$; it follows that $\check{X}$ cannot be quasi regular along $\check{S}$. But $\check{X}$ is quasi regular because it is even a regular variety . Contradiction .

In the general case we start with the very good resolution of X : (°) $X_n \xrightarrow{p(n)} X_{n-1} \longrightarrow \ldots\ldots \xrightarrow{p(1)} X_0 = X$ where X_n is a smooth manifold and each X_i is (a "better" A-space than X_{i-1}) obtained by topologically blowing up (with nice spine S_i) X_{i-1} as described in sect. 1 . The "<u>algebraic tower</u>" construction of [7] says that we can"approximate" the above resolution by a tower :

$$(°°) \qquad \begin{array}{cccccc} \check{X}_n & \xrightarrow{\bar{p}(n)} & \check{X}_{n-1} & \longrightarrow \ldots & \xrightarrow{\bar{p}(1)} & \check{X}_0 = \hat{X} \\ \uparrow g_n & & \uparrow g_{n-1} & & & \uparrow \\ X_n & \xrightarrow{p(n)} & X_{n-1} & \longrightarrow \ldots & \xrightarrow{p(1)} & X_0 = X \end{array}$$

where : (i) each $\check{X}_i$ satisfies the statement of theorem 1.1 with respect to X_i ; (ii)each vertical arrow is a smooth homeomor-phism preserving the strata and such that $g_i(S_i) = \check{S}_i$ is a closed subvariety of $\check{X}_i$; (iii) all $\bar{p}(i)$ are regular algebraic morphisms such that the diagram (°°) is commutative and each $\bar{p}(i)|\, \check{X}_i \setminus \check{S}_i$ is an algebraic isomorphism onto $\check{X}_{i-1} \setminus \bar{p}_i(\check{S}_i)$.

Thus , by applying the above kind of argument step by step , we obtain that : $\check{X}_n$ (regular) quasi regular $\Longrightarrow \check{X}_{n-1}$ quasi regular $\Longrightarrow \ldots \Longrightarrow \check{X}_0 = \hat{X}$ quasi regular .

Furthermore as the resolution (°) filters to the skeltons of each X_i as the algebraic tower filters to the singularities of $\check{X}_i$, $\mathrm{Sing}(\check{X}_i)$. It follows that also $\mathrm{Sing}^k(\check{X}_i)$ is quasi regular for each k . Thus the theorem is proved actually getting more that each X_i is strongly coherent . To develop completely this proof we have to consider the details of the algebraic tower constructions which is long and needs often to be careful .

4.3 COROLLARY <u>Every closed compact P.L. manifold is homeomor-</u>

phic to a strongly coherent variety .

At last we prove a sort of converse of theorem 3.3 .

6.3 THEOREM *Let* X *be a compact analytic subset of* R^n. *Assume all the sets* X , Sing(X) ,..., $Sing^k(X)$,... *to be real coherent analytic spaces . Then* X *admits a structure of closed* $\tilde{A}$*-space* .

Proof.(Sketch) Let V be a compact smooth stratified space (which we assume to be of pure dimension d for the sake of simplicity) such that each stratum of X has a system of locally trivial "tubular neighbourhoods" (see [16]) . The properties (i) and (ii) of the definition of (closed) $\hat{A}$-spaces (see sect. 1) can be reformulated in an essentially equivalent manner as follows : there exists a topological resolution of singularities of V :

$$V_k \xrightarrow{p_k} V_{k-1} \longrightarrow \dots \xrightarrow{p_1} V_0 = V$$

such that : (1) V_k is a closed smooth manifold ; (2) Each V_i is a compact smooth stratified space with locally trivial tubular neighbourhoods for every stratum and of pure dimension d ; (3) Each p_i is an *onto* continuous map obtained by a finite number of topological blow ups of V_{i-1} along the closed submanifolds M^i_1 ,...., $M^i_{h_i}$ (in fact closed strata of V_{i-1}) that is : (a) $p_i | V_i \setminus p_i^{-1}(\bigcup_j N^i_j)$ is an isomorphism (of stratified spaces) onto $V_i \setminus (\bigcup_j N^i_j)$, the $N^i_1,\dots,N^i_{h_i}$ being disjoint tubular neighbourhoods of $M^i_1,\dots,M^i_{h_i}$; (b) p_i is a smooth morphism of stratified spaces , eventually by refining the stratification of V_i inside $p_i^{-1}(\bigcup_j N^i_j)$; (4) If we denote by cod V_i = max $\{$ codimension of V^i_s in V_i , V^i_s stratum of $V_i \}$

then cod $V_i < \text{cod } V_{i-1}$.
By using this further characterization of $\hat{A}$-spaces's properties (i) and (ii) , it follows from the Hironaka's first desingularization theorem for real analytic spaces (as stated in [4] pag. 159) that X can be stratified to obtain a space V' satisfying those properties . To see this fact one needs to remark : (I) The analytic resolution of singularities works by means of a succession of (admissible) monoidal transformations (or generalized blow ups , see [4]) with regular centers , which are in particular "topological blow ups" (that is satisfy the above (a) and (b) - with respect to suitable stratifications -) ; (II) The succession can be chosen such that : only a finite number of monoidal transformations occur and if D_r , D_s are both centers of such transformations q_r, q_s , if q_r appears before q_s , then

$$\text{cod}(D_r \text{ in } X) \geq \text{cod}(D_s \text{ in } X)$$

; (III) If D is the center of an admissible monoidal transformation of X (X is "normally flat" along D ; see [4]) then X is smoothly locally trivial along D ; (IV) If X' is obtained from X by an admissible monoidal transformation p , then it satisfies the hypotheses of theorem 6.3 ; (V) (last but crucial remark) If $X' \xrightarrow{p} X$ is as in (IV) then p is a proper onto analytic map . (Note that (I) and (II) are general properties of the resolution of singularities , while (III),(IV) and (V) depend strictly on our coherence assumptions) .
By means of fairly standard arguments of stratification of analytic spaces it is not so hard to obtain now the required V' . It remains to see that X can be stratified to satisfy also

property (iii) of $\tilde{A}$-spaces (the existence of nice spines) ; but this follows soon from the so-called Hironaka's "normal crossing" second desingularization theorem ([4] pag. 160) . The theorem is proved .

7.3 COROLLARY {Compact coherent real analytic spaces of dim.=2} "=" "=" {Compact strongly coherent algebraic varieties of dim.=2}"=" "=" $\mathfrak{F}$. (Where "=" means up to homeomorphisms) .Moreover a similar statement holds for germs .

REFERENCES

1. D.SULLIVAN , Combinatorial invariants of analytic spaces Proc.of Liverpool Sing.Symposium I , Lecture Notes n° 192, Springer Verlag (1971) .
2. E.AKIN , Stiefel-Whitney homology classes and bordism , T.A.M.S. Vol. 205 , 1975 .
3. C.MCCRORY , Euler singularities and homology operations , Proc.of Symposia in Pure Math . Vol. 27 , 1975 .
4. H.HIRONAKA , Resolution of singularities of an algebraic variety over a field of characteristic zero I and II , Ann. of Math. 79 (1964) .
5. M.KATO , Topological resolution of singularities , Topology Vol. 12 (1973) .
6. S.BUONCRISTIANO,M.DEDO' , Local blow-up of stratified sets up to bordism , to appear on T.A.M.S.
7. S.AKBULUT,H.C.KING , A topological characterization of of real algebraic varieties , B.A.M.S. (new series) vol.2, 1, (1980) .
8. S.AKBULUT,L.TAYLOR , A topological resolution theorem , B.A.M.S. (new series) Vol.2 , 1, (1980) .

9. N.H.KUIPER , Algebraic equations for non smoothable 8-manifolds , Publ.Math. I.H.E.S. n° 33 (1967) .

10. R.BENEDETTI,M.DEDO' , The topology of two-dimensional real algebraic varieties , to appear on Ann.Mat.Pura App.

11. M.GALBIATI , Stratifications et ensemble de non-cohérence d'un espace analytique réel , Inventiones Math. 34 (1976).

12. A.TOGNOLI , Proprietà globali degli spazi analitici reali Ann.Mat.Pura App. 75 (4)

13 H.CARTAN , Variétés analitiques réelles et variétés analitiques complexes , Bull.Soc.Math.France , 85 (1957) .

14. A.TOGNOLI , Algebraic geometry and Nash functions , Institutiones Math. , Vol.III , Academic Press , London and New York (1978) .

15. R.BENEDETTI,A.TOGNOLI , Theoremes d'approximation en géometrie algebrique réelle , in Séminaire sur la Géom.Alg.Réelle (Risler) ,Publ.Math. Univ. Paris VII ,1980.

16. J.MATHER , Notes on topological stability , (mimeo.), Harward , 1970 .

REMARKS AND COUNTEREXAMPLES IN THE THEORY OF REAL ALGEBRAIC VECTOR BUNDLES AND CYCLES

R.Benedetti (Pisa)-A.Tognoli (Tours)

Introduction

Let X be a compact non singular affine real algebraic variety of dimension m (we shall call X shortly a compact algebraic manifold) . For each natural number k , associate to X

$V^k_{alg}(X)$ = the set of isomorphism classes of algebraic k-vector bundles over X (that is bundles defined by regular rational cocycles) .

$V^k_{s\text{-}alg}(X)$ = the subset of $V^k_{alg}(X)$ of those classes having strongly algebraic representatives , that is bundles F of the type $F = g^*(F_{nk})$ $g: X \longrightarrow G_{nk}$ being a regular rational map from X to a suitable Grassmann manifold G_{nk} and $F_{nk} \longrightarrow G_{nk}$ being the tautological bundle over G_{nk} .

$H^{alg}_{m-k}(X)$ = the subgroup of $H_{m-k}(X) = H_{m-k}(X,\mathbb{Z}_2)$ generated by the set of algebraic subvarieties of X of codimension k (see [1]) .

$H^k_{alg}(X)$ = the subgroup of $H^k(X) = H^k(X,\mathbb{Z}_2)$ corresponding to $H^{alg}_{m-k}(X)$ by the Poincaré duality .

There are several questions concerning these objects associated to X which naturally arise , for example , in the study of the topology of real algebraic varieties or the global analytic equivalence between analytic and algebraic varieties (see [2] , [3] , [4]) . In [2] we proved : if we denote for a compact manifold M by $V^k(M)$ the set of continuous isomorphism classes of k-vector bundles over M , then there exists an algebraic manifold $\overline{M}$ diffeomorphic to M such that the natural map $f_k: V^k_{s\text{-}alg}(\overline{M}) \longrightarrow V^k(\overline{M})$ is onto ; on the other hand we produced an example of an irreducible algebraic manifold Y of dimension 2 with two connected components both homeomorphic to the torus T_2 such that the above f_1 is not onto .

The main goal of this note is to improve that example by showing in fact that : (see section 5)

" for each $d \geq 3$ there exist connected algebraic manifolds of dimension d (say Y) such that even the other natural map $\overline{f}_1: V^1_{alg}(Y) \longrightarrow V^1(Y)$ is not onto"(and , hence , $H^{alg}_{d-1}(Y) \neq H_{d-1}(Y)$; this answers to a question of [3] and [4]).

To construct the above examples we need the following result proved in section 3 :

" for every algebraic vector bundle F over X there exists a subvariety S of X of codimension ≥ 2 such that the restriction of F over $X \setminus S$ is actually strongly algebraic."

Furthermore we shall give in section 2 a simple proof of the following result which is essentially "well known" (see[1]):

" $H^*_{alg}(X)$ is a subring of $H^*(X)$ which contains all the Stiefel Whitney classes of every strongly algebraic vector bundle over X ".

We would like to thank N.H.Kuiper and J.Bochnak for their useful comments about the topics of this note .

1. An useful lemma .

Let V be an affine variety in $\mathbb{R}^n \times \mathbb{R}^q$ and p the natural projection onto $\mathbb{R}^q$. Assume that V is irreducible and $\dim p(V) = \dim V$. For each s consider the natural inclusions $\mathbb{R}^s \subset \mathbb{C}^s \subset \mathbb{P}^s(\mathbb{C})$. Let $\overline{V}$ be the projective closure of V in $\mathbb{P}^n(\mathbb{C}) \times \mathbb{P}^q(\mathbb{C})$; if $\overline{p}$ is the projection onto the second factor $\mathbb{P}^q(\mathbb{C})$, one has :

1. $\overline{p}(\overline{V})$ is a projective complex subvariety of $\mathbb{P}^q(\mathbb{C})$;
2. $\overline{V} \cap (\mathbb{R}^n \times \mathbb{R}^q) = V$, $\dim_{\mathbb{C}} \overline{V} = \dim V$;
3. $\dot{V} = \overline{p}(\overline{V}) \cap \mathbb{R}^q$ is the affine closure of p(V) in $\mathbb{R}^q$;
4. $\overline{V}$ and $\overline{p}(\overline{V})$ are irreducible .

Since $\dim V = \dim p(V)$ there exists a closed subset $\overline{S}$ of $\overline{V}$ (in the Zariski topology) such that $\overline{p}: \overline{V} \setminus \overline{S} \longrightarrow \overline{p}(\overline{V} \setminus \overline{S})$ is a covering with (well defined) degree d ; moreover V is not contained in $\overline{S}$ by dimensional reasons .

1.1 LEMMA <u>If d is odd , then $\dim (\dot{V} \setminus p(V)) < \dim V$.</u>

<u>Proof</u>. Assume that $\dim (\dot{V} \setminus p(V)) = \dim V = \dim p(V)$. Then there would exist $x \in (\dot{V} \setminus p(V)) \cap \overline{p}(\overline{V} \setminus \overline{S})$. But , in this case d must be an even number , because $\overline{p}^{-1}(x)$ consists of pairs of conjugate points in $\overline{V}$.

2. Some results about $H^k_{alg}(\cdot)$.

Let X be an affine real algebraic variety. There are various equivalent ways to define strongly algebraic vector bundles over X. Here we recall only that an algebraic vector bundle over X is strongly algebraic if and only if it admits an algebraic complexification (see [2] for other definitions). The following proposition states some remarkable properties of strongly algebraic vector bundles (for the proof see [2])

PROPOSITION For each compact affine variety X and each k the map $f_k: V^k_{s\text{-}alg}(X) \longrightarrow V^k(X)$ is injective.

For any affine variety X, continuous (C^q, smooth...) sections of a strongly algebraic vector bundle over X can be approximated (with respect to the usual topologies) by algebraic sections on every compact subset of X.

Moreover recall that $V^q_{alg}(-)$ and $V^q_{s\text{-}alg}(-)$ can be different (see [5] pag. 40).

A purpose of this section is to give a simple proof of the fact that all characteristic classes of any strongly algebraic vector bundle over an algebraic manifold X belong to $H^*_{alg}(X)$. For each number k define $T_k(X)$ to be the subgroup of $H_k(X)$ of all classes $\bar{s}$ admitting as a representative an algebraic pair (V,f) V being a compact algebraic manifold and $f: V \longrightarrow X$ a rational regular map. If we assume that X is compact let us denote by $T^*(X)$ the Poincaré dual of $T_*(X)$.

2.2 THEOREM Let X be a compact algebraic manifold; then:

(a) for each regular rational map $g: X \longrightarrow W$ between X and a compact algebraic manifold W and for each $\bar{s} \in T^k(W)$ the pullback $g^*(\bar{s})$ belongs to $T^k(X)$.

(b) If $\bar{s}$ and $\bar{t}$ are in $T^*(X)$ then also $\bar{s}\cup\bar{t}$ belongs to $T^*(X)$.

Proof. Let us denote by D the Poincaré duality. Assume that the algebraic pair (V,f) is a representative for $D(\bar{s})$. It is well known that to get a representative for $D(g^*(\bar{s}))$ it is enough to find smooth maps f' and g' such that f' is trasverse to g' and f' (g') approximates f (g) . Then take $(g'^*(V,f'),p)$ where p is the projection into X . We need to do it without changing g . Note that g(X) is a compact semialgebraic subset of W , hence it can be stratified in such a way that the Whitney conditions are satisfied . Standard arguments of transversality show that we are able to find a smooth map f' close to f which is transverse to all strata of g(X) (hence transverse to g itself) and (smoothly) homotopic to f (hence (V,f') and (V,f) are cobordant pairs) . In these hypotheses we can apply the proposition 4.1 of [2] (see also the sect. e of [7]) to obtain a diffeomorphism $h: V \longrightarrow \bar{V}$ between V and another algebraic manifold $\bar{V}$ and a regular rational map $\bar{f}: \bar{V} \longrightarrow W$ such that $\bar{f} \circ h$ is close to f' . Then the usual pull-back of $(\bar{V},\bar{f})$ by means of g gives an algebraic representative of $D(g^*(\bar{s}))$.

Let (V_s,f_s) and (V_t,f_t) be now algebraic representatives of $D(\bar{s})$ and $D(\bar{t})$ respectively . For any pair of transverse smooth maps f'_s and f'_t close to f_s and f_t $(f'^*_s(V_t,f'_t),f'_s \circ p)$ is a representative of $D(\bar{s}\cup\bar{t})$. By applying as before the results of [2] to both f_s and f_t we construct ,for r = s,t , diffeomorphisms $h_r: V_r \longrightarrow \bar{V}_r$ and regular rational maps $\bar{f}_r: \bar{V} \longrightarrow X$ such that f'_r is close to $\bar{f}_r \circ h_r$.

The algebraic pair $(\bar{f}_s^*(\bar{V}_t,\bar{f}_t),\bar{f}_s op)$ is a representative for $D(\bar{s}\cup\bar{t})$.

Now we want to strengthen the above proposition as follows :

2.3 PROPOSITION <u>Let X be as before . Then for each k</u>

<u>$T_k(X) = H_k^{alg}(X)$.</u>

<u>Proof</u>. First we shall show that $T_k(X)$ is contained in $H_k^{alg}(X)$. Let $\bar{a} \in T_k(X)$ and (V,f) be an algebraic representative of $\bar{a}$. If $\bar{a} = 0$ the proposition is obvious. Assume $\bar{a} \neq 0$. Thus $\dim f(V) = \dim V = k$. We may suppose that V is irreducible , otherwise we should work component by component; moreover consider V realized in some $\mathbb{R}^n$, X in $\mathbb{R}^q$. Apply the content of section 1 to the graph of f (which we call V again) and to the projection onto $\mathbb{R}^q$, say p .

<u>Claim</u> If the degree d of the covering is even, then $\bar{a} = 0$. In fact , consider the mapping cylinder of $p : V \longrightarrow p(V)$,

$$C = V \times [0,1] \coprod p(V)_{/\,(x,1)\sim p(x)}$$

and define the continuous map $F: C \longrightarrow X$, $F(x,t) = p(x)$ for $t \neq 1$, $F(x,1) = F(x,p(x)) = p(x)$. The couple (C,F) produces an explicit homology between (V,p) and zero; in fact we can triangulate C in such a way that $S = p(V) \cap \bar{p}(\bar{S})$ is a subcomplex and , since d is even , every k simplex in $C \setminus V \times \{0\}$ is face of an even number of k + 1 simplexes .

Then d is necessarily odd . By applying lemma 1.1 it is clear that p(V) carries a fundamental class (the same of $\dot{V}$) which equals $\bar{a}$,as one can easily see by the mapping cylinder argument again . The converse follows immediately from the Hironaka desingularization theorem .

As an immediate corollary of 2.2 , 2.3 and the well known fact that for every Grassmann manifold G_{nk} $H^*(G_{nk}) = H^*_{alg}(G_{nk})$ we have :

2.4 THEOREM <u>$H^*_{alg}(X)$ is a subring of $H^*(X)$ which contains all the Stiefel Whitney classes of every strongly algebraic vector bundle over X</u> .

The above result becomes more precise for $H^1_{alg}(X)$.

2.5 PROPOSITION <u>For every class $\bar{a}$ in $H^1_{alg}(X)$ the dual $D(\bar{a})$ can be represented by algebraic submanifolds of X</u> .

<u>Proof</u>. Let Z be an algebraic subvariety of X of codimension 1 representing $D(\bar{a})$. The line bundle F_Z associated to Z is of course strongly algebraic (take complexifications of X and Z to obtain one of F_Z itself) . By the proposition at the beginning of this section , there exists an algebraic global section $\underline{z}$ of F_Z transverse to its zero section ; $Y = \{ \underline{z} = 0 \}$ is a regular algebraic hypersurface of X . Actually by using again the proposition recalled before and the fact that the first Stiefel Whitney classes classify the (continuous) line bundles over X it is not hard to prove that $Z \xrightarrow{i} F_Z \xrightarrow{j} Y$ defines in fact $H^{alg}_{m-1}(X) \xrightarrow{i} V^1_{s\text{-}alg}(X) \xrightarrow{j} H^{alg}_{m-1}(X)$ such that $j \circ i = id$ and i is onto . The proposition is proved.

2.6 COROLLARY (Theorem 4 of [3]) <u>If X is unorientable then $H^1_{alg}(X) \neq 0$</u> .

<u>Proof</u>. The tangent bundle of X is strongly algebraic and its first Stiefel Whitney class is just a complete obstruction to get an orientation of X .

3. Algebraic bundles are strongly algebraic out of subvarieties of codimension greater than two .

3.1 THEOREM Let X be an affine variety and F an algebraic vector bundle over X . Then there exists a closed subvariety S of X such that : (i) $\dim S \leqslant \dim X - 2$; (ii) The restriction of F over $X \setminus S$ is strongly algebraic .

Proof. We may assume that X is irreducible . Fix a complexification $\bar{X}$ of X and let $g_{ik}: U_i \cap U_k \longrightarrow GL(n, \mathbb{R})$ be a regular rational cocycle defining F with respect to the open covering of X (in the Zariski topology) $\mathcal{U} = \{U_j = X \setminus T_j\}$. We can find (see [5] , sect. 3 and 4) open neighbourhoods $\bar{U}_{ik}$ of $U_{ik} = U_i \cap U_k$ in $\bar{X}$, and regular rational maps $\bar{g}_{ik} : \bar{U}_{ik} \longrightarrow GL(n,\mathbb{C})$ extending the g_{ik}s . Since X ,and hence $\bar{X}$, is irreducible the $\bar{g}_{ik}$s satisfy the cocycle conditions (whenever they are defined) . To get a complexification of F it would be enough to find open neighbourhoods $\bar{U}_j$ of U_j in $\bar{X}$ (for each j) in such a way that $\bar{U}_i \cap \bar{U}_k = \bar{U}_{ik}$. In general this is impossible . However we shall construct a closed subvariety S of X , satisfying (i) , such that $\dot{g}_{ik} = g_{ik} | U_{ik} \setminus S$ can be complexified in the sense just explained . We note that it is enough to prove that for each fixed pair (i,k) of indexes there exists a closed subvarieties S_{ik} of X and two open sets $\dot{U}_i, \dot{U}_k$ of $\bar{X}$ such that : $\dim S_{ik} \leq \dim X - 2$, $\dot{U}_i \cap \dot{U}_k$ contains $U_{ik} \setminus S_{ik}$, $\dot{U}_i \cap \dot{U}_k$ is contained in $\bar{U}_{ik}$; in fact in this case take $S = \bigcup_{ik} S_{ik}$. Let $d_{ik} = \det g_{ik}$; we may write $d_{ik} = h/f$ where h and f are regular rational functions on X , $h \neq 0$ and $f \neq 0$ on U_{ik} . The ring of regular rational functions R(X) is Noetherian , hence we can write h and f as products of irreducible elements of R(X) : say $h = h_1 \dots h_r$, $f = f_1 \dots f_t$.

Claim Let q be an irreducible element of R(X) and $Y_q = \{q = 0\}$; if Y_q is reducible then codim $Y_q \geq 2$.

In fact , take an extension $\bar{q}$ of q to $\bar{X}$. $Z = \{\bar{q} = 0\}$ is an algebraic hypersurface of $\bar{X}$. Assume codim $Y_q = 1$. In this case there exists an irreducible component of Z , say $\bar{Y}$, which is the complexification of $\bar{Y} \cap X$. More-over we are able to find an equation g for $\bar{Y}$ with real coefficients , so that g divides q . Since q is irreducible Y_q must coincide with $\bar{Y}$ and , hence , is irreducible .

Take as S_{ik} the union of the varieties of the form $\{h_j = 0\}$ or $\{f_p = 0\}$ which are reducible . By the claim codim $S_{ik} \geq 2$. Moreover , since all the subvarieties of $X \setminus S_{ik}$ with equations h_j or f_p are irreducible they are contained either in $T'_i = X \setminus (S_{ik} \cup U_i)$ or in $T'_k = X \setminus (S_{ik} \cup U_k)$. Set $U'_j = U_j \setminus S_{ik}$ for $j = i,k$. It is clear by the construction that the cocycle $g_{ik} \big| U'_i \cap U'_k$ can be complexified to an open covering of the form : $\{\bar{X} \setminus \bar{T}'_j \cup \bar{S}_{ik}\}$ where $\bar{T}'_j$ and $\bar{S}_{ik}$ are complexifications in $\bar{X}$ of T'_j and S_{ik} respectively . The theorem is proved .

4. Basic example of dimension two

Let C be the plane curve defined by the equation :

$$X^4 + Y^4 - 2aX^2 + b = 0 \quad , \text{ where } a^2 > b > 0 \ .$$

C is a compact irreducible non singular curve with two connected components $C = C_0 \cup C_1$. Let S^1 be the unit circle in $\mathbb{R}^2$, $W = W_0 \cup W_1 = C \times S^1$, $W_i = C_i \times S^1$.
Choose a point x_0 in S^1 and let $D = C_0 \times \{x_0\} \subset W_0$. Let us denote by F_D the (smooth) line bundle over W associated to D and by a_D the related class in $H_1(W)$.

4.1 THEOREM (Essentially contained in [2]pag.98) (1) <u>F_D is not isomorphic to any strongly algebraic bundle over W</u>; (2) <u>a_D doesn't belong to $H_1^{alg}(W)$</u> ;(3) <u>D cannot be approximated in W by algebraic submanifolds of W</u> .

<u>Proof</u>. The same argument as in the proof of proposition 2.5 shows that (1),(2) and (3) are equivalent conditions .
Assume that (3) is true and let V be an algebraic submanifold of W close to D . Applying the lemma 1.1 to V and to p = "the projection of W onto C " (the restriction of p to V is an analytic diffeomorphism onto C_0 , so the degree of the covering is odd) , one would obtain that there exists a plane curve $\dot{C}$ containing C_0 and such that dim $\dot{C} \setminus C_0 = 0$. Clearly this is absurd .

The following theorem shows that the above F_D is not even isomorphic to any algebraic bundle :

4.2 THEOREM <u>F_D is not isomorphic to any algebraic vector bundle over W</u> .

<u>Proof</u>. Assume that there exists a line bundle over W algebraic and isomorphic to F_D . Denote this bundle F_D again .
By 3.1 there is a finite set $A = \{a_1 \dots a_k\}$ in W such that the restriction of F_D over $W \setminus A$ is strongly algebraic .
Choose "small" open balls B_i about a_i $i = 1,\dots,k$.

By the proposition at the beginning of section 2 one constructs an algebraic section $\underline{z}$ of F_D over $W \setminus A$ such that $\{\underline{z} = 0\} \cap (W \setminus (\bigcup_j B_j)) = D'$ is an analytic regular curve close to D . The smallest algebraic subvariety of W containing D' is an irreducible curve Z such that $Z \setminus D' = \bigcup_j Z_j$, each Z_j contained in B_j . Apply the lemma 1.1 to Z and to the projection p of Z to C . As in 4.1 the degree is odd and clearly for each Z_j either : or $Z_j \subset \overline{S}$ (see 1.1) or $p(Z_j) \subset C_o$. Hence $\dim p(\bigcup_j Z_j) \cap C_1 = 0$; thus we can conclude again that there would exists a plane curve E containig C_o and such that $E \setminus C_o = \{$finite set$\}$. Contradiction .

4.3 REMARK It is rather easy to construct an algebraic variety X homeomorphic to $T_2 \vee T_2$ ($T_2 = S^1 \times S^1$) , having a line bundle without any algebraic structure : let C ,W be as before and fix two points y_o and y_1 in W_o and W_1 respectively . There exists a regular rational map $q : W \longrightarrow X$ such that: X is an affine variety ; $q(y_i) = z_o \in X$; q gives an algebraic isomorphism between $W \setminus \{y_o, y_1\}$ and $X \setminus \{z_o\}$(see for instance the proposition 3.4 of [6]) . The line bundle over X associated to $q(D)$ works (by assuming $y_o \notin D$) .

5. The examples of dimension ≥ 3

We shall produce two examples of dimension 4 and one of dimension 3 . Of course our method can be used to get further examples or , perhaps , more general statements . On the other hand we are not able to construct a connected example of dimension two (the case of curves is trivial) . (See the paper of Risler "sur l'homologie des surfaces réelles", this volume).

Let :

$$Y_1 = S^1 \times S^1 \times S^1 \times S^1 = T_4$$

$$Y_2 = S^3 \times S^1 \quad (S^3 \text{ being the 3-sphere})$$

$$Y_3 = S^1 \times (T_2 \# T_2) \quad (\# \text{ denoting the "connected sum"}).$$

For $i = 1,2,3$ set $m(i) = \dim Y_i$ and $d(i) = 2m(i) + 1$.

5.1 THEOREM <u>For each $i = 1,2,3$, there exists an algebraic manifold $X_i \subset \mathbb{R}^{d(i)}$, such that (a) X_i is diffeomorphic to Y_i ; (b) $\bar{f}_1: V^1_{alg}(X_i) \longrightarrow V^1(X_i)$ is not onto and , a fortiori , $H^1_{alg}(X_i) \neq H^1(X_i)$.</u>

<u>Proof</u>. <u>$i = 1$</u> Set $V_0 = \{t_0\} \times \{x_0\} \times S^1 \times S^1$,

$$V_1 = \{t_1\} \times S^1 \times S^1 \times \{z_0\}$$

where $V_j \subset Y_1$, $t_0 \neq t_1$. Let $p: Y_1 \longrightarrow S^1$ be the last factor projection .

<u>$i = 2$</u> Fix S and T to be respectively (smooth) copies of S^1 and T_2 embedded inside S^3, $S \cap T = \emptyset$. Set : $V_0 = S \times S^1$ and $V_1 = T \times \{x_0\}$, $V_j \subset Y_2$. Let $p: Y_2 \longrightarrow S^1$ be the natural projection .

<u>$i = 3$</u> Fix three copies of S^1 , say S_i $i= 1,2,3$ and maps :

$$f_j: S_j \longrightarrow S^1 \quad \deg f_j = 1 \quad j = 1,2 ,$$

$$f_3: S_3 \longrightarrow S^1 \quad \deg f_3 = 2 .$$

Clearly we can construct a smooth manifold M , homeomorphic to the one obtained by removing three disjoint open discs from the 2-sphere S^2 , such that (a) the boundary of M consists of the union $\bigcup_i S_i$, (b) there exists a smooth map $F': M \longrightarrow S^1$ extendig f_i $i = 1,2,3$.

Take the smooth double of M , say T , which is diffeomorphic to $T_2 \# T_2$, and let $F: T \longrightarrow S^1$ be a smooth map which extends F' . We may consider $Y_3 = S^1 \times T$. Set $V_o = S^1 \times S_1$ and $V_1 = S^1 \times S_3$. Let $p = F \circ q : Y_3 \longrightarrow S^1$ where q is the second factor projection of Y_3 onto T .

Take W as in section 4 ; in each case fix diffeomorphisms $g_j: V_j \longrightarrow W_j$ j = 0,1 (and assume that they preserve the order of the factors) . Consider $\mathbb{R}^4 \times \{0\}$ naturally included inside $\mathbb{R}^{d(i)}$ and let $g: Y_i \longrightarrow \mathbb{R}^{d(i)}$ be an embedding extending g_j for i = 1,2,3 and j = 0,1 . Now we can find an algebraic submanifold X_i of $\mathbb{R}^{d(i)}$ that is diffeomorphic to Y_i , is close to $g(Y_i)$ and contains W . This is possible because the normal bundle of W in $g(Y_i)$ is in each case trivial so we can apply the relative approximation theorem 3.2 of [2]. Let $h_i: X_i \longrightarrow Y_i$ be a diffeomorphism such that $h_i | W_j = g_j^{-1}$ for j = 0,1 . At last if we denote by L the non trivial line bundle over S^1 , then $(p \circ h_i)^*L$ is a line bundle over X_i i = 1,2,3 which is not isomorphic to any algebraic bundle; in fact it is immediate to see that the restriction of this bundle over W is just isomorphic to F_D of sect.4 ; the theorem is proved .

5.2 REMARK In [3] and [4] is stated the following problem : Let $X \subset \mathbb{R}^n$ be a non singular , compact , connected algebraic variety . Suppose that $H^1(X) \neq 0$. Is $H^1_{alg}(X) \neq 0$? The above X_2 produces a counterexample . In fact $H^1(X_2) = \mathbb{Z}_2$ and , by 5.1 , $H^1_{alg}(X_2) = 0$. Moreover the same X_2 shows that in general the "duality" $H^1(-) = H^{m-1}(-)$ is not longer true for $H^{alg}(-)$. In fact it is easy to see that $H_1^{alg}(X_2) = \mathbb{Z}_2$.

REFERENCES

[1] BOREL A. and HAEFLIGER A. La classe d'homologie fondamentale d'un espace analytique Bull.Soc.Math. France 89 (1961) pp.461-513 .

[2] BENEDETTI R. and TOGNOLI A. On real algebraic vector bundles Bull.Sc.math. 2e série ,104, 89-112 (1980)

[3] BOCHNAK J.,KUCHARZ W. and SHIOTA M. The divisor class groups of global real analytic,Nash or rational regular function . This volume.

[4] BOCHNAK J. Topology of real algebraic sets-some open problems. This volume.

[5] TOGNOLI A. Algebraic geometry and Nash functions Institutiones mathematicae , Vol.III , Acad.Press 1978 .

[6] BENEDETTI R. and DEDO' M. The topology of two dimensional real algebraic varieties Ann.Mat.Pura Appl. (IV) vol. CXXVII (1981) pp. 141-171 .

[7] TOGNOLI A. Algebraic approximation of manifolds and spaces Sém.Bourbaki , 32 éme année (1979)n° 548 .

Riccardo Benedetti , Istituto Matematico Univ. di Pisa

Alberto Tognoli , Istituto Matematico Univ. di Ferrara

Inst. Math. Univ. de Tours.

TOPOLOGY OF REAL ANALYTIC SETS - SOME OPEN PROBLEMS

by J. BOCHNAK

Let X be a compact non singular real algebraic subset of $\mathbb{R}^n$ of dimension m (briefly : X is a C.N.A.S) and let $\mathcal{R}(X)$ be the ring of regular rational functions on X.

1. REALIZATION OF HOMOLOGY CLASSES BY ALGEBRAIC SUBSETS. VANISHING ALGEBRAIC CYCLES.

Let $H^{alg}_{m-k}(X)$ be the subgroup of $H_{m-k}(X,\mathbb{Z}_2)$ of homology classes represented by algebraic subsets of X ; $H^{alg}_*(X) = \sum_0^m H^{alg}_i(X)$.

Problem 1. Let M be a compact smooth manifold. Does there exist a C.N.A.S. X diffeomorphic to M with $H^{alg}_*(X) = H_*(X, \mathbb{Z}_2)$?

This problem is probably very hard. Two positive results toward its solution are known.

THEOREM 1 [2], [6]. *Given a compact smooth manifold* M *of dim* m, *there is a C.N.A.S.* X *diffeomorphic to* M *with* $H^{alg}_{m-1}(X) = H_{m-1}(X, \mathbb{Z}_2)$.

The second result needs a definition. We say [1] that a real vector bundle $F \to X$ is a strongly algebraic vector bundle if there is a regular rational map (i.e. rational, smooth, everywhere defined) $g : X \to G$ such that $F = g^*(\gamma)$. Here G is a suitable Grassmann manifold with its standard algebraic structure and γ is the tautological bundle on G. (Several equivalent definition are given in [1]).

Example : the tangent bundle τ_X is a strongly algebraic vector bundle.

THEOREM 2 [2], [6], [8]. *If* $F \to X$ *is a strongly algebraic vector bundle, then every homology class which is dual to some Stiefel - Whitney class of* F *is in* $H^{alg}_*(X)$.
[2] *If the duals of two cohomology classes* $\alpha, \beta \in H^*(X, \mathbb{Z}_2)$ *are in* $H^{alg}_*(X)$, then

the dual of $\alpha \cup \beta$ *is also in* $H_*^{alg}(X)$.

On the other hand we have :

THEOREM 3 [2]. *Given a compact smooth m-manifold* M, $m \geq 3$, *there is a C.N.A.S* X *diffeomorphic to* $M \times S^1$ *with* $H_m^{alg}(X) \neq H_m(X, \mathbb{Z}_2)$.

A proof of this theorem for $M = S^3$ is given in [2] ; the general case is the same. Examples of 3 - dimensional C.N.A.S. X with $H_2^{alg}(X) \neq H_2(X, \mathbb{Z}_2)$ are also known [2], [6].

Problem 2 . Does there exist a 2 - dimensional connected C.N.A.S. X with $H_1^{alg}(X) \neq H_1(X, \mathbb{Z}_2)$?

Let $\mathcal{A}_X$ = {the family of vector bundles obtained from the tangent bundle τ_X by constructions involving a finite number of operations such as Λ^k (the k-th exterior power), Hom (, $\mathbb{R}$) (the dual vector bundle), $\otimes$ (the tensor product), $\oplus$ (the Whitney sum)}. Let $\mathcal{L}^X$ be the subring of $H^*(X, \mathbb{Z}_2)$ generated by the Stiefel-Whitney classes of elements of $\mathcal{A}_X$ and the set of classes representable by submanifolds of X and let $\mathcal{L}_X$ be the subgroup of $H_*(X, \mathbb{Z}_2)$ which is Poincaré dual to $\mathcal{L}^X$. It follows from the theory described above that $\mathcal{L}_X \subset H_*^{alg}(X)$. In some sense $\mathcal{L}_X$ is the biggest possible groups always contained in $H_*^{alg}(X)$.

Problem 3. Given a compact smooth m - manifold M, is it possible to find a C.N.A.S. X, diffeomorphic to M, with $\mathcal{L}_X = H_*^{alg}(X)$?

Is it possible to find a C.N.A.S. X diffeomorphic to M, with $H_{m-1}^{alg}(X) = 0$ (if M orientable) and $H_{m-1}^{alg}(X) = \mathbb{Z}_2$ (if M non orientable)?

Problem 4. Characterize the class of smooth compact manifolds M, satisfying the following property : for any C.N.A.S. X diffeomorphic to M, $H_*(X, \mathbb{Z}_2) = H_*^{alg}(X)$.

It is known that $P^{2k}(\mathbb{R})$ is in this class (proof : for any X diffeomorphic to $P^{2k}(\mathbb{R})$, $\mathcal{L}_X = H_*^{alg}(X)$). Does $P^3(\mathbb{R})$ is also in this class?

2. ALGEBRAIC COVERINGS

DEFINITION. *Let* $\pi : \tilde{X} \to X$ *be a topological covering of* X, $X, \tilde{X}$ *connected. We say that* π *is an algebraic covering if* $\tilde{X}$ *is a C.N.A.S. and* π *is a regular rational map.*

Almost nothing seems to be known about algebraic coverings of C.N.A.S.

Problem 5. Let $\pi: X' \to X$ be a finite topological covering of X. Does there exist an equivalent algebraic covering of X?

The simplest open case : let X be diffeomorphic to $P^3(\mathbb{R})$. Does there exist an algebraic connected two-fold covering of X?

Remark. If X is non orientable, then it has an orientable algebraic two-fold covering [5].

Problem 6. Let $\pi : \tilde{X} \to X$ be an algebraic covering of X and let G be the group of covering transformations of $\tilde{X}$. Does any element of G is an algebraic morphism?

Let $\mathcal{R}(\tilde{X})^G$ be the subring of $\mathcal{R}(\tilde{X})$ of G-invariant rational regular functions on $\tilde{X}$. Describe the size of $\pi^*\mathcal{R}(X)$ in $\mathcal{R}(\tilde{X})^G$. When $\pi^* \mathcal{R}(X) = \mathcal{R}(X)^G$?

3. ALGEBRAICITY OF REAL ANALYTIC SETS

DEFINITION. *Let* $Y \subset X$ *be a real analytic subset of* X. *We say that* Y *is analytically isomorphic (in* X*) to an algebraic set if there is a* C^ω *diffeomorphism* $\sigma : X \to X$ *such that the set* $\sigma(Y)$ *is algebraic.*

Similary we may define the analogous local notion for a germ of an analytic set.

Conjecture [4]. Let $Y \subset X$ be a coherent analytic hypersurface and let $H^{alg}_{m-1}(X) = H_{m-1}(X,\mathbb{Z}_2)$. Then the following conditions are equivalent :

a) Y is analytically isomorphic (in X) to an algebraic set ;

b) at each point $x \in Y$, the germ Y_x is analytically isomorphic to an algebraic set A^x (A^x depends on x).

The conjecture is known to be true in some particular cases [3], [4], [7]. If the set of singular points of Y is finite, the conjecture is related to the following.

Problem 7. Let $V_q \subset X$ be a germ at $q \in X$ of an analytic coherent hypersurface. Assume that V_q has an isolated singular point at q and that V_q is analytically isomorphic to an algebraic germ. Let U be a neighborhood of q in X. Does there exist an algebraic hypersuface $W \subset X$ such that

(i) $W \subset U$;

(ii) the germs V_q and W_q are C^{ω} isomorphic (in a neighborhood of q) ;

(iii) any point of W, different from q, is non singular.

Problem 7 is a particular case of a general problem of constructing algebraic sets with given singularities.

4. EXISTENCE OF RATIONAL MORPHISMS

Problem 8. (Algebraic automorphisms of C.N.A.S.). Let A(X) be the group of algebraic automorphisms of X. Let M be a smooth compact connected manifold. Does there exist C.N.A.S. X and Y, both diffeomorphic to M, with A(X) finite and A(Y) infinite?

Problem 9. For which algebraic manifolds X the group A(X) of automorphisms acts transitively on X? S^n, Grassmannian manifolds and their products are obvious examples. Are there others? Is it possible to describe all of them?

Problem 10. Given a neighborhood U of a point $q \in X$, is it possible to find a rational morphism $\varphi: X \to X$ with $\varphi(X) \subset U$, $\varphi \neq$ constant?

Problem 11. We say that a compact algebraic manifold X is of class $\mathfrak{D}$ if there is a Zariski-open covering $\{U_i\}$ of X, with each U_i algebraically isomorphic to a Zariski open subset V_i of $\mathbb{R}^m$.

Describe the elements of $\mathfrak{D}$. (Obviously S^m, the Grassmannians and their products are in $\mathfrak{D}$). The family $\mathfrak{D}$ is probably very small.

REFERENCES

[1] BENEDETTI R., TOGNOLI A., On real algebraic vector bundles, Bull. Sc. Math. II serie 104 (1980), 89-112.

[2] BENEDETTI R., TOGNOLI A., Remarks and counterexamples in the theory of real algebraic vector bundles and cycles. This volume.

[3] BOCHNAK J., KUCHARZ W., SHIOTA M., On equivalence of ideals of real global analytic functions and the 17-th Hilbert problem, Invent. Math. 63 (3) (1981).

[4] BOCHNAK J., SHIOTA M., On algebraicity of global real analytic functions, preprint, University of Amsterdam (Vrije) (1981

[5] BOCHNAK J. KUCHARZ W., SHIOTA M., The divisor class groups of some rings of global real analytic, Nash, or rational regular functions. This volume.

[6] KING H., A letter, April 1981.

[7] KUCHARZ W., A letter 1981.

[8] SHIOTA M., Real Algebraic Realization of Characteristic Classes, preprint, Kyoto University 1981.

Note added in proof. During the months after the Rennes Conference, some problems stated above has been solved. In particular

a) The answer on problem 1 is negative. It was shown by R. Benedetti and M. Dedo (Counterexamples to representing homology classes by real algebraic subvarieties up to homeomorphism, preprint University of Pisa 1982), that for any integer $k \geq 11$ there is a smooth compact, connected K dimensional manifold M, such that for any C.N.A.S. X diffeomorphic to M, one has

$$H_{k-2}(X,\mathbb{Z}_2) \neq H^{alg}_{k-2}(X,\mathbb{Z}_2) .$$

b) The answer on problem 2 is affirmative. See the papers of Risler and Silhol in this volume.

c) We suggest the reader to consult the following papers of Akbulut and King concerning the topology of real algebraic sets :

1. Real algebraic structures on topological spaces, Publ. I.H.E.S., 53 (1981), 79-162 ;

2. The topology of real algebraic sets with isolated singularities, Ann. of Math. 113(3) (1981), 425-446 ;

3. A relative Nash theorem, Trans. A.M.S. 267(2) (1981), 465-481.

Also several informations about the topology of real algebraic sets can be found in an excellent survey

A. Tognoli - Algebraic approximation of manifold and spaces, Séminaire Bourbaki (1979), exposé 548.

Vrije Universiteit
Department of Mathematics
P.O. Box 7161
1007 MC AMSTERDAM

THE DIVISOR CLASS GROUPS OF SOME RINGS OF GLOBAL REAL ANALYTIC, NASH OR RATIONAL REGULAR FUNCTIONS

Jacek BOCHNAK, Wojciech KUCHARZ, Masahiro SHIOTA

1. Introduction. Let A be a normal, noetherian commutative ring and $K = A_{(0)}$ its quotient field. A *fractional ideal* I is an A-submodule of K for which there exists an element $\alpha \in A$, $\alpha \neq 0$, such that $\alpha I \subset A$. A fractional ideal is called a *principal ideal* if it is generated by one element. We say that I is *divisorial* if $I \neq 0$ and if I is an intersection of principal ideals. The set D(A) of divisorials ideals has the structure of a commutative group given by the composition law $(I,J) \to I*J$, where $I*J = A:(A:IJ)$; A is the unit element of D(A). We call a *local ring of* A any ring of the form A_M, where M is a maximal ideal of A. A fractional ideal I is called *locally principal* if IA_M is principal for all maximal ideals M of A. Any locally principal ideal is divisorial, and if I is locally principal, then for any $J \in D(A)$, we have $I*J = IJ$.

Recall that a *factorial ring* is a commutative integral domain in which every non-zero element can be factorized into irreducible ones in an essentially unique way. Every factorial ring is normal and a normal noetherian ring is factorial if and only if every divisorial ideal is principal. A ring is said to be *locally factorial* if all its local rings are factorial. Every factorial ring is locally factorial, but the converse is not true in general (a criterion of factoriality of locally factorial rings is given in [3]). If A is locally factorial, then any divisorial ideal of A is locally principal.

Let us denote by P(A) the subgroup of D(A) of principal (fractional) ideals. To each noetherian, normal domain A we associate the group D(A)/P(A) which is called the *divisor class group* of A and is denoted by C(A). This group mesures to some extent the factoriality of A. In particular A is factorial if and only if C(A) = 0. Of course all these notions and properties are well known [7], [9].

In this paper we shall compute the divisor class group of the rings of global real analytic functions, Nash functions and real regular rational functions on a large class of real analytic or algebraic sets. Some of our main results are listed below.

Theorem 1. Let X be a compact, coherent, irreducible real analytic space. Suppose that the ring $\mathcal{O}(X)$ of real analytic functions on X is normal. Then the following conditions are equivalent:

(a) The groups $C(\mathcal{O}(X))$ and $H^1(X,\mathbb{Z}_2)$ are isomorphic;

(b) The ring $\mathcal{O}(X)$ is locally factorial.

Theorem 2. Let $X \subset \mathbb{R}^n$ be a real algebraic set. Suppose that X is coherent (as an analytic set) and that the ring $N(X)$ of Nash functions on X is an integral, locally factorial domain. Then

(a) There is a monomorphism $\psi\colon C(N(X)) \to H^1(X,\mathbb{Z}_2)$.

(b) If X is compact, then $C(N(X))$ and $H^1(X,\mathbb{Z}_2)$ are isomorphic.

Corollary 1. Let X be a compact, coherent real analytic (resp. algebraic) set. Then the ring of real analytic functions $\mathcal{O}(X)$ (resp. Nash functions $N(X)$) is factorial if and only if it is locally factorial and $H^1(X,\mathbb{Z}_2) = 0$.

Let $X \subset \mathbb{R}^n$ be an algebraic set and let $P[X] = \mathbb{R}[Y_1,\dots,Y_n]/(\text{ideal of } X)$ be the *ring of polynomials* on X. Consider the ring $R(X)$ of *regular rational functions* on X i.e.

$$R(X) = \{f/g \,:\, f,g \in P[X],\ g^{-1}(0) = \emptyset\}.$$

Theorem 3. Let $X \subset \mathbb{R}^n$ be a compact, irreducible real algebraic set.

Suppose that the ring $\mathcal{R}(X)$ is locally factorial. Then there is a monomorphism

$$\psi : C(\mathcal{R}(X)) \longrightarrow H^1(X,\mathbb{Z}_2)\ .$$

In particular, if $H^1(X,\mathbb{Z}_2) = 0$, then (X) is factorial.

This theorem, for X non singular, has been proved in L. Bröcker, "Reelle Divisoren" Arch. der Math. 35 (1980), 140-143, and in R.G. Swan, "Topological examples of projective modules" Trans. A.M.S. 230 (1977), 201-234.

Corollary 2. If V is a compact non-singular algebraic set and $H^1(V,\mathbb{Z}_2) = 0$, then the ring of polynomials $\mathcal{P}[V]$ on V is factorial if and only if each strictly positive, irreducible polynomial in $\mathcal{P}[V]$ is prime.

In general, the groups $H^1(X,\mathbb{Z}_2)$ and $C(\mathcal{R}(X))$ are not isomorphic, even if X is a non-singular, compact, connected algebraic set. Indeed it follows from Theorem 3 that, at least for X non-singular, the group $C(\mathcal{R}(X))$ is isomorphic to the subgroup $H^{alg}_{m-1}(X,\mathbb{Z}_2)$ of $H_{m-1}(X,\mathbb{Z}_2)$ generated by codim 1 algebraic cycles; dim X = m. Recall [6], that any codim 1 algebraic subset of X defines an element of $H_{m-1}(X,\mathbb{Z}_2)$; the set of all such elements form the subgroup $H^{alg}_{m-1}(X,\mathbb{Z}_2)$. The following striking result has been communicated to us by R. Benedetti and A. Tognoli.

Example 1. [27]. There exists a non-singular 4-dimensional algebraic set $X \subset \mathbb{R}^n$, diffeomorphic to $S^3 \times S^1$, such that $H^{alg}_3(X,\mathbb{Z}_2) = 0$. In particular $\mathcal{R}(X)$ is factorial, while $\mathcal{R}(S^3 \times S^1)$ is not.

More information concerning the realization of homology classes of an algebraic manifold by its algebraic subsets can be found in [27],[28],[29]. However our knowledge of the structure of $H^{alg}_{m-1}(X,\mathbb{Z}_2)$ is still very incomplete.

Open problems. Let M be a compact, connected, orientable, smooth m-manifold. Is there a non-singular algebraic set $X \subset \mathbb{R}^n$ diffeomorphic to M with $H^{alg}_{m-1}(X,\mathbb{Z}_2) = 0$?

Is there a 2-dimensional compact, connected, non-singular algebraic set with $H_1(X,\mathbb{Z}_2) \neq H^{alg}_1(X,\mathbb{Z}_2)$? [See the papers of Risler, "sur l'homologie des surfaces réelles" and Silhol, "A bound of the order of $H^{(a)}_{n-1}(X,\mathbb{Z}/2)$ in this volume].

We are able to prove the following.

Theorem 4. Let $X \subset \mathbb{R}^n$ be a non-singular, m-dimensional algebraic set. Then the element of $H_{m-1}(X,\mathbb{Z}_2)$ corresponding to the first Stiefel-Whitney class of X is in $H^{alg}_{m-1}(X,\mathbb{Z}_2)$. Moreover, it can be realized by a non-singular algebraic hypersurface of X.

Corollary 3. If $X \subset \mathbb{R}^n$ is a non-singular, nonorientable m-dimensional algebraic set, then $H^{alg}_{m-1}(X,\mathbb{Z}_2) \neq 0$.

Crollary 4. If $X \subset \mathbb{R}^n$ is as in Corollary 3 and $H_{m-1}(X,\mathbb{Z}_2) = \mathbb{Z}_2$, then $H^{alg}_{m-1}(X,\mathbb{Z}_2)$ is a topological invariant of X.

Using Theorem 4 we may also prove

Theorem 5. If the set of regular points of an algebraic set $X \subset \mathbb{R}^n$ is nonorientable, then $R(X)$ (and a fortiori $P[X]$) is not factorial.

The paper is organized as follows. Theorem 1,2 and 3 are proven in sections 2,4 and 5, respectively. Section 5 also contains the proof of Theorem 5. Theorem 4 is proven in section 6. Corollary 2 is a consequence of a more general statement given in section 7. Finally section 8 contains a method for constructing singular algebraic sets with $O(X)$, $N(X)$ and $R(X)$ locally factorial.

2. The ring of real analytic functions.

Let X be a coherent real analytic space. Let us denote by

O the sheaf of germs of real analytic functions on X;

$\mathcal{O}^*$ the subsheaf of $\mathcal{O}$ of germs of nowhere vanishing analytic functions on X;

$\mathcal{M}$ the sheaf of germs $(f/g)_x$, where $f_x \in \mathcal{O}_x$ and $g_x \in \mathcal{O}_x \setminus S_x$, where S_x is the set of divisors of zero in $\mathcal{O}_x$;

$\mathcal{M}^*$ the subsheaf of $\mathcal{M}$ of invertible elements in $\mathcal{M}$;

$\mathcal{D} = \mathcal{M}^*/\mathcal{O}^*$ the sheaf of real divisors on X;

$\Gamma(X,F)$ the set of sections of a sheaf F;

$\alpha: \Gamma(X,\mathcal{M}^*) \to \Gamma(X,\mathcal{D})$ the canonical homomorphism.

Let $\mathcal{O}(X)$ be the ring of analytic functions on X. It is easy to see (using Theorem A of Cartan) that if a function $f \in \mathcal{O}(X)$ is not a divisor of zero in $\mathcal{O}(X)$, then for each $x \in X$, the germ f_x of f at $x \in X$ is is not in S_x.

Now assume that the set X is (globally) irreducible; then $\mathcal{O}(X)$ is an integral domain. Using the previous remark we may define a map

$$\sigma: \mathcal{O}(X)_{(0)} \to \Gamma(X,\mathcal{M})$$

from the field of fractions $\mathcal{O}(X)_{(0)}$ of $\mathcal{O}(X)$ into $\Gamma(X,\mathcal{M})$, by the formula $(\sigma(f/g))_x = f_x/g_x$. Using again Theorem A of Cartan we obtain

Lemma 1. σ is an isomorphism.

Throughout this section we assume that X is a *compact, coherent, irreducible, real analytic space*. Then $\mathcal{O}(X)$ is a noetherian ring [22], and any maximal ideal of $\mathcal{O}(X)$ is of the form $M_a = \{f \in \mathcal{O}(X): f(a) = 0\}$, for some $a \in X$. We suppose also that $\mathcal{O}(X)$ is a *normal* ring.
In order to investigate the divisor class group of $\mathcal{O}(X)$, we shall construct a homomorphism of $\Gamma(X,\mathcal{D})$ into $D(\mathcal{O}(X))$.

<u>Construction of a homomorphism</u> $\psi: \Gamma(X,\mathcal{D}) \to D(\mathcal{O}(X))$.

Let $d \in \Gamma(X,\mathcal{D})$. Then for any $x \in X$ there is an element $m_x \in \mathcal{M}^*_x$ such that $d_x = m_x\mathcal{O}^*_x$. Define a coherent sheaf of $\mathcal{O}$-modules $\mathcal{K}$, taking $\mathcal{K}_x = m_x\mathcal{O}_x$, for $x \in X$. Put $I = \Gamma(X,\mathcal{K}) \subset \Gamma(X,\mathcal{M}) = \mathcal{O}(X)_{(0)}$ and observe that I is a non-zero, finitely generated $\mathcal{O}(X)$-submodule of $\mathcal{O}(X)_{(0)}$ i.e. a fractional ideal of $\mathcal{O}(X)$. Moreover, we shall show that I is locally principal and hence $I \in D(\mathcal{O}(X))$. Let $I = (m_1,\dots,m_k)\mathcal{O}(X)$. Then $\mathcal{K}_x = (m_1,\dots,m_k)\mathcal{O}_x = m_x\mathcal{O}_x$ and

$$m_i = \alpha^i_x m_x \quad , \quad m_x = \sum_{i=1}^{k} \beta^i_x m_i \; ,$$

for some $\alpha^i_x, \beta^i_x \in \mathcal{O}_x$. From this it follows that $m_x(1-\sum_{i=1}^{k} \beta^i_x\alpha^i_x) = 0$. But since $m_x \notin S_x$, we get $\sum_{i=1}^{k} \beta^i_x \alpha^i_x = 1$. In particular for some i_x, $\alpha^{i_x}_x(x) \neq 0$ and $m_{i_x}\mathcal{O}_x = I\mathcal{O}_x$. This, together with the inclusion $m_{i_x}\mathcal{O}(X)_{M_x} \subset I\mathcal{O}(X)_{M_x}$ and the flatness of $\mathcal{O}_x$ over $\mathcal{O}_{M_x}$ imply that $m_{i_x}\mathcal{O}(X)_{M_x} = I\mathcal{O}(X)_{M_x}$.

Now define $\psi: \Gamma(X,\mathcal{D}) \to D(\mathcal{O}(X))$ taking $\psi(d) = I$. Since for each $d \in \Gamma(X,\mathcal{D})$ the ideal $\psi(d)$ is locally principal, we have $\psi(d_1d_2) = \psi(d_1)\psi(d_2) = \psi(d_1)*\psi(d_2)$, i.e. ψ is a homomorphism.

<u>Proposition 1.</u>

(a) The homomorphism $\psi: \Gamma(X,\mathcal{D}) \to D(\mathcal{O}(X))$ is injective.

(b) Its image $\psi(\Gamma(X,\mathcal{D}))$ is precisely the set of locally principal fractional ideals of $\mathcal{O}(X)$.

(c) $\psi(\alpha(\Gamma(X,\mathcal{M}^*)) = P(\mathcal{O}(X))$.

<u>Proof.</u>

(a) Suppose that for some $d \in \Gamma(X,\mathcal{D})$, we have $\psi(d) = \mathcal{O}(X)$. Let $d_x = m_x\mathcal{O}^*_x$, $m_x \in \mathcal{M}^*_x$, $x \in X$. Then $m_x\mathcal{O}_x = \mathcal{O}_x$, i.e. $m_x \in \mathcal{O}^*_x$ and $d_x = 1$.

(b) By construction $\psi(\Gamma(X,\mathcal{D}))$ is contained in the set of locally principal fractional ideals. To show the converse let's pick a locally principal ideal I of $\mathcal{O}(X)$ and prove that $\psi(d) = I$, for some $d \in \Gamma(X,\mathcal{D})$. For each $x \in X$ choose an element $m^x \in I$ such that $I\mathcal{O}(X)_{M_x} = m^x\mathcal{O}(X)_{M_x}$ and define a section $d \in \Gamma(X,\mathcal{D})$ by $d_x = (m^x)_x\mathcal{O}^*_x$, $x \in X$. Clearly such a definition is correct and $\psi(d) = I$.

(c) Follows from the construction of ψ.

Corollary 5. Let X be a compact, coherent, irreducible real analytic space and let $\mathcal{O}(X)$ be a normal ring. Then the following conditions are equivalent:

(a) The homomorphism $\psi\colon \Gamma(X,\mathcal{D}) \to D(\mathcal{O}(X))$ (defined above) is an isomorphism;

(b) The induced homomorphism $\bar{\psi}\colon \Gamma(X,\mathcal{D})/\alpha(\Gamma(X,\mathcal{M}^*)) \to C(\mathcal{O}(X))$ is an isomorphism;

(c) $\mathcal{O}(X)$ is locally factorial.

Proof of Theorem 1. Follows from Proposition 1 and the following lemma (which is "well known").

Lemma 2. Let X be a coherent real analytic space. Then $H^1(X,\mathbb{Z}_2)$ is isomorphic to $\Gamma(X,\mathcal{D})/\alpha(\Gamma(X,\mathcal{M}^*))$.

Proof. From the exact sequence of sheaves

$$0 \to \mathcal{O} \xrightarrow{\exp} \mathcal{O}^* \to \mathbb{Z}_2 \to 0$$

we obtain an isomorphism $\delta\colon H^1(X,\mathcal{O}^*) \to H^1(X,\mathbb{Z}_2)$. From the exact sequence

$$1 \to \mathcal{O}^* \to \mathcal{M}^* \to \mathcal{D} \to 1$$

we obtain a monomorphism

$$\beta: \Gamma(X,\mathcal{D})/\alpha(\Gamma(X,\mathcal{M}^*)) \to H^1(X,\mathcal{O}^*)$$

and it remains to show that β is surjective. Let $u \in H^1(X,\mathcal{O}^*)$ and suppose that a 1-cocycle $\{g_{ij}\}$, where $g_{ij} \in \mathcal{O}^*(U_i \cap U_j)$ and $(U_j)_{j\in J}$ is an open covering of X, represents u. Let (E,X,π) be an analytic 1-vector bundle over X with the transition functions $\{g_{ij}\}$. The sheaf of germs of analytic sections of E is coherent, which implies (Theorem A of Cartan) the existence of a global non-trivial analytic section s of E. Such a section induces on each U_j a function $f_j \in \mathcal{O}(U_j)$, not identically equal to zero, and such that $f_i = f_j g_{ij}$ on $U_i \cap U_j$. The family $\{f_j\}_{j\in J}$ defines a section d of $\mathcal{D}$. Clearly if $\bar{d}$ is its image in $\Gamma(X,\mathcal{D})/\alpha(\Gamma(X,\mathcal{M}^*))$, then $\beta(\bar{d}) = u$.

Corollary 6. Let X be as in Theorem 1. Suppose that at each point $x \in X$, the ring $\mathcal{O}_x$ of germs of real analytic functions at x is factorial. Then $H^1(X,\mathbb{Z}_2)$ and $C(\mathcal{O}(X))$ are isomorphic.

Proof. The ring $\mathcal{O}(X)$ is then locally factorial (cf. [9], Cor.6.11 p. 35), so we may apply Theorem 1.

Example 2. If $\mathcal{O}(X)$ is locally factorial, then the ring of germs $\mathcal{O}_x$ need not be necessarily factorial at each point of X. Indeed, let's consider an algebraic compact, connected subset $X \subset \mathbb{R}^3$, $\dim X = 2$, $H^1(X,\mathbb{Z}_2) = \mathbb{Z}_2$. Suppose that X has only one singular point, say $0 \subset \mathbb{R}^3$, of the form $(x^2 + y^2 + (z-1)^2-1)(x^2 + y^2 + (z+1)^2-1) = 0$, in a suitable analytic coordinates system around 0. Such a set exists (cf. section 8) and is completely determined analytically by these conditions. It is rather easy to see that $\mathcal{O}(X)$ is locally factorial, but the ring $\mathcal{O}_0(X)$ of germs at 0 is not even an integral domain.

3. The ring of germs of analytic functions on a compact subset of an analytic manifold.

Let M be a real analytic (resp. complex Stein) manifold and let $K \subset M$ be a compact, connected subanalytic set of M[11]. Let denote by $\mathcal{O}_K = \Gamma(K,\mathcal{O})$ the ring of germs of real analytic (resp. holomorphic) functions at K. It is known, that under these assumptions the ring $\mathcal{O}_K$ is noetherian [22] and normal. Using the method of section 2, we may prove

Theorem 1' The divisor class group $C(\mathcal{O}_K)$ of $\mathcal{O}_K$ is isomorphic to $H^1(K,\mathbb{Z}_2)$ in the real case, and to $H^2(K,\mathbb{Z})$ in the complex case.

4. The ring of Nash functions.

Let X be a real algebraic subset of $\mathbb{R}^n$ endowed with its canonical analytic structure and let $U \subset X$ be an open semi-algebraic subset of X. Recall [5], that an analytic function $f\colon U \to \mathbb{R}$ is said to be a *Nash function*, if the graph of f is a semi-algebraic subset of $\mathbb{R}^n \times \mathbb{R}$, (or equivalently, if there exists a polynomial $p\colon X \times \mathbb{R} \to \mathbb{R}$, $p \in P[X][z]$, $p \not\equiv 0$, such that $p(x,f(x)) \equiv 0$ on U). It is known that if the ring $N(U)$ of Nash functions on U is normal (which happens for example if X is a normal analytic space), then $N(U)$ is noetherian (for a proof in the non-singular case see [16] or [5]; the general case is analogous). Any maximal ideal of $N(U)$ is of the form $M_a =$ $= \{\phi \in N(U): \phi(a) = 0\}$, for some $a \in U$; [5],[16].

The main goal of this section is to prove Theorem 2 of section 1. Before beginning the proof of this result, some preparation is necessary. Let $\mathcal{O}$, $\mathcal{O}^*$, M, M^*, $\mathcal{D}$, Γ etc..., has the same meaning as in section 2 and assume that $N(X)$ is an *integral, locally factorial domain.* We shall construct a homomorphism

$$\Phi: D(N(X)) \to \Gamma(X,\mathcal{D}).$$

Let $I \in D(N(X))$. For each $x \in X$ choose an element $m^x \in I$ such that $IN(X)_{M_x} = m^x N(X)_{M_x}$ and define a section $d = d(I) \in \Gamma(X,\mathcal{D})$ by

$$d_x = (m^x)_x \mathcal{O}^*_x \quad , \quad x \in X,$$

where, as usual, $(m^x)_x$ denotes the germ of m^x at x. Such a definition is correct, since if $m = f/g$, $f, g \in N(X)\setminus\{0\}$, then f_x and g_x are not zero divisors in $\mathcal{O}_x$ (due to the flatness of $\mathcal{O}_x$ over N_{M_x}). Moreover $IN(X)_{M_y} = m^x N(X)_{M_y}$ for all y near x (I is a finitely generated $N(X)$-submodule of $N(X)_{(0)}$).

Now let us put $\Phi(I) = d(I)$. Obviously Φ is a group homomorphism; by assumption we have $I*J = IJ$ for $I, J \in D(N(X))$, so $\Phi(I)\Phi(J) = \Phi(I*J)$. Finally observe that $\Phi(P(N(X)) \subset \alpha(\Gamma(X,M^*))$, so Φ induces a homomorphism

$$\bar{\Phi}: C(N(X)) \to \Gamma(X,\mathcal{D})/\alpha(\Gamma(X,M^*)).$$

<u>Proof of (a) of Theorem 2.</u> We shall show that $\bar{\Phi}$, defined above, is a monomorphism. Suppose that $\Phi(I) \in \alpha(\Gamma(X,M^*))$, for some $I \in D(N(X))$.

We shall prove that $I \in P(\mathcal{N}(X))$. Without loss of generality we may assume that I is an integral ideal, i.e. $I \subset \mathcal{N}(X)$. Let $g_1, \ldots, g_p$ be a set of generators of I. We know by assumption, that there is a meromorphic function $f \in \Gamma(X, \mathcal{M}^*)$ such that $I\mathcal{O}_x = f\mathcal{O}_x$ for all $x \in X$. Since I is integral, f is in $\mathcal{O}(X)$. This shows that $(\sum_{i=1}^{p} g_i^2)\mathcal{O}(X) = f^2\mathcal{O}(X)$. In particular, there is a function $v \in \mathcal{O}(X)$, $v^{-1}(0) = \emptyset$, such that $\sum_{i=1}^{p} g_i^2 = (vf)^2$. Then $\phi = fv$ is in $\mathcal{N}(X)$ and ϕ obviously generates I (flatness of $\mathcal{O}_x$ over $\mathcal{N}(X)_{M_x}$). Therefore $\bar{\Phi}$ is a monomorphism. Now we define $\psi = \delta \circ \beta \circ \bar{\Phi}$, where β and δ are isomorphisms defined in section 2. The proof of (a) is complete.

Before the proof of (b) of Theorem 2, it is necessary first to establish some notions.

Let $\xi = (E,B,\pi)$ be a continuous k-vector bundle over a set B. We say ([2],[25]) that ξ is a *strong Nash* (resp. *strong algebraic) vector bundle* if the following conditions are satisfied:

(α) E and B are Nash subsets of an open semi-algebraic subset of $\mathbb{R}^s$ (resp. E and B are algebraic subsets of $\mathbb{R}^s$), for some $s \in \mathbb{N}$;

(β) the projection $\pi: E \to B$ is a Nash (resp. regular) map;

(c) Each point $x \in B$ has an open semi-algebraic (resp. Zariski open) neighborhood $W \subset B$, such that $\xi | W$ is a trivial Nash (resp. trivial algebraic) vector bundle.

Any strong algebraic vector bundle is obviously also a strong Nash vector bundle.

<u>Example 3</u>. Let's consider the Grassmann manifold $G_k(\mathbb{R}^m)$ embedded as a non-singular algebraic set of the vector space $L(\mathbb{R}^m)$ of all linear endomorphisms of $\mathbb{R}^m$. The universal k-vector bundle $\gamma_{m,k}$ over $G_k(\mathbb{R}^m)$ can be

canonically embedded in $L(\mathbb{R}^m) \times \mathbb{R}^m \cong \mathbb{R}^{m^3}$ as a non-singular algebraic set. More precisely, $\gamma_{m,k}$ is a subbundle of the trivial vector bundle $G_k(\mathbb{R}^m) \times \mathbb{R}^m$; the total space of $\gamma_{m,k}$ is a non-singular algebraic subset of $G_k(\mathbb{R}^m) \times \mathbb{R}^m$ [15]. Clearly $\gamma_{m,k}$ is canonically endowed with a strong algebraic structure and inherits an algebraic riemannian structure from $G_k(\mathbb{R}^m) \times \mathbb{R}^m$.

<u>Lemma 3.</u> Let $X \subset \mathbb{R}^n$ be a compact algebraic set and let $\xi = (E,X,\pi)$ be a continuous k-vector bundle over X. Then there is an open semi-algebrai neighborhood U of X in $\mathbb{R}^n$ and a strong Nash vector bundle $\eta = (F,U,p)$ such that

(a) The restriction $\eta|X$ is C^0-isomorphic to ξ;

(b) There is a Nash riemannian structure <,> on η;

(c) There are Nash sections $s_1,\dots,s_\ell$ of η, such that $s_1(x),\dots,s_\ell(x)$ generate $F_x = p^{-1}(x)$, for each $x \in U$.

<u>Proof.</u> Let $f: X \to G_k(\mathbb{R}^m)$ be a classifying map for ξ. Let W be a semi-algebraic open neighborhood of $G_k(\mathbb{R}^m)$ in $L(\mathbb{R}^m)$ and let $\rho: W \to G_k(\mathbb{R}^m)$ be a Nash retraction of W onto $G_k(\mathbb{R}^m)$ [15]. Choose a continuous map $\tilde{f}_1: \mathbb{R}^n \to L(\mathbb{R}^m)$ with $\tilde{f}_1|X = f$ and let $\tilde{f}_2: \mathbb{R}^n \to L(\mathbb{R}^m)$ be a Nash mapping which is C^0-close to $\tilde{f}_1$ in a neighborhood of X. Choose a semi-algebraic open neighborhood U_0 of X, $U_0 \subset \tilde{f}_1^{-1}(W) \cap \tilde{f}_2^{-1}(W)$ and define $f_i = \rho\circ\tilde{f}_i|U_0$: $U_0 \to G_k(\mathbb{R}^m)$, $i = 1,2$. Clearly f_2 is a Nash map, and we may assume (taking $\tilde{f}_2$ sufficiently close) that f_1 and f_2 are homotopic. Then the vector bundles $\xi_i = (E_i,U_0,\pi_i)$, where $\xi_i = f_i^*(\gamma_{m,k})$, are C^0-isomorphic. Moreover, ξ_2 has a strong Nash vector bundle structure and a Nash riemannian structure $<,>: E_2 \oplus E_2 \to \mathbb{R}$. Let $t_1,\dots,t_\ell: U_0 \to E_2$ be a

continuous sections of ξ_2 such that $t_1(x),\dots,t_\ell(x)$ generate $E_{2,x}$, for all $x \in U_0$. Since E_2 is a Nash closed submanifold of an open semi-algebraic subset of $\mathbb{R}^\lambda$, for some $\lambda \in \mathbb{N}$, we may choose an open semi-algebraic neighborhood W' of E_2 in $\mathbb{R}^\lambda$ and a Nash retraction $\sigma\colon W' \to E_2$. Choose $t_i'\colon U_0 \to \mathbb{R}^\lambda$, $i = 1,\dots,\ell$, a Nash mappings, with t_i' close to t_i in a neighborhood of X, and choose a neighborhood U_0' of X in U_0 with $U_0' \subset \bigcap_{i=1}^{\ell} t_i'^{-1}(W')$. Then, for any $i = 1,\dots,\ell$, $\pi_2\circ\sigma\circ t_i'|U_0'\colon U_0' \to \mathbb{R}^n$ is a Nash map, close to the identity in a neighborhood of X. So we may choose open, semi-algebraic neighborhoods $U, U_1,\dots,U_\ell$ of X in U_0' such that $\pi_2\circ\sigma\circ t_i'|U_i\colon U_i \to U$ is a Nash diffeomorphism of U_i onto U, for $i = 1,\dots,\ell$. Define $\eta = \xi_2|U = (F,U,p)$ and $s_i = (\sigma\circ t_i')\circ(\pi_2\circ\sigma\circ t_i'|U_i)^{-1}$, $i = 1,\dots,\ell$. Then each s_i is a Nash section of η, and taking if necessary U smaller, we may assume that $s_1(x),\dots,s_\ell(x)$ generate F_x, for each $x \in U$. Clearly $\xi = f_1^*(\gamma_{m,k})|X \cong f_2^*(\gamma_{m,k})|X = \eta|X$.

<u>Proof of (b) of Theorem 2.</u> We must only show that the map $\Phi' = \beta\circ\bar{\Phi}\colon C(N(X)) \to H^1(X,\mathcal{O}^*)$ is surjective. Let $v \in H^1(X,\mathcal{O}^*)$ and let $\xi = (E,X,\pi)$ be an analytic 1-vector bundle over X, which is defined by a cocycle corresponding to v. For ξ let's choose a strong Nash vector bundle $\eta = (F,U,p)$ and a Nash sections $s_1,\dots,s_\ell$ of η satisfying the conditions of Lemma 3. Since vector bundles ξ and $\eta|X$ are C^0-isomorphic, they are also analytically isomorphic. In particular, any system of transition functions for $\eta|X$ is a 1-cocycle on X defining the element v (see [12] p.41). Let $s\colon U \to F$ be any Nash section of η, which is not identically zero on X. Then $f_i\colon U \to \mathbb{R}$, where $f_i(x) = \langle s(x), s_i(x)\rangle$ for $x \in U$, $i = 1,\dots,\ell$, are in $N(U)$. Observe that if we define $U_i = \{x \in U\colon s_i(x) \neq 0\}$, then $\{g_{ij}\}$, where $g_{ij} = f_i/f_j \in \mathcal{O}^*(U_i \cap U_j)$, is a system of transition functions for η. Put $I = (f_1,\dots,f_\ell)N(X)$ and $J_x = (f_1,\dots f_\ell)\mathcal{O}_x$. Observe that J_x is a principal ideal of $\mathcal{O}_x$; indeed $J_x = f_i\mathcal{O}_x$, where i is such

that $s_i(x) \neq 0$. Since the natural injection $\mathcal{N}(X)_{M_x} \to \mathcal{O}_x$ is flat, the ideal $I\mathcal{N}(X)_{M_x}$ is also principal, for all $x \in X$. Hence I is locally principal non zero ideal and therefore $I \in D(\mathcal{N}(X))$. Clearly $\Phi'([I]) = v$, where $[I]$ is the image of I in $C(\mathcal{N}(X))$.

Remarks. (1) It can be shown that in Theorem 2(a), the assumption X coherent, can be replaced by the assumption X compact, and in Theorem 2(b) the assumption X compact can be replaced by X smooth.

(2) Theorem 2(a) holds true for a large class of subrings of $\mathcal{O}(X)$. More precisely we have

Theorem 2'. Let X be a coherent real analytic space and let A(X) be a subring of $\mathcal{O}(X)$. Suppose that

(a) A(X) is noetherian and normal;

(b) If $f \in A(X)$ and $f^{-1}(0) = \emptyset$, then $1/f \in A(X)$;

(c) $\forall x \in X$, the canonical injection $A(X)_{M_x} \to \mathcal{O}_x$ is flat;

(d) $\forall g \in \mathcal{O}(X)$, if $g^2 \in A(X)$ then $g \in A(X)$.

Then there is a monomorphism $\psi: C(A(X)) \to H^1(X,\mathbb{Z}_2)$.

The proof of this theorem is quite similar to that given above for $A(X) = \mathcal{N}(X)$ and will be omitted. A large class of so-called semi-algebraic rings [5] satisfies the assumption of Theorem 2'.

5. The ring of regular functions.

Proof of Theorem 3. Let $X \subset \mathbb{R}^n$ be a *compact irreducible* real algebraic set and $V \subset X$ a Zariski open subset of X. Recall that the ring of regular functions on V is the ring

$$R(V) = \{f/g: f,g \in P[X],\ g^{-1}(0) \cap V = \emptyset\}.$$

Suppose that $R(X)$ is locally factorial. We shall prove Theorem 3 of section 1 i.e. we shall construct a monomorphism

$$\psi: C(R(X)) \to H^1(X,\mathbb{Z}_2).$$

Let I be a divisorial ideal of $R(X)$ and let $f_1,\dots,f_k \in R(X)_{(0)}\setminus\{0\} = P[X]_{(0)}\setminus\{0\}$ be a set of generators of I. Define

$$U_j = \{x \in X: IR(X)_{M_x} = f_j R(X)_{M_x}\},$$

$j = 1,\dots,k$. Observe that each U_j is Zariski open and for each $1 \le i \le k$ there is a regular function $h_{ij}: U_j \to \mathbb{R}$ such that $f_i = h_{ij} f_j$ on U_j. The family of nowhere vanishing regular functions $g_{ij} = h_{ij}|U_i \cap U_j \in R(U_i \cap U_j)$ defines a 1-cocycle i.e. $\{g_{ij}\} \in Z^1(\{U_j\}, R^*)$. Given $I \in D(R(X))$, we may associate to I an element $v(I) \in H^1(X,\mathbb{Z}_2)$, corresponding to a cocycle $\{g_{ij}/|g_{ij}|\} \in Z^1(\{U_j\},\mathbb{Z}_2)$. It is easy to see that $v(I)$ does not depend on the choice of the generators f_i. Moreover, if I is a principal divisorial ideal, then $v(I)$ is the unit element of $H^1(X,\mathbb{Z}_2)$. The above allows us to define a map

$$\psi: C(R(X)) \to H^1(X,\mathbb{Z}_2)$$

by taking $\psi([I]) = v(I)$. Using the fact that $I*J = IJ$, we conclude that ψ is a group homomorphism. Now we shall show that ψ is injective. Let $I \in D(R(X))$ and assume that $v(I)$ is the unit element in $H^1(X,\mathbb{Z}_2)$. Let f_i, U_j, h_{ij}, g_{ij} have the same meaning as above. We shall construct a strong algebraic 1-vector bundle $\xi = (E,X,\pi)$ for which $\{g_{ij}\}$ is a system of the transition functions. Let us define

$$E = \{(x,e_1,\dots,e_k) \in X \times \mathbb{R}^k:\ \text{if } x \in U_j,\ \text{then } e_i = h_{ij}(x)e_j \text{ for each } i = 1,\dots,k\}.$$

Note that E is an algebraic subset of $X \times \mathbb{R}^k$, the map $\pi: E \to X$, $\pi(x,e) = x$, is regular and $\pi^{-1}(x) \subset \{x\} \times \mathbb{R}^k$ is a vector subspace of $\{x\} \times \mathbb{R}^k \cong \mathbb{R}^k$. We claim that $\xi = (E,X,\pi)$ is the required strong algebraic vector bundle. On each U_j we have $\xi|U_j$ algebraically trivial, with an algebraic trivialization given by $\phi_j: \xi|U_j \to U_j \times \mathbb{R}$, $\phi_j(x,e_1,\dots,e_k) = (x,e_j)$. By construction, ξ is a C^0 trivial vector bundle ([12] p.41). Using the property that ξ is a strong algebraic vector bundle, we deduce that ξ is in fact an algebraicly trivial vector bundle [2],[24]. In particular there exist a regular, nowhere zero section $s: X \to E$. The section s is of the form $s = (s_1,\dots,s_k)$, where $s_i \in \mathcal{R}(X)$ and $s_i(x) = h_{ij}(x)s_j(x)$. We shall show that $f = \sum_{i=1}^{k} s_i f_i$ generates I (and hence that ψ is injective). Indeed, on U_j we have $f = \sum_{i=1}^{k} h_{ij}s_j h_{ij} f_j = (\sum_{i=1}^{k} h_{ij}{}^2)s_j f_j$. The function $(\sum_i h_{ij}{}^2)s_j$ is regular and nowhere zero on U_j. Hence $f\mathcal{R}(X)_{M_x} = f_j\mathcal{R}(X)_{M_x} = I\mathcal{R}(X)_{M_x}$ for all $x \in U_j$, which implies that $f\mathcal{R}(X) = I$. Thus we have proven Theorem 3.

<u>Remark (3).</u> In general, an algebraic vector bundle, which is trivial as a C^0 vector bundle, is not algebraically trivial [2],[24]. A theorem of Benedetti-Tognoli used above insures that such a situation cannot occur for a strong algebraic vector bundle. In our construction the structure of the bundle $\xi = (E,X,\pi)$ is particularly simple. We may avoid the theorem of Benedetti-Tognoli and deduce directly the existence of an algebraic nowhere zero section s of ξ as follows. Let

$$E' = \{(x,c) \in X \times \mathbb{R}^k:\ \sum_{i=1}^{k} c_i e_i = 0 \text{ for all } e = (e_1,\dots,e_k) \in \pi^{-1}(x)\}.$$

Then $E \oplus E' = X \times \mathbb{R}^k$ and there is a natural " orthogonal" retraction of $X \times \mathbb{R}^k$ onto E, "parallel" to E', $\rho: X \times \mathbb{R}^k \to E$. In particular $\rho(x,r) \in \pi^{-1}(x)$. Since E and E' are algebraic sets, the map ρ is regular. Now let $s': X \to E$ be a continuous, nowhere zero section of E and let $s_1: X \to X \times \mathbb{R}^k$ be an algebraic section of $X \times \mathbb{R}^k$ which is close to s'. Then $s = \rho \circ s_1$ is a nowhere vanishing algebraic section of E.

Now we shall study the image of the monomorphism $\psi: C(\mathcal{R}(X)) \to H^1(X,\mathbb{Z}_2)$. Let $X \subset \mathbb{R}^n$ be an irreducible algebraic set, $\mathcal{R}^*$ the sheaf (over X) of nowhere vanishing regular rational functions and $\mathcal{U} = \{U_i\}_{i \in J}$ a given finite Zariski open covering of X. Then we may consider the multiplicative group of algebraic 1-cocycles $Z^1(\mathcal{U},\mathcal{R}^*)$. It is convenient to introduce a subgroup $\tilde{Z}^1(\mathcal{U},\mathcal{R}^*)$ of $Z^1(\mathcal{U},\mathcal{R}^*)$ of *algebraic divisorial 1-cocycles* which we define as follows: $\{g_{ij}: U_i \cap U_j \to \mathbb{R}^*\}_{ij}$ is in $\tilde{Z}^1(\mathcal{U},\mathcal{R}^*)$ if there is a family $f_i: X \to \mathbb{R}$ of regular rational functions, $f_i \neq 0$, $i \in J$ such that $f_i|U_j/f_j|U_j \in \mathcal{R}(U_j)$ and $f_i = g_{ij}f_j$ on $U_i \cap U_j$. Let us denote by $H^1(\mathcal{U},\mathcal{R}^*)$, $\tilde{H}^1(\mathcal{U},\mathcal{R}^*)$, $H^1(X,\mathcal{R}^*) = \varinjlim H^1(\mathcal{U},\mathcal{R}^*)$, $\tilde{H}^1(X,\mathcal{R}^*) = \varinjlim \tilde{H}^1(\mathcal{U},\mathcal{R}^*)$ the corresponding cohomology groups. We consider $\tilde{H}^1(X,\mathcal{R}^*)$ as a subgroup of $H^1(X,\mathcal{R}^*)$. We denote by $h: H^1(X,\mathcal{R}^*) \to H^1(X,\mathbb{Z}_2)$ the natural homomorphism and we define $H^1_{Alg}(X,\mathbb{Z}_2) = h(H^1(X,\mathcal{R}^*))$, $H^1_{alg}(X,\mathbb{Z}_2) = h(\tilde{H}^1(X,\mathcal{R}^*))$.

Corollary 7 For X as in Theorem 3, the groups $C(\mathcal{R}(X))$ and $H^1_{alg}(X,\mathbb{Z}_2)$ are isomorphic. More precisely $\psi(C(\mathcal{R}(X))) = H^1_{alg}(X,\mathbb{Z}_2)$.

Proof. Follows directly from the proof of Theorem 3.

□.

It is rather easy to see that for a compact, non-singular, m-dimensional algebraic set X, the groups $H^1_{alg}(X,\mathbb{Z}_2)$ and $H^{alg}_{m-1}(X,\mathbb{Z}_2)$ are isomorphic.

We assume from now on that $X \subset \mathbb{R}^n$ is a non-singular, compact, irreducible, m-dimensional algebraic set and we shall compare the various cohomology groups on X. Is known that, $H^1(X,\mathcal{R}^*) \neq \tilde{H}^1(X,\mathcal{R}^*)$, for all X, dim X > 1 [26]. As it was pointed out in the introduction, there exists a 4-dimensional compact, connected non-singular algebraic set $X \subset \mathbb{R}^n$, with $H^1_{alg}(X,\mathbb{Z}_2) = H^{alg}_3(X,\mathbb{Z}_2) \neq H_3(X,\mathbb{Z}_2)$ [27]. Also it can be shown [27], that for the same set $H^1_{Alg}(X,\mathbb{Z}_2) \neq H^1(X,\mathbb{Z}_2)$. In fact it is known that $H^1_{alg}(X,\mathbb{Z}_2)$ and $H^1_{Alg}(X,\mathbb{Z}_2)$ are isomorphic. We may summarize the above results on the following diagrams.

$$\begin{array}{ccccc}
\tilde{H}^1(X,R^*) & \xrightarrow{\alpha_1} & H^1_{alg}(X,\mathbb{Z}_2) & \xrightarrow{\alpha_7} & H^{alg}_{m-1}(X,\mathbb{Z}_2) \\
\downarrow{\scriptstyle \alpha_3} & & \downarrow{\scriptstyle \alpha_4} & \searrow{\scriptstyle \alpha_5} & \\
H^1(X,R^*) & \xrightarrow{\alpha_2} & H^1_{Alg}(X,\mathbb{Z}_2) & \xrightarrow{\alpha_6} & H^1(X,\mathbb{Z}_2)
\end{array}$$

where α_i is the inclusion map for $3 \le i \le 6$, α_1 and α_2 are induced by h, and α_7 (the cohomology class of $\{g_{ij}\}$) = the homology class of the algebraic set determined by $\{f_i\}$, where $\{g_{ij}\} \in \tilde{Z}^1(U,R^*)$ and $f_i = g_{ij}f_j$.

Now let us consider the following table

	α_1	α_2	α_3	α_4	α_5	α_6	α_7
surjective	+	+	-	+	-	-	+
injective	+	-	+	+	+	+	+

where the symbol (+) (resp. (-)) means that a given map has (resp. does not have) the corresponding property for all (resp. for some) compact, irreducible, non-singular algebraic sets of dimension m.

We shall study the group $H_{m-1}^{alg}(X,\mathbb{Z}_2)$ in the next section. We conclude this section with the proof of Theorem 5 (assuming Theorem 4, which will be proved in section 6).

<u>Theorem 5.</u> If the set of regular points of an algebraic set $X \subset \mathbb{R}^n$ is nonorientable, then $R(X)$ (and a fortiori $P[X]$) is not factorial.

<u>Proof of Theorem 5.</u> Let Σ be the set of singular points of an m-dimensional algebraic subset $X \subset \mathbb{R}^n$. We shall show that if the set of regular points $X\setminus\Sigma$ is nonorientable, then the ring $R(X)$ is not factorial.

Suppose first that $\Sigma = \emptyset$. Using Theorem 4 we may choose an algebraic, non-singular, irreducible hypersurface $Y \subset X$, which defines a nontrivial element of $H_{m-1}(X,\mathbb{Z}_2)$. Let f be any irreducible polynomial on X, vanishing on Y, and such that grad $f(y) \neq 0$ for some $y \in Y$, and let $g_1,\dots,g_k$ be a set of generators of the ideal of polynomials on X vanishing on Y. Then $g = \sum_{i=1}^{k} g_i^2$ is irreducible in $R(X)$ and $f^2 = gh$ for some $h \in P[X]$ with $Y \not\subset h^{-1}(0) \neq \emptyset$. But f itself is not divisible neither by g nor by h, so $R(X)$ cannot be factorial.

Now let us consider the general case. Let S be a multiplicative set of $R(X)$ defined by $S = \{f \in R(X): f^{-1}(0) \subset \Sigma\}$. Then $R(X\setminus\Sigma) = R(X)_S$ the localization of $R(X)$ with respect to S. As above, the ring $R(X\setminus\Sigma)$ is not factorial. Hence $R(X)$ is not factorial either.

Remark (4) It can be shown that $\mathcal{R}(X)$ is not factorial even if we only assume that set of C^∞ regular points of maximal dimension of X is nonorientable.

6. The algebraic cycles of codimension one.

In this section let $X \subset \mathbb{R}^n$ be a non-singular, m-dimensional (not necessarily compact) algebraic set. We shall investigate the subgroup $H^{alg}_{m-1}(X,\mathbb{Z}_2)$ of $H_{m-1}(X,\mathbb{Z}_2)$ generated by algebraic cycles.

First we shall show that "generically" we have $H^{alg}_{m-1}(X,\mathbb{Z}_2) = H_{m-1}(X,\mathbb{Z}_2)$, at least if X is compact. This is virtually contained in the following statement.

Theorem [1],[24]. Let V be a smooth compact manifold and let $K_1,\dots,K_s$ be a family of codim 1 closed submanifolds of V, which are in general position. Then there exists a non-singular algebraic subset $V' \subset \mathbb{R}^p$, a family of algebraic sets $K_1',\dots,K_s'$ in V' and a C^∞ diffeomorphism $u\colon V \to V'$ such that $u(K_i) = K_i'$, $i = 1,\dots,s$.
Moreover if $V \subset \mathbb{R}^p$, $p \geq 2\dim V + 1$, and $\varepsilon > 0$, then we may choose u such that $|u(x)-x| \leq \varepsilon$ for all $x \in V$.

Corollary 8. Let V be a smooth, compact m-dimensional manifold. Then there exists a non-singular algebraic set V' diffeomorphic to V, such that $H^{alg}_{m-1}(V',\mathbb{Z}_2) = H_{m-1}(V',\mathbb{Z}_2)$.
Moreover, if $V \subset \mathbb{R}^p$, $p \geq 2m+1$, and $\varepsilon > 0$, then we may assume $V' \subset \mathbb{R}^p$ and that there is a diffeomorphism $u\colon V \to V'$ satisfying $|u(x)-x| \leq \varepsilon$ for all $x \in V$.

Corollary 9. Given a compact, connected smooth manifold V, there always exist a non-singular real algebraic set V', such that

(a) V is diffeomorphic to V';

(b) The divisor class group of $R(V')$ is isomorphic to $H^1(V',\mathbb{Z}_2)$.

The main goal of this section is to prove Theorem 4 stated in the introduction.

Proof of Theorem 4. It is sufficient to show the theorem for X compact. Indeed, if X is not compact, we consider $\hat{X}$ the desingularization of Hironaka of the algebraic closure of X in $P^n(\mathbb{R})$. By asssumption $X \subset \hat{X}$, so the restriction to X of a realization of the dual of the first Stiefel-Whitney class of $\hat{X}$ is that of X. Hence we may assume X compact. Also without loss of generality, we may suppose that X is connected. Now let $W \subset X$ be a smooth, closed, connected, codim 1 submanifold of X representing, by the Poincaré duality theorem, the first Stiefel-Whitney class of X. Then Theorem 4 is included in the following statement.

Theorem 4'. The submanifold W is homologous to a connected, non-singular algebraic hypersurface of X.

Before the proof of Theorem 4', we need several lemmas. Given a smooth codim 1 submanifold Y of X, we note by [Y] the image of its $\mathbb{Z}_2$-fundamental class in $H_{m-1}(X,\mathbb{Z}_2)$. Given two algebraic varieties over $\mathbb{R}$ (in the sense of Serre [18]), (A,R_A) and (B,R_B), with the structural sheaves R_A and R_B respectively, we say (as usual) that a map $f\colon A \to B$ is an *algebraic morphism*, if $f^*R_B \subset R_A$.
We note $S^k = \{x \in \mathbb{R}^{k+1} : |x| = 1\}$ and $q\colon S^k \to P^k(\mathbb{R})$ the standard covering map; q is an algebraic morphism, when S^k and $P^k(\mathbb{R})$ are equipped with its

standard algebraic structures.

Lemma 4. Let X be a compact, connected, non-singular m-dimensional real algebraic set. Assume that there exists an algebraic morphism $\phi: X \to P^k(\mathbb{R})$ of X into a k-dimensional projective space $P^k(\mathbb{R})$, such that the set

$$\tilde{X} = \{(x,y) \in X \times S^k : \phi(x) = q(y)\}$$

is connected. Let $\pi: \tilde{X} \to X$ be the natural 2-fold covering map, $\pi(x,y) = x$, and let $Y \subset X$ be a smooth, closed, codim 1 submanifold, such that the homology class $[\pi^{-1}(Y)]$ is zero in $H_{m-1}(\tilde{X},\mathbb{Z}_2)$. Then Y is homologous to a connected, non-singular algebraic hypersurface of X. In particular, if $[Y] \neq 0$, then $H^{alg}_{m-1}(X,\mathbb{Z}_2) \neq 0$.

Proof. Without loss of generality we may assume that Y is connected and $[Y] \neq 0$. Then $X\setminus Y$ is also connected and the assumption $[\tilde{Y}] = 0$, where $\tilde{Y} = \pi^{-1}(Y)$, implies that $\tilde{X}\setminus\tilde{Y}$ has precisely two connected components, say X_1 and X_2. Let $\sigma: \tilde{X} \to \tilde{X}$, $\sigma(x,y) = (x,-y)$, be the natural involution of $\tilde{X}$. Obviously σ is an algebraic morphism and $\sigma(X_1) = X_2$. Let $f: \tilde{X} \to \mathbb{R}$ be a smooth function with $f^{-1}(0) = \tilde{Y}$ and which is regular on $\tilde{Y}$. We may also assume, taking $f\circ\sigma - f$ instead of f, that $f\circ\sigma = -f$. Now choose a polynomial $F \in \mathcal{P}[X]$ sufficiently C^1-close to f, such that $(\tilde{X},\tilde{Y})$ is diffeomorphic to $(\tilde{X},F^{-1}(0))$. Since $\frac{1}{2}(F-F\circ\sigma)$ is also C^1-close to f, we may suppose that $F\circ\sigma = -F$. Hence $\sigma(F^{-1}(0)) = F^{-1}(0)$ and $Z = \pi(F^{-1}(0))$ is a smooth submanifold of X homologous to Y. We shall show that Z is an algebraic set. Observe that F may be chosen in $\mathcal{P}[X] \otimes \mathcal{P}[S^k]$ i.e. we may asssume that F^2 is of the form

$$F^2(x,y) = \sum_i f_i(x)g_i(y) \quad , \quad (x,y) \in \tilde{X} \subset X \times S^k,$$

where $f_i \in P[X]$ and $g_i \in P[S^k]$. Since $F^2 \circ \sigma = F^2$ and $\sigma(x,y) = (x,-y)$, it follows that $F^2(x,y) = \sum_i f_i(x)\gamma_i(y)$, with $\gamma_i(y) = \frac{1}{2}(g_i(y) + g_i(-y))$. Then $\gamma_i(y) = \gamma_i(-y)$ and each morphism $\gamma_i \in P[S^k]$ induces a unique algebraic morphism $g_i'\colon P^k(\mathbb{R}) \to \mathbb{R}$ satisfying $\gamma_i = g_i' \circ q$. We have

$$F^2(x,y) = \sum_i f_i(x)g_i'(q(y)) = \sum_i f_i(x)g_i'(\phi(x)).$$

Letting $H = \sum_i f_i(g_i' \circ \phi)$, we obtain an element $H \in R(X)$ satisfying $F^2 = H \circ \pi$. Hence $Z = \pi(F^{-1}(0)) = H^{-1}(0)$ is an algebraic subset of X. Obviously Z is a non-singular hypersurface. This completes the proof of Lemma 4. □

Let $G_m(\mathbb{R}^n)$ (resp. $\tilde{G}_m(\mathbb{R}^n)$) be the Grassmann manifold of m-dimensional linear subspaces of $\mathbb{R}^n$ (resp. oriented m-dimensional subspaces of $\mathbb{R}^n$), and let $V_m(\mathbb{R}^n)$ be the Stiefel manifold of orthonormal m-frames at the origin of $\mathbb{R}^n$. We consider $G_m(\mathbb{R}^n)$ and $V_m(\mathbb{R}^n)$ as a non-singular algebraic subsets of an Euclidean space [15]. More precisely, $V_m(\mathbb{R}^n)$ is considered as a subset of the vector space of $n \times m$ real matrices $M(n,m)$

$$V_m(\mathbb{R}^n) = \{(v_1,\ldots,v_m) \in M(n,m): \langle v_i, v_j\rangle = \delta_{ij}\},$$

where δ_{ij} is the Kronecker symbol, and $G_m(\mathbb{R}^n)$ is embedded in the linear space $L(\mathbb{R}^n)$ of all linear endomorphisms of $\mathbb{R}^n$, by identifying a space $W \in G_m(\mathbb{R}^n)$ with the orthogonal projection of $\mathbb{R}^n$ onto W. There is an obvious algebraic fibration $p\colon V_m(\mathbb{R}^n) \to G_m(\mathbb{R}^n)$.

<u>Lemma 5.</u> There is an algebraic morphism $h\colon G_m(\mathbb{R}^n) \to \mathbb{P}^k(\mathbb{R})$, $k = \binom{n}{m} - 1$,

such that the set

$$h^*(S^k) = \{x,y) \in G_m(\mathbb{R}^n) \times S^k \quad : h(x) = q(y)\}$$

is diffeomorphic to $\tilde{G}_m(\mathbb{R}^n)$.

Proof. Put $J = \{(\alpha_1,\dots,\alpha_m) \in \mathbb{Z}^m : 1 \le \alpha_1 < \dots < \alpha_m \le n\}$, and let $M'(n,m)$ be a Zariski open subset of $M(n,m)$ of the matrices of maximal rank. Given $v = (v_1,\dots,v_m) \in M'(n,m)$ and $\alpha \in J$, we define $g_\alpha(v)$ to be the determinant of the $m \times m$ matrix, which consists of α_i - th rows of v, $i = 1,\dots,m$. The map $\tilde{g}: M'(n,m) \to \mathbb{R}^{k+1}\setminus\{0\}$, given by $\tilde{g}(v) = (g_\alpha(v))_{\alpha \in J}$ is a well defined algebraic morphism satisfying $\tilde{g}(vz) = |z|\tilde{g}(v)$, for any $v \in M'(n,m)$ and any non-singular $m\times m$ matrix z, $|z| = \det z$. Taking the restriction $g = \tilde{g}|V_m(\mathbb{R}^n)$, we may define a unique smooth map $h: G_m(\mathbb{R}^n) \to P^k(\mathbb{R})$ which makes the diagram

$$\begin{array}{ccc} V_m(\mathbb{R}^n) & \xrightarrow{\ g\ } & \mathbb{R}^{k+1}\setminus\{0\} \\ p\downarrow & & \downarrow q \\ G_m(\mathbb{R}^n) & \xrightarrow{\ h\ } & P^k(\mathbb{R}) \end{array} \qquad (1)$$

commutative. We shall show that h is an algebraic morphism. Let us fix an element $v \in V_m(\mathbb{R}^n)$ and define a morphism

$$\tau_v : L(\mathbb{R}^n) \to M(n,m)$$

by $\tau_v(\phi) = (\phi(v_1),\dots,\phi(v_m))$. Denote by $\tilde{p}: M'(n,m) \to G_m(\mathbb{R}^n)$ the natural extension of p and by $L_v(\mathbb{R}^n) = \tau_v^{-1}(M'(n,m))$ a Zariski open subset of $L(\mathbb{R}^n)$. Observe that $\tilde{p}\circ\tau_v$ is the identity map on $G_m(\mathbb{R}^n) \cap L_v(\mathbb{R}^n)$, and that given two matrices $u, w \in M'(n,m)$ with $\tilde{p}(u) = \tilde{p}(w)$, we have $q(\tilde{g}(u)) = q(\tilde{g}(w))$ (because $u = vz$ for some non-singular $m \times m$ matrix z). Hence for all $\phi \in G_m(\mathbb{R}^n) \cap L_v(\mathbb{R}^n)$ we have

$$h(\phi) = h(\tilde{p}(\tau_V(\phi))) = h(p(\overline{\tau_V(\phi)})) = q\circ g(\overline{\tau_V(\phi)}) = q\circ\tilde{g}(\tau_V(\phi)),$$

where $\overline{\tau_V(\phi)}$ is the orthonormalization of $\tau_V(\phi)$, i.e. $h = q\circ\tilde{g}\circ\tau_V$ on $G_m(\mathbb{R}^n) \cap L_V(\mathbb{R}^n)$. Since the sets of the form $G_m(\mathbb{R}^n) \cap L_V(\mathbb{R}^n)$ constitute a Zariski open covering of $G_m(\mathbb{R}^n)$, it follows that h is an algebraic morphism.

Now let p': $V_m(\mathbb{R}^n) \to \tilde{G}_m(\mathbb{R}^n)$ and r: $\tilde{G}_m(\mathbb{R}^n) \to G_m(\mathbb{R}^n)$ be the natural surjection maps. Evidently $r\circ p' = p$. Let $v = (v_1,\ldots,v_m) \in V_m(\mathbb{R}^n)$ and $v' = (-v_1,v_2,\ldots,v_m)$. Then $g(v) = -g(v')$, $r\circ p'(v) = r\circ p'(v')$ and $p'(v) \neq$ $\neq p'(v')$. Using the commutativity of the diagram (1) above, we deduce that the map

$$\tilde{G}_m(\mathbb{R}^n) \ni p'(v) \to (r\circ p'(v),\ g(v)/|g(v)|) \in h^*(S^k)$$

is a well defined diffeomorphism of $\tilde{G}_m(\mathbb{R}^n)$ onto $h^*(S^k)$.

Lemma 6. Let $X \subset \mathbb{R}^n$ be a compact, connected, non-singular and nonorientable algebraic set; $\dim X = m$. Then there is an algebraic morphism ϕ: $X \to P^k(\mathbb{R})$, $k = \binom{n}{m} - 1$, such that the algebraic set

$$\tilde{X} = \{(x,y) \in X \times S^k\colon \phi(x) = q(y)\}$$

is orientable. In particular $\tilde{X}$ is a connected, algebraic, 2-fold covering space of X.

Proof. Consider the map

$$\theta\colon X \ni x \to T_xX \in G_m(\mathbb{R}^n),$$

where T_xX is the tangent space to X at x. This is an algebraic morphism. Indeed, let us take a point $\alpha \in X$ and choose a system of $n - m$ polynomials $f_1,\ldots,f_{n-m} \in P[\mathbb{R}^n]$ such that $X \subset \bigcap_{i=1}^{n-m} f_i^{-1}(0)$ and grad $f_1(\alpha),\ldots,$grad $f_{n-m}(\alpha)$ are linearly independent. Let

$$U = \{x \in X: \text{grad } f_1(x),\dots,\text{grad } f_{n-m}(x) \text{ are linearly independent}\}.$$

Obviously U is a Zariski open neighborhood of α. On U, the map θ is given by the formula

$$\theta(x) = \eta\left(\sum_{i=1}^{n-m} \mathbb{R}\,\text{grad } f_i(x)\right)$$

where $\eta: G_{n-m}(\mathbb{R}^n) \ni W \to W^{\perp} \in G_m(\mathbb{R}^n)$, so evidently θ is algebraic.

Now we define $\phi = h\circ\theta$, where h is the map defined in Lemma 5.

We shall show that for such a ϕ, the set $\tilde{X}$ is orientable. First we note that

$$\tilde{X} = (h\circ\theta)^*(S^k) = \theta^*(h^*(S^k)) = \theta^*(\tilde{G}_m(\mathbb{R}^n)).$$

If $\theta^*(\tilde{G}_m(\mathbb{R}^n))$ were not orientable, then we could choose a Jordan curve

$$\phi = (\phi_1,\phi_2): [0,1] \to \theta^*(\tilde{G}_m(\mathbb{R}^n)) \subset X \times \tilde{G}_m(\mathbb{R}^n),$$

such that any neighborhood of $\phi([0,1])$ in $\theta^*(\tilde{G}_m(\mathbb{R}^n))$ is nonorientable. This allows us to find a continuous map $\psi: [0,1] \to V_m(\mathbb{R}^n)$, such that $\theta\circ\phi_1 = p\circ\psi$, $\psi(0) = (v_1,\dots,v_m)$, $\psi(1) = (-v_1,v_2,\dots,v_m)$ and $p'\circ\psi(0) = \phi_2(0)$. Then $p'\circ\psi(0) \neq p'\circ\psi(1)$.

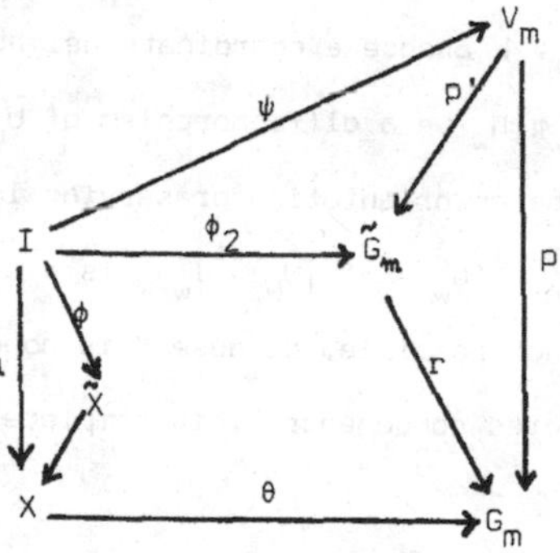

Since $r\circ\phi_2 = \theta\circ\phi_1$, we have $r\circ\phi_2 = r\circ p'\circ\psi$. Using unique-lifting property of covering space we deduce that $\phi_2 = p'\circ\psi$. But $\phi_2(0) = \phi_2(1)$, contradicting the property $p'\circ\psi(0) \neq p'\circ\psi(1)$. Hence $\tilde{X}$ must be orientable.

Now we are ready to finish the proof of Theorem 4'.

<u>Proof of Theorem 4'.</u> Let $W \subset X$ be as in Theorem 4'. We shall apply Lemma 4 with $Y = W$ and $\phi: X \to P^k(\mathbb{R})$ defined in Lemma 6. So let $\pi: \tilde{X} \to X$ be the 2-fold orientable algebraic covering of X, $\tilde{X} = \phi^*(S^k)$. We only must show that $[\tilde{W}] = 0$ in $H_{m-1}(\tilde{X},\mathbb{Z}_2)$, where $\tilde{W} = \pi^{-1}(W)$. Since $\tilde{X}\setminus\tilde{W}$ has at most two connected components (because $X\setminus W$ is connected), we only need prove that $\tilde{X}\setminus\tilde{W}$ is not connected. Then by a simple general topology argument involving the involution σ of $\tilde{X}$, $\pi\circ\sigma = \pi$, we deduce that $\tilde{W}$ must be the boundary of each of these components, and hence $[\tilde{W}] = 0$. Since W is a submanifold realizing the dual of the first Stiefel-Whitney class of X, the set $X\setminus W$ is orientable. Suppose that $\tilde{X}\setminus\tilde{W}$ is connected and given an orientation on $\tilde{X}$, choose an orientation of $X\setminus W$ such that $\pi: \tilde{X}\setminus\tilde{W} \to X\setminus W$ be an orientation preserving local diffeomorphism. It would allows us to extend the orientation on the whole X. Indeed, let $(U_\alpha, g_\alpha)_{\alpha\in A}$ be a covering of $X\setminus W$ by coherently orientec coordinate neighborhoods. For each $w \in W$ choose a coordinate neighborhood $\tilde{U}_w$ of a point $\tilde{w} \in \pi^{-1}(w)$, such that $\pi|\tilde{U}_w$ be a diffeomorphism of $\tilde{U}_w$ onto a neighborhood U_w of w in X. Let $\tilde{g}_w$ be an orientation preserving local coordinate map on $\tilde{U}_w$. Then $(U_\alpha, g_\alpha)_{\alpha\in A} \cup (U_w, \tilde{g}_w\circ(\pi|\tilde{U}_w)^{-1})_{w\in W}$ is a coherently oriented atlas of X. This is not possible, because X is nonorientable. Hence $\tilde{X}\setminus\tilde{W}$ must have two connected components. This completes the proof of Theorem 4'.

7. The ring of polynomial functions.

The results of sections 2-6 are basically *geometric*. The question of factoriality of the ring of polynomials $P[X]$ is essentially an *arithmetic* problem. The example of an ovaloid $X = \{x^4 + y^4 + z^2 = 1\} \subset \mathbb{R}^3$ given in [21] is significant: this simple surface (trivially diffeomorphic to S^2, for which $P[S^2]$ is factorial) has the ring $P[X]$ not factorial (indeed, we have $x^4 + y^4 = (1-z)(1+z) = (x^2 + \sqrt{2}xy + y^2)(x^2 - \sqrt{2}xy + y^2)$ in $P[X]$).

The arithmetical nature of the problem of factoriality of polynomial rings becomes more clear if we observe that essentially only the set of strictly positive polynomials is responsible for an eventual lack of factoriality of $P[X]$ (if $H^1(X,\mathbb{Z}_2) = 0$). More precisely we have

Theorem 6. Let $X \subset \mathbb{R}^n$ be a compact algebraic set. Then the ring $P[X]$ of polynomials on X is factorial if and only if the following three conditions are satisfied:

(α) The local ring $P[X]_{M_x}$ at each point $x \in X$ is factorial;

(β) The subgroup $H^1_{alg}(X,\mathbb{Z}_2)$ of $H^1(X,\mathbb{Z}_2)$ generated by algebraic divisorial cocyles is trivial.

(γ) Each strictly positive, irreducible element of $P[X]$ is prime.

Proof. Follows from Theorem 3 and a theorem of Nagata ([17] p. 31 or [9]).

8. Construction of non-trivial examples of algebraic sets X with $O(X)$, $N(X)$ and $R(X)$ locally factorial.

It is well known that for a compact, connected, non-singular algebraic set V, each of the rings $O(V)$, $N(V)$ and $R(V)$ is regular and hence locally factorial. The following theorem enables us to construct several examples of algebraic sets with singularities, whose corresponding rings are locally factorial.

Theorem [24] Let $U_1,\dots,U_s$ be a family of germs of real algebraic sets in $\mathbb{R}^k$ (U_i is a germ at $a_i \in \mathbb{R}^k$, $a_i \neq a_j$ for $i \neq j$). Suppose that codim $U_i = 1$ and each U_i has an isolated singularity at a_i. Then there exists a compact, connected, irreducible algebraic set $X \subset \mathbb{R}^k$ such that

(α) Sing $X \subset \{a_1,\dots,a_s\}$;

(β) for each $i = 1,\dots,s$, there exists a local analytic diffeomorphism $\sigma_i: (\mathbb{R}^k,a_i) \to (\mathbb{R}^k,a_i)$ with $\sigma_i(U_i) = X_{a_i}$, where X_{a_i} is a germ of X at a_i.

Now if we suppose that each ring of germs of analytic functions $\mathcal{O}_{a_i}(\mathbb{R}^k)/(\text{ideal of } U_i)$ is factorial, then applying the theorem above we obtain an algebraic set X with $\mathcal{O}(X)$, $\mathcal{N}(X)$ and $\mathcal{R}(X)$ locally factorial.

8. Bibliographical Note. The equivalence $\{\mathcal{O}(V)$ factorial$\} \Leftrightarrow \{H^1(V,\mathbb{Z}_2) = 0$, V compact, connected$\}$ has been proved (for V an analytic manifold), by a different method in [4],[19]. The equivalence $\{\mathcal{N}(U)$ factorial$\} \Leftrightarrow$ $\Leftrightarrow \{H^1(U,\mathbb{Z}_2) = 0\}$ for U a connected, semi-algebraic, smooth submanifold of $\mathbb{R}^n$, was proved in [5],[20]. The theorem $C(\mathcal{N}(U)) \cong H^1(U,\mathbb{Z}_2)$, for $U \subset \mathbb{R}^2$ open, semi-algebraic, has been proved in [8]. The implication $\{H^1(V,\mathbb{Z}_2) = 0$, V a compact, irreducible, non-singular algebraic set$\} \Rightarrow \{\mathcal{R}(V)$ factorial$\}$ is proved by a different method in [21], which also contains more results concerning the factoriality of $\mathcal{R}(V)$. Theorem 3 is related, at least for X non-singular, to a work of Silhol [23]. More informations concerning the theory of Nash vector bundles and algebraic vector bundles can be found in [2], [13], [24], [25], [26],[27].

Aknowledgement. We would like to thank R. Benedetti, N. Habegger, M. Kervaire and A. Tognoli for valuable remarks during the preparation of this paper.

References.

1. Benedetti, R., Tognoli A.: Approximation theorems in real algebraic geometry, Seminaire Risler, Université Paris VII, 1980.
2. Benedetti R., Tognoli A.: On real algebraic vector bundles, Bull.Sc. Math., 104, 89-112, (1980).
3. Bochnak J.: Un critère de factorialité des anneaux globaux réguliers, C.R.A.S. Paris, 283, 285-286, (1976).
4. Bochnak J.: Sur la factorialité des anneaux de fonctions analytiques, C.R.A.S. Paris, 283, 269-273, (1974).
5. Bochnak J., Efroymson G.: Real Algebraic Geometry and the 17th Hilbert Problem, Math. Ann. 251, 213-241 (1980).
6. Borel A., Haefliger A.: La class d'homologie fondamentale d'un espace analytique, Bull.Soc.Math. France 89, 461-513, (1961).
7. Bourbaki N.: Algèbre Commutative, Ch. VII, Paris 1965.
8. Efroymson G.: Nash rings on planar domains, Trans.Amer.Math.Soc. 249(2), 435-445, (1979).
9. Fossum R.: The Divisor Class Group of a Krull Domain, Springer Verlag 1973.
10. Hironaka H.: Introduction to real analytic sets and real analytic maps, Instituto Matematico "L.Tonelli" dell'Universita di Pisa 1973.
11. Hironaka H.: Subanalytic sets, Volume in honor of Y.Akizuki, Number theory, algebraic geometry and commutative algebra, 453-493, Tokyo 1973.
12. Hirzebruch F.: Topological methods in Algebraic Geometry, Springer Verlag, 1966.
13. Hubbard J.: On the cohomology of Nash sheaves, Topology 11, 265-270, (1974).
14. Kucharz W.: On analytic sets with given singularities, to appear.
15. Palais R.: Equivariant, real algebraic differential topology, Brandeis University, (preprint), 1972.
16. Risler J.J.: Sur l'anneau des fonctions de Nash globales, Ann.Sc. de l'Ecole Nor.Sup. 8 (3), 365-378, (1975).
17. Samuel P.: Anneaux Factoriels, Bol.Soc.Math., São Paulo, 1964.
18. Serre J.P.: Faisceaux algébriques cohérents, Ann. of Math., 81, 197-278, (1955).
19. Shiota M.: Sur la factorialité de l'anneau des fonctions analytiques, C.R.A.S. Paris 285, 253-255, (1977).

20. Shiota M.: On the unique factoriality of the ring of Nash functions Publ. R.I.M.S., Kyoto University, to appear.

21. Shiota M.: Sur la factorialité de l'anneau des fonctions lisses rationnelles, C.R.A.S., Paris. 292, 67-70 (1981).

22. Siu Y.: Noetheriannes of ring of holomorphic functions, Proc.Amer. Math. Soc. 21, (1969).

23. Silhol R.: Etude cohomologique des variètés algebriques réelles, preprint, University of Ferrara, (1980).

24. Tognoli A.: Algebraic approximation of manifolds and spaces, Seminaire Bourbaki, November 1979.

25. Tognoli A.: Une remarque sur les fibrés vectoriels analytiques et de Nash, C.R.A.S. Paris 290, 321-324 (1980).

26. Tognoli A.: Algebraic geometry and Nash functions, Institutiones mathematicae, Vol. III, Acad. Press, London-New York, 1978.

27. Benedetti R., Tognoli A.: Remarks and counterexamples in real algebrai vector bundles and cycles, This volume.

28. Shiota M.: Real algebraic realization of characteristic classes, preprint, Kyoto University (1981).

29. Akbulut S., King H.: article in preparation, University of Maryland.

Jacek BOCHNAK, Vrije Universiteit, Department of Mathematics, P.O.Box 7161, 1007 MC Amsterdam, The Netherlands.
Wojciech KUCHARZ, University of Katowice, Department of Mathematics, Bankowa 14, Katowice, Poland.
Masahiro SHIOTA, Kyoto University, Research Institute for Mathematical Sciences, Kyoto, Japan.

Real Spectra and Distributions of Signatures

by

Ludwig Bröcker

0. Introduction

Let V be an algebraic integral variety over a real closed field R. In this article we consider quadratic-forms over the function field R(V) and their signs on the set V(R) of the closed real points. For such a form ρ the $\mathbb{Z}$-valued function sign(ρ) is defined up to a Zariski-closed subset of V(R); it is constant on semialgebraic subsets.

Conversely divide V(R) into semialgebraic countries $S_1,\ldots,S_k$ (up to sets of lower dimension) and provide each country with a number $D_i \in \mathbb{Z}$. Does there exist a quadratic form ρ over R(V) with sign(ρ) = D_i on S_i? The given datum $(S_1,\ldots,S_k,D)$ is called distribution of signatures, and we say, that the above form ρ would solve that distribution. To get criteria for the solvability, we follow the fundamental idea in E. Artins work on Hilberts 17. problem [A]; there the study of signs on points is reduced to the study of signs on orderings. The main results of this article are known to me since two years or longer, but I changed the way writing them down several times: Once, when Brumfiel's book [Bru 1] appeared, where I found much of my base work; then, after I was convinced by the usefulness of the real spectrum, which was invented by Coste and Coste - Roy [CR], [C-CR]. This article includes a rapid introduction to real spectra, which is based on the notion of the restricted topological space [D-K2]; it contains some new results on real spectra of schemes and perhaps some new simplifications.

I want to thank M. Knebusch for stimulating conversations and in particular for a simplification to the description of X(V) by ultrafilters. I thank also H. Delfs, F. Ischebeck and H.W. Schülting for interesting and helpful discussions.

1. Restricted topological spaces

1. Definition. A restricted topological space $(X, \mathcal{B})$ is a pair, consisting of a topological space X together with a base $\mathcal{B}$ for the open sets of X, such that $\emptyset \in \mathcal{B}$, $X \in \mathcal{B}$ and $\mathcal{B}$ is closed under $\cap$ and $\cup$ (finite operations)

So $\mathcal{B}$ forms a lattice. Let $k(\mathcal{B})$ be the lattice, which is generated by $\mathcal{B}$ and the sets $X \setminus B$ for $B \in \mathcal{B}$. The elements of $k(\mathcal{B})$ are called constructible, the elements of $\mathcal{B}$ are called open-constructible. Note, that open constructible sets must not be open-constructible in general. A set $D \in k(\mathcal{B})$ can be expressed as $D = \bigcup_1^n (A_i \cap B_i)$ and also as $D = \bigcap_1^m (A_i' \cup B_i')$ with A_i, A_i' of the form $X \setminus B$ for some $B \in \mathcal{B}$ and B_i, $B_i' \in \mathcal{B}$. Thus there exist $B_1, \ldots, B_m \in \mathcal{B}$, $B_i \neq \emptyset$, such that $D \supset B_1 \cap \ldots \cap B_m$ or $D \subset X \setminus B$ for some $B \in \mathcal{B}$, $B \neq \emptyset$. If X is a Noetherian space $\mathcal{B}$ consists of all open sets of X. We are more interested in the case where X is not Noetherian and not irreducible.

2. Definition. A morphism $f : (X, \mathcal{B}) \to (X', \mathcal{B}')$ of restricted topological spaces is a map $f : X \to X'$ such that $f^{-1}(B') \in \mathcal{B}$ for all $B' \in \mathcal{B}'$. Thus f is continuous. We denote by $\mathcal{T}$ the category of the restricted topological spaces.

3. Main example. Let $(k, >)$ be an ordered field, $p_o = \{a \in k \mid a \geq 0\}$, $(R, >)$ a real closure of k and A a finitely generated k-algebra. For $U = \operatorname{spec}(A)$ we have $U \times_k R = \operatorname{spec}(A \otimes_k R)$. Now set $U_R := (U \times_k R)(R)$. We provide U_R with a lattice of open sets: So for $a \in A$ set $S(a) := \{x \in U_R \mid a(x) > 0\}$ where $a(x)$ is the residue-class of a at x. Let $\mathcal{S}$ be the lattice, which is generated by all $S(a)$ for $a \in A$. The space $(U_R, \mathcal{S}) \in \mathcal{T}$ is called the semialgebraic space of A. The elements of $k(\mathcal{S})$ are called semialgebraic, those of $\mathcal{S}$ open-semialgebraic. If k is real closed, it can be shown, that open-semialgebraic is just open and semialgebraic [D], [C-CR]. [1)]

---

1) Compare the remark at the end of this article.

More general, let V be a k-variety.[1)] Again we set $V_R := (V \times_k R)(R)$ and call a set $S \subset V_R$ open-semialgebraic, if for all open affine k-subvarieties $U \subset V$ the set $U_R \cap S$ is open-semialgebraic in U_R. For the lattice S of all open-semialgebraic sets we get the <u>semi-algebraic space</u> (V_R, S) of V. In fact $\sigma\colon \mathcal{V} \to \mathcal{T}$; $V \to (V_R, S)$ is a functor for the category $\mathcal{V}$ of k-varieties.

4. <u>Subspaces</u>. For a space $(X,\mathcal{B}) \in \mathcal{T}$ let $Y \subset X$ be a subset. One provides Y with the induced lattice $\mathcal{B}|Y = \{B \cap Y \mid B \in \mathcal{B}\}$. $(Y, \mathcal{B}|Y)$ is called a <u>subspace</u> of $(X,\mathcal{B})$. In particular, we denote by $(\underline{X}, \mathcal{B}|\underline{X})$ the subspace of all closed points of $(X,\mathcal{B})$.

5. <u>Site</u>. We provide $(X,\mathcal{B}) \in \mathcal{T}$ with the following site: "Open sets" are the elements of $\mathcal{B}$ "morphisms" are the inclusion maps and "coverings" are the <u>finite</u> coverings in the usual sense. Thus $(X,\mathcal{B})$ is called <u>connected</u>, if it cannot be decomposed into disjoint sets $B_1, B_2 \in \mathcal{B}$, $B_i \neq \emptyset$.

6. <u>Specialization</u>. For $(X,\mathcal{B}) \in \mathcal{T}$ let be $x_1, x_2 \in X$. We say x_1 <u>specializes</u> x_2 if x_1 lies in the closure of x_2, and we say x_2 <u>generalizes</u> x_1.

7. <u>Dimension</u>. Let D be a constructible set in the space $(X,\mathcal{B}) \in \mathcal{T}$. We call D thin in X if D admits no interior points. We denote this by $D \prec X$. More general, for $D_1, D_2 \in k(\mathcal{B})$ with $D_1 \subset D_2$ and D_1 thin in $(D_2, \mathcal{B}(D_2))$ we write $D_1 \prec D_2$. Now we define the <u>dimension</u> $\dim(X,\mathcal{B})$ to be the maximal integer n such that there exists a chain $\emptyset \neq D_0 \prec D_1 \prec \ldots \prec D_n = X$ of sets $D_i \in k(\mathcal{B})$. If this number does not exist, we set $\dim(X,\mathcal{B}) = \infty$.

It can be shown, that in the above example the dimension of a semi-algebraic set S is just the dimension of the Zariski-closure of S

---

1) More precisely $(V, \mathcal{O}_V)$ is a k-variety; but in this note we do not use structural sheaves explicitly.

in $V \times_k R$, which does not depend upon the question, whether we take the k- or the R- topology.

8. Canonical completion. For $(X,\mathcal{B}) \in \mathcal{T}$ let $\hat{X}$ be the set of all ultrafilters of $k(\mathcal{B})$ and more general set $\hat{D} := \{F \in \hat{X} \mid D \in F\}$. Then we have: $\widehat{D_1 \cup D_2} = \hat{D}_1 \cup \hat{D}_2$, $\widehat{D_1 \cap D_2} = \hat{D}_1 \cap \hat{D}_2$ and $\widehat{D_1 \setminus D_2} = \hat{D}_1 \setminus \hat{D}_2$. So $\hat{\mathcal{B}} := \{\hat{B} \mid B \in \mathcal{B}\}$ is a lattice, and $k(\hat{\mathcal{B}}) = \{\hat{D} \mid D \in k(\mathcal{B})\}$. We call $(\hat{X},\hat{\mathcal{B}})$ canonical completion of $(X,\mathcal{B})$. We regard "$\wedge$" as a functor : $\mathcal{T} \to \mathcal{T}$. In fact, for a morphism $f : (X,\mathcal{B}) \to (Y,\mathcal{C})$ the map $\hat{f} : (\hat{X},\hat{\mathcal{B}}) \to (\hat{Y},\hat{\mathcal{C}})$ is defined by $\hat{f}(F) = \{D \in k(\mathcal{C}) \mid f^{-1}(D) \in k(\mathcal{B})\}$.

9. Properties of "$\wedge$". 1) $(\hat{X},\hat{\mathcal{B}})$ is quasicompact.

2) $\wedge : k(\mathcal{B}) \to k(\hat{\mathcal{B}})$; $D \to \hat{D}$ is an isomorphism of lattices, its restriction to $\mathcal{B}$ is an isomorphism : $\mathcal{B} \to \hat{\mathcal{B}}$.

3) $D \in k(\mathcal{B})$ belongs to $\mathcal{B}$, iff $\hat{D}$ is open in $(\hat{X},\hat{\mathcal{B}})$.

4) For $F_1,F_2 \in \hat{X}$ the filter F_1 specializes F_2, iff $\mathcal{B} \cap F_1 \subset F_2$.

5) For $x \in X$ denote by $F(x)$ the principal filter generated by x. The map $x \to F(x)$ defines a morphism : $(X,\mathcal{B}) \to (\hat{X},\hat{\mathcal{B}})$, and its image is dense in $\hat{X}$. Moreover, if for each pair $x,y \in X$ with $x \neq y$ there exists a $D \in k(\mathcal{B})$ with $x \in D$, $y \notin D$, then the above map is injective.

6) The following objects are in one to one correspondence:
- Finite decompositions of X into disjoint sets of $\mathcal{B}$.
- Decompositions of $\hat{X}$ into disjoint open sets.

In particular $(X,\mathcal{B})$ is connected, iff $(\hat{X},\hat{\mathcal{B}})$ is connected in the usual sense.

10. Example. Set $X = \mathbb{Q}$, provided with the lattice $\mathcal{B}$, which is generated by all open intervals with boundary points in $\mathbb{Q}$. Then

$\hat{\underline{X}} \simeq -\infty \cup \mathbb{R} \cup \infty$.

There are many other concepts, which can be handled on the abstract level of restricted topological spaces, for instance local dimension, proper maps and completeness, but we must not enter in this here.

2. The real spectrum

Let $\mathcal{A}$ be the category of all commutative rings with unit and $\mathcal{S}$ the category of all quasicompact separated schemes. To begin with we fix a Ring $A \in \mathcal{A}$.

1. <u>Definition</u>. A set $P \subset A$ is called positive-cone of A , if the following holds : $P + P \subset P$, $P \circ P \subset P$, $A = P \cup -P$ and $-1 \notin P$. P is called <u>prime</u>, if moreover $P \cap -P$ is a prime ideal in A . We denote by $X(A)$ the set of all prime positive-cones of A

2. <u>Definition</u>. A set $T \subset A$ is called precone of A , if $T + T \subset T$, $T \cdot T \subset T$, $A^2 \subset T$ and $-1 \notin T$. Set $X/T := \{P \in X(A) \mid P \supset T\}$ The set $Q(A)$ of all sums of squares in A is a precone, iff $-1 \notin Q(A)$, in which event A is called <u>formally real</u>.

3. <u>Proposition</u> [Br 4]. Let T be a precone in A . Then

$$\bigcap_{P \in X/T} P = \{a \in A \mid a^{2n+1} + at \in T \text{ for some } n \in \mathbb{N} \text{ and } t \in T\} \neq A .$$

4. <u>Definition</u>. For $a \in A$ let be $B(a) := \{P \in X(A) \mid -a \notin P\}$ and $\mathcal{B}(A) = \mathcal{B}$ the lattice which is generated by all sets $B(a)$ for $a \in A$. $(X(A), \mathcal{B}(A))$ is called the <u>real spectrum</u> of A .

For $A, A' \in \mathcal{A}$ and a homomorphism $f : A \to A'$ we define $Xf : (X(A'), \mathcal{B}(A')) \to (X(A), \mathcal{B}(A))$ by $Xf(P') := f^{-1}(P)$ for $P' \in X(A')$. Thus we regard X as a contravariant functor : $\mathcal{A} \to \mathcal{T}$.

Again we fix a Ring $A \in \mathcal{A}$.

5. Remark. For $P, P' \in X(A)$ the following properties are equivalent:

a) P' specializes P

b) $P' \supset P$.

Proof: obvious.

6. Remark. Specializations of a fixed $P \in X(A)$ form a chain, more general, if P and $P' \in X(A)$ are incomparable, then P and P' admit disjoint neighbourhoods.

Proof: We set $p := P \cap -P$ and $p' := P' \cap -P'$ and consider three cases:

i) $p = p'$: choose $a \notin p$ with $a \in P \cap -P'$. Then $P \in B(a)$ and $P' \in B(-a)$.

ii) $p \subset p'$. Then $P \setminus p \not\subset P' \setminus p'$, hence there exists $a \in P \setminus p$ with $-a \in P' \setminus p'$, thus $P \in B(a)$ and $P' \in B(-a)$.

iii) There exist $a \in p \setminus p'$ and $b \in p' \setminus p$: Wo.l.g. $a \in P', b \in P$. Then $P \in B(b-a)$ and $P' \in B(a-b)$.

7. Propósition [C-CR], [CR]. Let $D \subset X(A)$ be constructible. Then

a) D is quasicompact

b) $\underline{D}$ is compact,

where $\underline{D}$ is the subspace of all closed points in $(D, \mathcal{B}(A) \mid D)$.

Proof a): Wo.l.g. D is of the form $\{P \in X(A) \mid a_1, \ldots, a_r \in P;$ $b_1, \ldots b_s \notin -P\}$.

By Alexanders subbasis lemma one must only show:

If D_λ, $\lambda \in \Lambda$ is a filter base of sets $D_\lambda = \{P \in D \mid a_\lambda \in P\}$ for certain elements $a_\lambda \in A$, then $\bigcap_{\lambda \in \Lambda} D_\lambda \neq \emptyset$.

Now consider the semiring $T \subset A$, which is generated by $A^2, a_1, a_1, \ldots, a_r$ $b_1, \ldots, b_s$ and all a_λ with $\lambda \in \Lambda$. Then $-1 \notin T$, hence T is a precone. For $b = b_1, \ldots b_s$ the set $T_b := \{\frac{t}{b^{2n}} \mid t \in T, n \in \mathbb{N}\}$ is even a precone in A_b by assumption. By prop. 1 T_b can be extended to an element $P' \in X(A_b)$. By the localization map $f : A \to A_b$ we get $P = f^{-1}(P') = Xf(P') \in X(A)$, which lies in $\bigcap_{\lambda \in \Lambda} D_\lambda$.

The proof of part b) uses the preceding remark and a similar argument.

8. Remark. For $b \in A$ let $f : A \to A_b$ be the localization. Then

$$Xf : X(A_b) \to B(b^2) \text{ is an isomorphism .}$$

Now we fix a scheme $S \in \mathfrak{S}$.

9. Definition. A prime positive cone P of S is a pair, consisting of a point $x = x(P) \in S$ and an ordering $P(x)$ of the residue class-field $k(x)$ of x. We denote by $X(S)$ the set of all these prime positive cones of S. For $S = \text{spec}(A)$ with $A \in \mathcal{A}$ we have $X(S) = X(A)$. We must provide $X(S)$ with a suitable lattice as base. So for $U \subset S$ we define $X(U) := \{P \in X(S) \mid x(P) \in U\}$.

10. Definition. A set $B \subset X(S)$ is called open-constructible, if for all open affine subschemes $U \subset S$ with $U = \text{spec}(A)$ the set $B \cap X(U)$ lies in $\mathcal{B}(A)$. $\mathcal{B}(S)$ is the lattice of all open-constructible sets in $X(S)$.

There are many elements in $\mathcal{B}(S)$, since we have

11. Proposition. Suppose, that $\text{spec}(A) = U$ is an open affine subscheme of S. Then the inclusion $(X(A), \mathcal{B}(A)) \to (X(S), \mathcal{B}(S))$ is an imbedding.

Proof. Let $\text{spec}(B) = V$ be another open affine subscheme of S. Then $\text{spec}(A) \cap \text{spec}(B)$ is of the form $\bigcup_{i=1}^{r} \text{spec}(B_{b_i})$. Therefore, by the preceding remark, $X(\text{spec}(A) \cap \text{spec}(B)) = \bigcup_{i=1}^{r} B(b_i^2)$, which is open constructible in $X(\text{spec}(B))$.

Corollary. $X(S)$ is quasicompact.

Again we regard X as a functor $\mathfrak{S} \to T$, which is now covariant. This functor is called real spectrum.

For $S \in \mathfrak{S}$ and $x \in S$ we will identify $X(k(x))$ with the subspace $\{P \in X(S) \mid x(P) = x\}$ of $X(S)$.

12. Definition. Let k be a field, $B \subset k$ a valuation ring with maximal ideal I and residue classfield $\overline{B}$. Moreover let $T \subset k$ be a precone (usually called preordering). Then T is named compatible with B, if $1+I \subset T$.

Then T induces a preordering $\overline{T}$ on $\overline{B}$. The valuations of k, which are compatible with a fixed ordering $P \subset k$, form a chain.

13. Proposition. For $S \in \mathfrak{S}$ and $P \in X(S)$ suppose, that B is a valuation ring of $x(P)$, which is compatible with $P(x)$ and admits the center $y \in \overline{x(P)}$. Then $P' := (y, \overline{P(x)} \mid k(y))$ specializes P.

This is rather obvious. Conversely one has the following result, which was proved by Brumfiel [Bru 1 p. 161] (For an easier proof see [Bru 2].) in the affine case.

14. Proposition (real place extension). For $S \in \mathfrak{S}$ and $P', P \in X(S)$, where P' specializes P, there is a valuation ring B of $k(x(P))$ with center $x(P')$, which is compatible with P, such that the residueclassfield $\overline{B}$ provided with the ordering $\overline{P}$ is an archimedean extension of the ordered field $(k(x(P')), P'(x))$.

15. Corollary. The specializations of a fixed element $P \in X(S)$ form a chain.

16. Corollary. If P and $P' \in X(S)$ are incomparable. then they admit disjoint neighbourhoods.

Proof. Set $x = x(P)$ and $x' = x(P')$. If x and x' lie in a common affine subscheme, we are through. Otherwise choose open affine subschemes U, U' with $x \in U$ and $x' \in U'$. Now take neighbourhoods of x and x' in $U = \mathrm{spec}(A)$ resp. $U' = \mathrm{spec}(B)$ of the form $B(a_1) \cap \ldots \cap B(a_n)$. If all these neighbourhoods mutually have a non void intersection, we get a filterbasis of open constructible sets and hence by proposition 7 an element $P_o \in X(U \cap U')$ such that P and P' specialize P_o. Contradiction.

17. <u>Corollary</u>. For $D \in k(\mathcal{B}(S))$ the space $\underline{D}$ is compact.

18. <u>Proposition</u>. All closed irreducible sets $I \subset X(A)$ are of the form $\overline{P_o}$ with $P_o \in X(S)$.

Proof. Suppose first, that $S = \mathrm{spec}(A)$ for $A \in \mathcal{A}$. Then we may regard the elements of $X(S) = X(A)$ as subsets of A. We set $P_o := \bigcap_{P \in I} P$ and get $I = \overline{P_o}$. [CR], [C-CR].

Now for an arbitrary scheme S let $S = U_1 \cup \ldots \cup U_n$ be a covering of S by open affine subschemes. Choose i such that $X(U_i) \cap I \not\subset X(U_j) \cap I$ or $X(U_i) \cap I = X(U_j) \cap I$, say $i = 1$. We claim, that $I \subset X(U_1)$: Suppose, that $I \not\subset X(U_1)$, say $P_2' \in X(U_2) \cap I$ with $P_2' \notin X(U_1)$. Now $I \cap X(U_1 \cap U_2) \neq \emptyset$ for otherwise we had a decomposition: $I \subset X(S \setminus U_1) \cup X(S \setminus U_2)$. So let $P \in I \cap X(U_1 \cap U_2)$. $I \cap X(U_i)$ is closed and irreducible in $X(U_i)$, hence $I \cap X(U_i) = \overline{P_i} \cap X(U_i)$ with $P_i \in X(U_i)$ for $i = 1, \ldots, n$. We have $P, P_2' \in \overline{P_2} \cap X(U_2)$ and thus two cases:

i) $P_2' \in \overline{P}$. We have $\overline{P} \subset \overline{P_1}$, hence $P_2' \in \overline{P_1}$ and thus $P_2' \in \overline{P'}$ for all $P' \in I \cap X(U_1)$, since otherwise $P' \in \overline{P_2'}$ for some $P' \in I \cap X(U_1)$ which yields $P_2' \in X(U_1)$; impossible. But now we have $I \cap X(U_1) \subset X(U_2)$. Contradiction.

ii) $P \subset \overline{P_2'}$. This implies $P_2' \in X(U_1)$. Contradiction.

19. Remark. Let k be a field with positivecone P_o , and let V be a k-variety with real spectrum $(X(V),\mathcal{B}(V))$. The subspace $(X(V/P_o),\mathcal{B}/P_o)$ of all elements $P \in X(V)$, which extend P_o , is closed. Therefore the preceding results hold for this subspace correspondingly.

3. Fans

In this section we recall some facts concerning the real spectrum of a field k . For a preordering $T \subset k$ one has

$$T = \bigcap_{P \in X/T} P \neq k \; .$$

1. Definition. T is called a fan, if for all $a \in K$, $a \notin -T$, one has $T+aT = T \cup aT$.

There are many equivalent conditions, which define fans. We refer to [B-K], [Br 2]. Examples of fans are orderings and intersections of two orderings. These fans are called trivial. One gets all fans by lifting trivial fans from residue fields of valuations. More precisely we have

2. First fan theorem [Br 2]. Let $T \subset k$ be a fan. Then there exists a valuation ring $B \subset k$, compatible with T , such that the induced preordering $\overline{T}$ is a trivial fan of $\overline{B}$.

It is easily seen, that the finest valuation ring, which is compatible with T , has again this property. It is called valuation ring of the fan T .

Now let $W(k)$ be the Wittring of all symmetric bilinear forms over k and $T \subset k$ a preordering. Let $I(T) \subset W(k)$ be the ideal, which is generated by the forms $<1,-t>$ for $t \in T$, $t \neq 0$. Then Pfisters principle says, that the sequence

$$0 \to I(T) \to W(k) \xrightarrow{\text{sign}} C(X/T,\mathbb{Z})$$

is exact [P], [B-K]. Here $C(X/T,\mathbb{Z})$ is the ring of all continuous functions: $X/T \to \mathbb{Z}$ ($\mathbb{Z}$ provided with the discrete topology), and sign is the total signature. $W(k/T) := W(k)/I(T)$ is called the mod. T <u>reduced</u> Wittring. It is characterized as a subring of $C(X/T,\mathbb{Z})$ by the following

3. <u>Second fan theorem</u> [B-Br]. For $f \in C(X/T,\mathbb{Z})$ there exists a form $\rho \in W(K)$ with $f = \text{sign}(\rho)$, iff $\sum_{P \in X/T'} f(P) \equiv 0 \mod \frac{1}{2}(k^*:T'^*)$ for all fans $T' \supset T$ with $(k^*:T'^*) < \infty$.

In view of this theorem it is interesting, how large this number $\frac{1}{2}(k^*:T'^*)$ can be in special cases. In [Br 2] it is shown, that the maximum of these numbers is just the reduced stability index of k. Hence by [Br 1] we get the

4. <u>Third fan theorem</u>. Let R be a real closed field and $K \supset R$ a formally real function field of transcendence degree d. Then there exist fans $T \subset K$ with $(K^*:T^*) = 2^{d+1}$ and this number is maximal with that property.

4. The ultrafiltertheorem

Again let (k,P_0) be an ordered field, R a real closure of (k,P_0) and V the category of all k-varieties. We have on the one hand the functor σ as in the main example of section 1, that is $\sigma: V \to T$; $V \to (V_R,S)$. On the other hand we have the real spectrum functor

$X\colon \mathcal{V} \to \mathcal{T}\ ;\ V \to (X(V/P_o), \mathcal{B}/P_o)$ and the canonical completion functor $\wedge\colon \mathcal{T} \to \mathcal{T}$.

These functors are related as follows (compare [Bru 1, p. 232]).

1. Ultrafiltertheorem. $X = \wedge \circ \sigma$, that means, for $V \in \mathcal{V}$ there are canonical mutually inverse isomorphisms

$$(\hat{V}_R, \hat{S}) \underset{F}{\overset{P}{\rightleftarrows}} (X(V/P_o)\ ,\ \mathcal{B}/P_o)\ .$$

Proof. For $F \in \hat{V}_R$ we look after a point $x(F) \in V$ and an ordering $P(F)(x)$ of $k(x)$ (which extends P_o). Let $d := \min\{\dim(S) \mid S \in F\}$ and $S \in F$ with $\dim S = d$. Let U be the Zariski k-closure of S in V with irreducible components $U = U_1 \cup \ldots \cup U_r$. Since F is an ultrafilter, we may assume, that $S \subset U = U_1$. Now for all $S \in F$ the set $S \cap U$ is k-dense in U . We define $x(F)$ to be the generic point of U . In order to get $P(F)(x)$ let W be an open affine neighbourhood of x . Then $U \subset W$ apart from a closed subset of lower dimension, so we may assume, that U is affine. Now for $f \in k[x]$ we define $f \in P(F)(x)$ if there exists $S \in F$ such that $f(y) \geq 0$ for all $y \in S \cap U$. This gives us in fact a unique ordering of $k[x]$; moreover the map P is injective, for if $F_1 \neq F$ and $x(P(F)) = x(P(F_1))$ we assume again, that V is affine.

There are $S, S_1 \in F$ and F_1 respectively with $S \cap S_1 = \emptyset$.

Wo.l.g. S is of the form $S(f_1) \cap \ldots \cap S(f_r)$. Then $S_1 \subset S(-f_1) \cup \ldots \cup S(-f_r)$, so we may assume, that $S_1 \subset S(-f_1)$ which yields, that $P \neq P_1$.

In order to construct the inverse map F we use the following

2. Lemma. Let U be an integral affine k-variety and P an ordering of its function field, which extends P_o . Then for $f_1,\ldots,f_n \in k[U] \cap P$, $f_i \neq 0$, the set $S(f_1) \cap \ldots \cap S(f_n)$ is k-dense in $U \underset{k}{\times} R$.

Now for $P \in X(V/P_o)$ and $x = x(P)$ let $W = \overline{x}$, which can assumed to be affine. Then by the lemma the sets $S(f)$ for $f \in P(x) \cap k[W]$ form a filterbasis F'. We define $F(P)$ to be the filter, which is generated F'. In fact $F(P)$ is an ultrafilter, for otherwise we had different ultrafilters F_1 and $F(P)$ over F', which contradicts the injectivity of the map P.

It is easily seen, that P and F are morphisms of restricted topological spaces.

Proof of the Lemma: Let $A := k[U][\sqrt{f_1},\ldots,\sqrt{f_n}]$ and $W = \text{spec}(A)$. Then P and thus P_o can be extended to the function field $k(W)$. By Artins theorem [A] W_R is k-dense in $W \times_k R$. Under the natural map $W \times_k R \to U \times_k R$ the set W_R is mapped to $S(f_1)\cap\ldots\cap S(f_n)$ (up to a set of lower dimension). Hence $S(f_1)\cap\ldots\cap S(f_n)$ must be k-dense in $W \times_k R$.

3. <u>Remark</u>. It can be shown, for instance by the Tarski-Seidenberg-principle [Bru 1, P. 268], that the semialgebraic sets of V_R, which are defined over R, can already be defined over k. Therefore, by the ultrafiltertheorem, the projection $\pi: V \times_k R \to V$ induces a homeomorphism.

$$X\pi: (X(V \times_k R),\mathcal{B}) \to (X(V/P_o),\mathcal{B}/P_o) \;;$$

but I do not know, whether open-semialgebraic sets over R can be defined as open-semialgebraic sets over k.

Henceforth we assume, that $k = R$ is real closed. Then, as already mentioned one can prove, that open semialgebraic sets in $V(R)$ are open-semialgebraic. By the properties of "$\wedge$" and the ultrafilter-theorem open semialgebraic sets of $V(R)$ correspond to open constructible sets of $X(V)$. This leads to the following

4. <u>Geometrical description of the specialization</u>. Let P, P' be $\in X(V)$ with corresponding ultrafilters $F(P)$, $F(P')$, generic points $x = x(P)$ and $x' = x(P')$ with closures W and W'. Then the following statements are equivalent:

a) P' specializes P

b) $F(P')$ specializes $F(P)$

c) The sets $S \cap W'(R)$; $S \in F(P)$, S closed, generate $F(P')$.

Moreover, if for a closed subvariety W'' of V and for all closed sets $S \in F(P)$ the set $S \cap W''(R)$ is Zariski-dense in W'', then the sets $S \cap W''(R)$; $S \in F(P)$, S closed, generate an ultrafilter, say $F(P'')$. Hence P'' specializes P and thus $W'' \subset W$.

We give, without proofs, an explicite description of $X(V) = \hat{V}(R)$ for the dimensions 1 and 2. We assume further, that V is complete.

5. <u>Dimension 1</u>. For $P \in X(V)$ we have $\dim x(P) = 0$ or $\dim x(P) = 1$. If $\dim x(P) = 0$, P corresponds to the closed point $x(P) \in V(R)$, that is, to the principalfilter $F(x(P)) \in \hat{V}(R)$.

So let $\dim x(P) = 1$, that is, $x(P)$ is an irreducible real component of V. Now we have two cases

1. P is specialized by an element $P' \in X(V)$. Then $\dim x(P') = 0$ and the point $x(P')$ lies in all closed sets of the filter $F(P)$. Therefore P corresponds to a real half-branch of the curve $V(R)$ at $x(P')$. That means, the real half branches at $x(P')$ correspond to the elements $P \in X(V)$, which generalize P'. The number of them is always even; it is 2 if $x(P')$ is smooth.

2. P is closed. Then P is called a gap.

Note, that <u>$X(V)$</u> $= V(R) \cup \{\text{gaps}\}$ is compact and consists of compact linear continua, which are patched together at discrete points (<u>$X(V)$</u> is a pseudograph).

If V is integral, by prop. 2.13 the gaps are just those orderings of $k(V)$, which are archimedean over R.

6. Dimension 2. Now for $P \in X(V)$ we may assume, that $\dim x(P) = 2$. One has the following cases

1. P is specialized by P_1 and this by P_0. Then P_0 corresponds to a point x_0, $x(P_1)$ is generic point of a real curve G through x_0. So P_1 corresponds to a real half-branch of G at x_0. Now by the geometrical description of the specialization P corresponds to one of the even number of real half-leafs of V at the above real half-branch of P_1. Cones P of this type can be described geometrically by symbols [K] like

point • , half-branch  , half-leaf 

as in the following picture

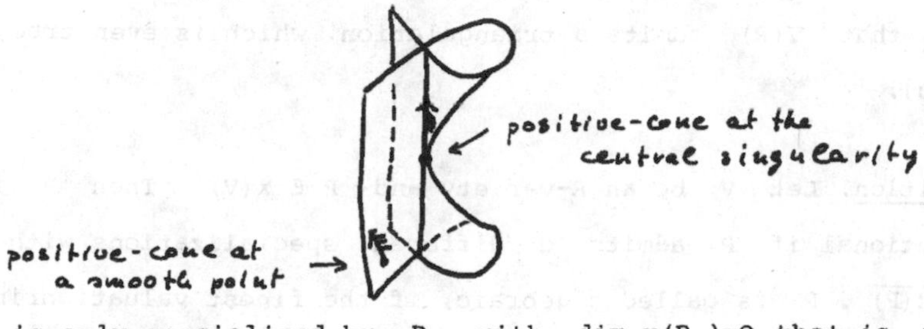

2. P is only specialized by P_0 with $\dim x(P_0)=0$,that is $x(P_0) = x_0$ is a closed point. If x_0 is smooth, then the real half-branches at x_0 of real curves through x_0 admit a natural circular ordering.
Now P corresponds to an open Dedekind-cut of this ordering. If x_0 is singular, then the set B of all real half-branches at x_0 of real curves through x_0 has the canonical structure of a finite union of circular ordered sets and an Eulerian graph, whose points are the singular real half-branches and whose edges are ordered.
For the above picture this graph at the origin looks like

Now again P corresponds to an open Dedekind-cut in one of these ordered sets.

3. P is specialized by P_1 and P_1 is closed. Then $x(P_1)$ is the generic point of a real curve G . P_1 corresponds to a gap on G(R) and P to a real half-leaf of V(R) at this gap.

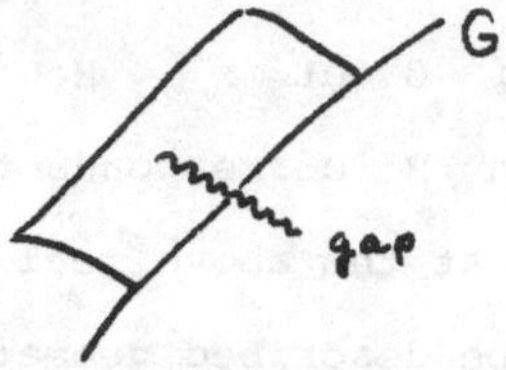

4. P is closed. Then P is a gap in V(R) . If V is integral, this happens, if and only if P is archimedean over the unique ordering of R .

Proofs for the above story are not difficult but rather long. One uses the fact, that V(R) admits a triangulation, which is even true, if $R \neq \mathbb{R}$ [D].

7. <u>Definition</u>. Let V be an R-variety and $P \in X(V)$. Then P is called rational if P admits d different specializations with $d = \dim \overline{x(P)}$. P is called algebraic, if the finest valuationring $B \subset R(x(P))$, which is compatible with P , is discrete of rang d .

Thus a rational P is always algebraic. Conversely if P is algebraic, by Zariskis local uniformization theorem there exists a model W of R(x(P)) such that P is rational in X(W) . In the above example P is rational just for the case 1 .

5. Distributions of signatures

Let R be a real closed field and V an integral R-variety. We assume V to be real, which means, that V(R) is dense in V or equivalently, that dim V(R) = dim V or that the function field R(V) is formally real. We set $d = \dim V$.

1. Definition. A distribution of signatures on V is represented by the following:

1. A partition of $V(R)$ into open semialgebraic sets $S_1,\dots,S_k$, each of dimension d, up to lower dimension.
This means $\dim(V(R) \setminus \bigcup_1^k S_i) < d$ and $S_1 \cap S_j = \emptyset$ for $i \neq j$.

2. An injective map $D : \{S_1,\dots,S_k\} \to \mathbb{Z}$.

Two such objects $(S_1,\dots,S_k,D)$ and $(S'_1,\dots,S'_k,D')$ are called equivalent, if $k = k'$, $S_i = S'_i$ up to a set of lower dimension in a suitable numeration $i = 1,\dots,k$ and $D(S_i) = D(S'_i)$. Now a distribution of signatures on V is an equivalence class of such objects.

Henceforth we consider semialgebraic sets only up to sets of dimension $< d$. So for partitions $V(R) = S_1 \cup \dots \cup S_k$ and $V(R) = S'_1 \cup \dots \cup S'_e$ one has a common refinement, and by this the set $\mathrm{Dis}(V)$ of all distributions of signatures on V has a natural structure of a commutative ring with unit. Now let ρ be a non degenerated quadratic form over $R(V)$. Then up to a Zariski-closed set of dimension $< d$ the signature of ρ is defined at the points of $V(R)$, and this is constant on semialgebraic sets. Therefore ρ defines a distribution of signatures on V. More precisely we get a map

$$\mathrm{Sign}: W(R(V)) \longrightarrow \mathrm{Dis}(V)$$

which is in fact a homomorphism of rings, where $W(R(V))$ is the Wittring of the function field $R(V))$.

We say, that a distribution of signatures on V can be solved, if it lies in the image of Sign. For instance it is easily seen, that the distribution on $\mathbb{R}^2$, which takes the value 2 on the first quadrant and the value 0 on the rest, cannot be solved.

Now by the ultrafiltertheorem we get

2. <u>Proposition</u>. There is a natural isomorphism
$f: \mathrm{Dis}(V) \longrightarrow C(X(R(V)), \mathbb{Z})$. Moreover the diagram

$$\begin{array}{ccc} & & \mathrm{Dis}(V) \\ & \overset{\mathrm{Sign}}{\nearrow} & \downarrow f \\ W(R(V)) & & \\ & \underset{\mathrm{Sign}}{\searrow} & \\ & & C(X, \mathbb{Z}) \end{array}$$

commutes.

In particular $\mathrm{Ker}(\mathrm{Sign}) = W(R(V))_{\mathrm{tor}}$.

In some sense by this proposition and the second fantheorem one has a criterion for the solvability of distributions of signatures. Unfortunately there is no practicable geometrical description in $V(R)$ of the valuations of $R(V)$. Nevertheless, the third fantheorem can immediately be translated to

3. <u>Corollary</u>. A distribution of signatures on V can be solved, if and only if it can be solved modulo 2^d .

4. <u>Remark</u>. Let $S \subset V(R)$ be a semialgebraic set of dimension d . Then one can as well define a distribution of signatures on S . Moreover a distribution of signatures on V by restriction induces one on S .

The next corollary, which corresponds to the first fantheorem, gives us some kind of local-global-principle for the solvability of distributions.

5. <u>Corollary</u>. Suppose, that $R = \mathbb{R}$ and V is complete (or at least real-complete [D-K 2], [C-CR]). Then a distribution $(S_1, \ldots S_k, D)$ of signatures on V is solvable, if and only if for each pair p, q

of points there exist semialgebraic neighbourhoods A and B of p and q respectively, such that the distribution restricted to $A \cup B$ is solvable.

Proof. By the proposition we have to show, that $f \circ D$ lies in the image of sign. According to the second fantheorem we restrict $f \circ D$ to $X(\mathbb{R}(V))/T$, where T is a fan. The valuationring B of T obviously contains $\mathbb{R}$ (at this point the special basefield $\mathbb{R}$ is needed). Since V is complete, B admits a centre W, which is a closed real subvariety of V. By the first fantheorem all orderings $P \in X(\mathbb{R}(V))/T$ are compatible with B and induce at most two different orderings $\overline{P_0}$ and $\overline{P_1}$ on the residue classfield $\overline{B}$. We restrict $\overline{P_0}$ and $\overline{P_1}$ to P'_0 and $P'_1 \in X(V)$, where $x = x(P'_0) = x(P'_1)$ is the generic point of W. Once more we need the special basefield $\mathbb{R}$ to conclude, that the ultrafilters $F(P'_0)$ and $F(P'_1)$, which by the ultrafiltertheorem belong to P'_0 and P'_1, converge respectively to points p and $q \in V(R)$. By proposition 2.13 all elements $P \in X(\mathbb{R}(V))/T$ generalize P'_0 or P'_1; so we learn from the geometric description of the specialization, that the corresponding ultrafilters converge also to p or q. But now by assumption $f \circ D$, restricted to $X(\mathbb{R}(V))/T$ lies in the image of sign. Since this is true for all fans of $X(\mathbb{R}(V))$ we get, that $f \circ D \in \text{sign}(W(\mathbb{R}(V)))$, hence by the proposition $(S_1,\ldots,S_k, D)$ can be solved.

6. <u>Corollary</u>. Suppose, that $R = \mathbb{R}$ and V is complete. Let $(S_1,\ldots,S_k, D)$ be a distribution of signatures, such that S_i is union of connected components for $i = 1,\ldots,k$. Then this distribution can be solved if (and only if) $D(S_i) \equiv D(S_j) \bmod 2$ for $0 \leq i,j \leq k$.

Proof. This is rather clear, if V is projective. For the general case, one may use nearly the same argument as in the preceding proof.

7. Remark. If in the preceding Corollary D takes only the values $+1$ and -1, the distribution can be solved by a 1-dimensional form.

Proof. Consider the discriminant of a solving form.

By an example of Schülting [Sch] the preceding three statements are no longer true, if $\mathbb{R}$ is replaced by an arbitrary real closed field.

8. Definition. Let $(S_1,\ldots,S_k, D)$ be a distribution of signatures on the real integral R-variety V. Then a point $p \in V(R)$ is called frontier-point of $(S_1,\ldots,S_k, D)$, if in each neighbourhood of P (With respect to the strong topology of $V(R)$) there are smooth points x and y with $x \in S_i$, $y \in S_j$ and $i \neq j$. (Apparently this property of p depends only on the equivalence-class of $(S_1,\ldots,S_k, D)$). A closed integral subvariety W of V is called frontier of $(S_1,\ldots,S_k, D)$ if the frontier points of $W(R)$ are Zariski-dense in W.

9. Definition. Let V be a real integral R-variety of dimension d. Then a fan T of $R(V)$ is called rational, if the following holds:

a) $|X(R(V))/T| = 2^d$

b) each $P \in X(R(V))/T$ is rational.

In the case, that V is smooth (or at least normal) and $d = 2$, the following objects are in 1-1 correspondence:

a) Rational fans T in $R(V)$.

b) Pairs of different real halfbranches on irreducible real curves W in V.

Here $X(R(V))/T$ consists of the two pairs of orderings, which belong to the pairs of real halfleafs at the two real halfbranches on W. With the symbol of example 4.5 $X(R(V))/T$ looks like

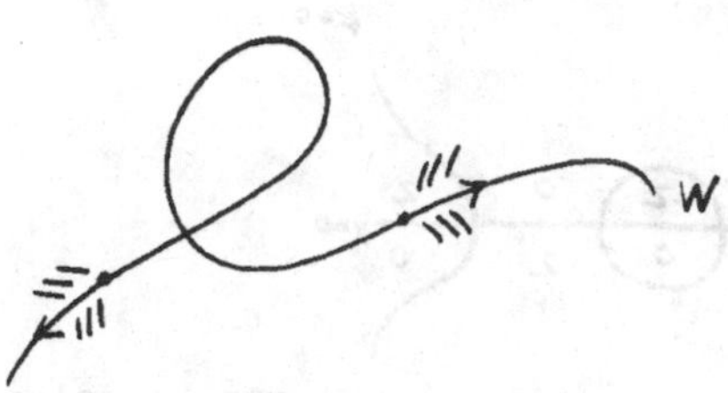

We call W the <u>curve</u> of the rational fan T .

Now we can state the following

10. <u>Rational-fan criterion</u>. Let V be a smooth complete integral real $\mathbb{R}$-variety of dimension 2 . Then a distribution $(S_1,\ldots,S_k, D)$ of signatures on V can be solved, if and only if the following holds:

a) $D(S_i) \equiv D(S_j) \bmod 2$ for $0 \leq i,j \leq k$.

b) For each rational fan T in $\mathbb{R}(V)$, such that the curve of T is a frontier of $(S_1,\ldots,S_k, D)$ one has $\sum_{P\in X(R(V))/T} D(P) \equiv 0 \bmod 4$.

Here D(P) has an obvious meaning, since there is a unique $i \in \{1,\ldots,k\}$ such that S_i belongs to the ultrafilter of P . For the above symbol that means, that the shaded real half leaf belongs to a unique country S_i for $i \in \{1,\ldots,k\}$.

Before we prove this criterion, let us give an

11. <u>Example</u>. In $V = P^2\mathbb{R}$ consider the elliptic curve W , which is given in affine coordinates by $f(x,y) = y^2-x(x-1)(x+1) = 0$ and looks like

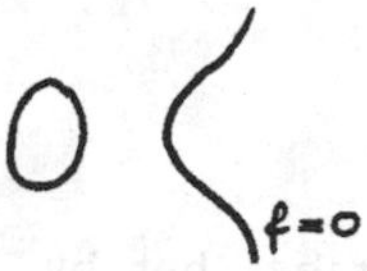

The distribution

cannot be solved, since

$$\sum_{P\in X(\mathbb{R}(V))/T} D(P) = 2 .$$

But the distribution 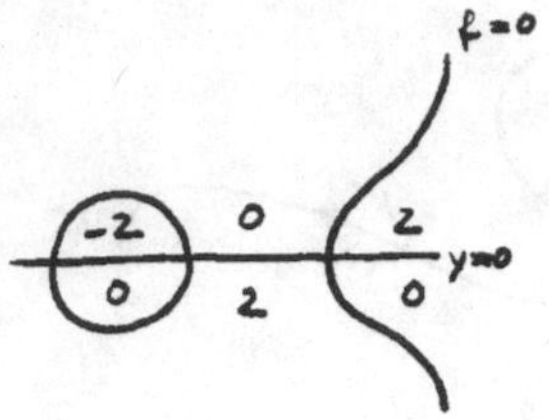

with frontiers $y = 0$, $f = 0$ can be solved. An explicite solution is given by the diagonalform $<x, -f, xf, -y, xy, xyf>$.

Proof of the criterion. It follows from the second fantheorem, that the conditions a) and b) are necessary. Now suppose, that these conditions hold. By corollary 3 we may assume, that D takes only the values $+1$ and -1 . Let $W_1, \ldots, W_r$ be the 1-dimensional frontiers of (S_1, S_2, D) . By condition b) apart from isolated points each point of $W_i(\mathbb{R})$ is frontierpoint for $i = 1,2$. Therefore $W_1(\mathbb{R}) \cup \ldots \cup W_r(\mathbb{R})$ is the boundary of $\overline{S_1}$ or $W_1(\mathbb{R}) + \ldots + W_r(\mathbb{R}) = \partial \overline{S_1}$ mod 2 for each triangulation of $\overline{S}$ (apart from isolated points). But then by [Br 3] the real divisor $W_1 + \ldots + W_r$ is principal [1], say it is real divisor of f for some $f \in \mathbb{R}(V)$. Now if $V(\mathbb{R})$ is connected, we are done, since $<f>$ or $<-f>$ solves the distribution. In the general situation by corollary 6 we find an element $g \in \mathbb{R}(V)$ such that $<gf>$ solves the distribution.

12. Remark. One can also prove this criterion without use of real divisors. One needs only the fantheorems and the fact, that V is locally factorial.

---

1) In [Br 3] this was only proved for projective varieties, but by a remark of Colliot Thélène [CT] the condition "projective" can be avoided. Compare also [EPT].

Post scriptum

I want to express my gratitude to M. Coste, who read this paper carefuly and pointed out to me some little errors. He also made the following interesting remarks.

1. As to the restricted topological spaces $(X,\mathcal{B})$ the ultrafilters of the lattice $k(\mathcal{B})$ correspond to the prime filters of the lattice $\mathcal{B}$. (A filter F is called prime, if $B_1 \in F$ or $B_2 \in F$ whenever $B_1 \cup B_2 \in F$.) Thus the ultrafilter completion $(\hat{X},\hat{\mathcal{B}})$ of $(X,\mathcal{B})$ may also be described as prime filter completion. The spaces $(X,\mathcal{B})$ for which $(X,\mathcal{B}) = (\hat{X},\hat{\mathcal{B}})$, are precisely the spectral spaces of M. Hochster (Prime ideal structure in commutative rings. Trans.Amer.Math.Soc. 142, 43-60(1969)). One defines the dimension of a spectral space (sometimes called combinatorial dimension) as the maximum length of chains of specializations. The combinatorial dimension is equal or greater than our dimension as defined in 1.7. Both dimensions coincide for real spectra of k-varieties but not for arbitrary spectral spaces.

2. In the main example 1.3 it can be shown, that open-semialgebraic is just open and semialgebraic, even in the case, where k is not real closed. Actually the semialgebraic space of A is isomorphic to the semialgebraic space of $A \otimes_k R$. Correspondingly the map $X\pi$ in remark 4.3 is in fact an isomorphism of restricted topological spaces. Coste proves this fact by use of the results in [C-CR], § 4.

References.

[A] Artin, E.: Über die Zerlegung definiter Funktionen in Quadrate. Abh. Math. Sem. Univ. Hamburg 3, 319-323 (1924)

[B-Br] Becker, E. and Bröcker, L.: On the description of the reduced Wittring. J. of Algebra, 328-346 (1978)

[B-K] Becker, E. und Köpping, E.: Reduzierte quadratische Formen und Semiordnungen reeller Körper. Abh. Math. Sem. Univ. Hamburg 46, 143-177 (1977).

[Br 1] Bröcker, L.: Zur Theorie der quadratischen Formen über formalreellen Körpern. Math. Ann. 210, 233-256 (1974)

[Br 2] Bröcker, L.: Characterizations of fans and hereditarily pythagorean fields. Math. Z. 152, 149-163 (1976)

[Br 3] Bröcker, L.: Reelle Divisoren. Arch. Math. 35, 140-143 (1980)

[Br 4] Bröcker, L.: Positivbereiche in kommutativen Ringen. to appear in Abh. Math. Sem. Univ. Hamburg

[Bru 1] Brumfiel, G.W.: Partial ordered rings and semi-algebraic geometry. Cambridge University Press (1979)

[Bru 2] Brumfiel, G.W.: Real valuation rings and ideals. Conference on real algebraic geometry and quadratic forms. Rennes (1981)

[CT] Colliot-Thélène, J.L.: Letter

[C-CR] Coste, M. et Coste-Roy, M.F.: La topologie du spectre réel. Manuscript. Paris-Nord 1980

[CR] Coste-Roy, M.F.: Spectre réel d'un anneau et topos étale réel. Thèse. Université Paris Nord (1980)

[D] Delfs, H.: Kohomologie affiner semialgebraischer Räume. Thesis, Regensburg (1980)

[D-K 2] Delfs, H. und Knebusch, M.: Semialgebraic topology over a real closed field II: Basic theory of semialgebraic spaces. Math. Z. 178, 175-213 (1981)

[EPT] Everybody proved that theorem.

[K] Knebusch, M.: On the local theory of signatures and reduced quadratic forms. Abh. Math. Sem. Univ. Hamburg 51 149-195 (1981)

[P] Pfister, A.: Quadratische Formen in beliebigen Körpern. Inventiones math. 1, 116-132 (1966)

[Sch] Schülting, H.W.: Real holomorphy rings in real algebraic geometry. Conference on real algebraic geometry and quadratic forms. Rennes (1981)

L. Bröcker
FB. Mathematik
Universität Münster
Einsteinstr. 64
D4400 Münster

TRANSVERSAL ZEROS AND POSITIVE SEMIDEFINITE FORMS

by

Man-Duen Choi*), Manfred Knebusch**),

Tsit-Yuen Lam***), Bruce Reznick***).

Introduction

For any natural number $n \geq 2$ and any even natural number $d \geq 2$ we consider the convex cone $P(n,d)$ consisting of the positive semidefinite (= psd) forms over $\mathbb{R}$ in n variables $x_1,\ldots,x_n$ of degree d, and the convex subcone $\Sigma(n,d)$ consisting of the finite sums of squares of forms of degree $d/2$ in the variables $x_1,\ldots,x_n$. As is well known $\Sigma(n,d) \neq P(n,d)$ except for very special pairs (n,d), namely the pairs with $n = 2$ or $d = 2$ or $(n,d) = (3,4)$ (Hilbert, cf. [CL] for an elementary proof).

In this paper we ask for relations between the sets $EP(n,d)$ and $E\Sigma(n,d)$ of extremal elements of the cones $P(n,d)$ and $\Sigma(n,d)$. Notice that, since our cones are closed (after adding the origin), every element in $P(n,d)$ resp. $\Sigma(n,d)$ is a finite sum of elements in $EP(n,d)$ resp. $E\Sigma(n,d)$. Thus the sets $EP(n,d)$ and $E\Sigma(n,d)$ deserve special attention.

Our main result, Theorem 6.1 in §6, is the determination of all pairs (n,d) such that $E\Sigma(n,d)$ is contained in $EP(n,d)$, which means $E\Sigma(n,d) = EP(n,d) \cap \Sigma(n,d)$. This answers Problem B in the survey article [CL_1].

In order to obtain the result a general observation turns out to be helpful:

a) Let H be an irreducible indefinite form in $\mathbb{R}[x_1,\ldots,x_n]$ of degree r. Then for any $F \in P(n,d)$

$$F \in EP(n,d) \iff FH^2 \in EP(n,d+2r),$$

$$F \in E\Sigma(n,d) \iff FH^2 \in E\Sigma(n,d+2r),$$

---

*) supported by NSERC of Canada

**) supported by DFG during a stay at Berkeley 1980

***) supported by NSF

cf. Theorem 5.1. We also feel that the following observation sheds light on the problem:

b) If $F \in E\mathcal{P}(n,d)$ then $F^2 \in E\Sigma(n,2d)$,

cf. Theorem 5.2.

Our "counterexamples" $G \in E\Sigma(n,d)$, $G \notin E\mathcal{P}(n,d)$ are of the form $G = H^2F^2$ with H a product of irreducible indefinite forms and F an irreducible psd form of some degree e which is not extremal in $\mathcal{P}(n,e)$. Basic counterexamples will be explicitly constructed in §6 for $(n,d) = (3,12)$ and $(n,d) = (4,8)$.

The observations a) and b) rely on the presence of "transversal zeros" for some forms coming up in the proofs. A transversal zero of a polynomial $F(x_1,\ldots,x_n)$ over $\mathbb{R}$ is a point $c \in \mathbb{R}^n$ such that F changes sign in every neighbourhood of c. If F has no multiple irreducible factors then a point c of the zero set $\mathcal{Z}(F) \subset \mathbb{R}^n$ turns out to be a transversal zero if and only if $\mathcal{Z}(F)$ has local dimension n-1 at c, cf. Theorem 3.4.

The first half of our paper is devoted to a geometric study of transversal zeros and to the question how far a polynomial is determined by its transversal zeros. We try to do all this on a natural level of generality. This leads us to study the set $|D|_{\mathbb{R}}$ of real points of an effective Weil divisor D on a normal algebraic variety X over $\mathbb{R}$. But for the applications of the theory of transversal zeros made in §5 and §6 it suffices to consider the case when X is a projective space $\mathbb{P}^{n-1}_{\mathbb{R}}$, or - if one wants to study also multiforms - a direct product of projective spaces.

We suspect that many of our considerations on transversal zeros are more or less "folklore", well known to the experts. However, to our knowledge, no coherent account of this useful theory seems to exist in the literature. Thus we feel that these Proceedings are a good place to explicate the basic facts.

In the whole paper we admit any real closed field R as ground field instead of the field $\mathbb{R}$ of real numbers. Using some standard results from semialgebraic topology, all contained in [DK] and §1 of the present paper, this does not cause additional difficulties. Thus we never need Tarski's principle to transfer elementary statements from $\mathbb{R}$ to other real closed fields.

§ 1 The pure dimensional parts of a semialgebraic set

We start with a variety X over a real closed field R, i.e. a reduced algebraic scheme over R. The set X(R) of rational points of X is a semialgebraic space in the sense of [DK], and we use freely the language of "semialgebraic topology" developed in that paper. In particular we make use of the dimension theory in [DK, §8].

Let N be a semialgebraic subset of X(R). For any point x of N the <u>local dimension</u> $\dim_x N$ of N at x is defined as the infimum of the dimensions of all semialgebraic neighbourhoods of x in N [DK, §13]. We introduce the sets (k = 0,1,2,...)

$$\Sigma_k(N) := \{x \in N \mid \dim_x N \geq k\}.$$

Of course $\Sigma_k(N)$ is empty if k exceeds the dimension d of N. It is clear from [DK, §8] that every $\Sigma_k(N)$ is a closed subset of N (in the strong topology, as always). We shall need some elementary facts about the sets $\Sigma_k(N)$ (actually only about $\Sigma_d(N)$), not covered by the paper [DK].

<u>Proposition</u> 1.1. $\Sigma_k(N)$ is semialgebraic for every $k \geq 0$.

It is trivial to verify this lemma using the theorem that every affine semialgebraic space can be triangulated [DK_1]. A more elementary proof, which also gives additional insight, runs as follows. Let d = dim(N). For k > d there is nothing to prove. We now deal with the case k = d. We may assume that X is affine. Let Y denote the Zariski closure of N in X, and let S denote the singular locus of Y. Then

$$N' := (Y(R) \setminus S(R)) \cap N$$

is an open semialgebraic subset of N and the complement in N, i.e. $N \cap S(R)$, has dimension at most d-1. Suppose we know already that $\Sigma_d(N')$ is semialgebraic. Let L be the closure of $\Sigma_d(N')$ in N. This is again a semialgebraic set. $N \setminus L$ is open in N and has dimension at most d-1. Thus $N \setminus L$ is disjoint from $\Sigma_d(N)$. On the other hand L is contained in $\Sigma_d(N)$, since $\Sigma_d(N)$ is closed and contains $\Sigma_d(N')$. Thus $\Sigma_d(N)$ coincides with the semialgebraic set L.

Replacing N by N' and X by $X \setminus S$ we assume now that Y is smooth. Let $Y_1,\ldots,Y_t$ denote the connected components of Y. The set $\Sigma_d(N)$ is the union of the sets $\Sigma_d(N \cap Y_i(R))$, and it suffices to prove that these sets are semialgebraic. $N \cap Y_i(R)$ is Zariski dense in Y_i. Replacing N by anyone of the sets $N \cap Y_i(R)$ we assume that in addition Y is connected, hence irreducible.

We have $N = N_1 \cup \ldots \cup N_r$ with non empty sets

$$N_i = \{x \in Y(R) \mid g_i(x) = 0,\ f_{ij}(x) > 0,\ j = 1,\ldots,s_i\},$$

where g_i, f_{ij} are functions in the affine ring $R[Y]$. If g_i is not zero then $\dim N_i \le n-1$. But if g_i is zero then N_i is open in $Y(R)$, hence $N_i \subset \Sigma_d(N)$, since Y is smooth and thus $Y(R)$ has local dimension d at every point [DK, §8]. It is now clear that $\Sigma_d(N)$ is the closure of the union of all N_i with $g_i = 0$ in the set N. Thus $\Sigma_d(N)$ is indeed semialgebraic.

Consider now the open semialgebraic subset $N_1 := N \setminus \Sigma_d(N)$ of N. Clearly

$$\Sigma_{d-1}(N) = \Sigma_d(N) \cup \Sigma_{d-1}(N_1).$$

We know from the proof already given that $\Sigma_{d-1}(N_1)$ and $\Sigma_d(N)$ are semialgebraic. Thus $\Sigma_{d-1}(N)$ is semialgebraic. Repeating this argument we see that all $\Sigma_k(N)$ are semialgebraic, and our lemma is proved.

<u>Proposition 1.2.</u> For every $k \ge 0$ the semialgebraic set

$$\Sigma_k^o(N) := \Sigma_k(N) \setminus \Sigma_{k+1}(N),$$

consisting of all points $x \in N$ with $\dim_x N = k$, is pure of dimension k, i.e. $\dim_x \Sigma_k^o(N) = k$ for every $x \in \Sigma_k^o(N)$.

<u>Proof.</u> Let x be a point of $\Sigma_k^o(N)$ and let U_o be an open semialgebraic neighbourhood of x in N with $\dim U_o = k$. For any open semialgebraic neighbourhood $U \subset U_o$ of x in N we then also have $\dim U = k$. Moreover for every such U there exists an open semialgebraic subset V of U which is semialgebraically isomorphic to an open non empty subset of R^k[DK, §8]. Clearly V is contained in $\Sigma_k^o(N) \cap U$. Thus $\dim(\Sigma_k^o(N) \cap U) = k$. Q.E.D.

We call $\Sigma_k^o(N)$ the <u>pure</u> <u>k-dimensional</u> <u>part</u> of N. More specifically, if $\dim N = d$ we call $\Sigma_d^o(N) = \Sigma_d(N)$ the <u>pure</u> <u>part</u> of N.

<u>Example 2.3.</u> If X is irreducible of dimension n, and if the set $X(R)_{reg}$ of regular points of X in X(R) is not empty, then the pure part $\Sigma_n(X(R))$ of X(R) is the closure of $X(R)_{reg}$ in X(R).

Indeed, $X(R)_{reg}$ is pure of dimension n, and X(R) has local dimension at most n-1 at every singular point which is not contained in the closure of $X(R)_{reg}$.

§ 2 Transversal zeros of algebraic functions

We assume in this section that the variety X over R is irreducible, that the set X(R) of real points is not empty, and that X is regular at every point of X(R). Then X(R) is an n-dimensional semialgebraic manifold [DK, §13] with n = dim X. We also assume that X is affine, and we denote the ring R[X] of regular functions on X by A. On the space X(R) every $f \in A$ takes values in R. We are interested in the zeros and the sign behaviour of the functions $f : X(R) \to R$.

Definition 2.1. Let L be a subset of X(R) on which f does not vanish everywhere. We say that f is positive semidefinite (resp. positive definite) on L if $f(x) \geq 0$ (resp. $f(x) > 0$) for all $x \in L$. In the same way we use the words "negative semidefinite" and "negative definite". If there exist points $x \in L$ and $y \in L$ with $f(x) > 0$ and $f(y) < 0$, then we call f indefinite on L.

Definition 2.2. Let f be a non zero element of A. A transversal zero of f is a point $x \in X(R)$ such that f is indefinite on every semialgebraic neighbourhood V of x in X(R). Notice that f cannot vanish everywhere on V since dim V = n.

We denote by $Z(f)$ the set of zeros of f on X(R) and by $Z_t(f)$ the set of transversal zeros of f. We finally denote by N(f) the closed reduced subscheme of all zeros of f on X. Thus $Z(f)$ is the set of real points of N(f) and $Z_t(f)$ is a subset of $Z(f)$. The set $Z(f)$ is closed and semialgebraic in X(R). The set $Z_t(f)$ is the intersection of the closure of the set of points of X(R) where f is positive with the closure of the set where f is negative. Thus $Z_t(f)$ is also closed and semialgebraic in X(R).

Proposition 2.3. For every non zero regular function f on X the set $Z_t(f)$ of transversal zeros is either empty or pure of dimension n-1.

Proof. Let a be a given point of $Z_t(f)$. We choose an open neighbourhood V of a in X(R) with a semialgebraic isomorphism $\varphi : V \xrightarrow{\sim} V'$ onto an open semialgebraic convex subset V' of R^n. (Recall that X(R) is a semialgebraic manifold.) We then choose a point $x_0 \in V$ with $f(x_o) > 0$

and an open semialgebraic subset $U \subset V$ such that $f(y) < 0$ for every $y \in U$ and such that $U' := \varphi(U)$ is convex in R^n. We finally choose a hyperplane H of R^n with $H \cap U' \neq \emptyset$ and not containing the point $x_o' := \varphi(x_o)$. Now consider the central projection

$$\pi \ : R^n \setminus \{x_o'\} \twoheadrightarrow H$$

onto H with center x_o'. We claim that

$$(*) \qquad \pi \circ \varphi(Z_t(f) \cap V) \supset H \cap U'.$$

Indeed, let $y' \in H \cap U'$ be given and let $\gamma' : [0,1] \to V'$ be the straight path from x_o' to y',

$$\gamma'(t) = (1-t)x_o' + ty'.$$

Then $\gamma := \varphi^{-1} \circ \gamma'$ is a semialgebraic path in V running from the point x_o to the preimage y of y'. Since $f(x_o) > 0$ and $f(y) < 0$ there exists some point $\tau \in]0,1[$ where the semialgebraic function $f \circ \gamma$ on $[0,1]$ changes sign. $\gamma(\tau)$ is clearly a transversal zero of f. The point $\gamma'(\tau)$ lies in $\varphi(Z_t(f) \cap V)$ and maps under π to the point y'. Thus the inclusion $(*)$ holds true. This implies that

$$\dim Z_t(f) \cap V \geq n-1,$$

since $\dim (H \cap U') = n-1$. But $Z(f) \cap V$ has dimension at most $n-1$ since this set is contained in $N(f)$. Thus $Z_t(f) \cap V$ has dimension $n-1$ for every open semialgebraic neighbourhood V of a.

Q.E.D.

<u>Corollary 2.4.</u> Let f and g be non zero regular functions on X. Let $a \in X(R)$ be a transversal zero of f and assume that $Z_t(f) \cap U$ is contained in $Z(g)$ for some neighbourhood U of a. Then f and g have a non trivial common factor in the regular local ring $\mathcal{O}_{X,a}$. {Recall that $\mathcal{O}_{X,a}$ is a unique factorization domain.}

<u>Proof.</u> For every affine Zariski neighbourhood W of a in X the semialgebraic set $W \cap U \cap Z_t(f)$ has dimension $n-1$ by Proposition 2.3 above. Our hypothesis implies that this set is contained in the intersection $N(f) \cap N(g) \cap W$ of the hypersurfaces $f = 0$ and $g = 0$ on W. Thus the (algebraic!) dimension of $N(f) \cap N(g) \cap W$ cannot be smaller than $n-1$ for any Zariski neighbourhood W of a. This implies that there exists some $h \in A$ which is a prime element in $\mathcal{O}_{X,a}$ and has the property that

$$N(h) \cap W \subset N(f) \cap N(g) \cap W$$

for small Zariski neighbourhoods W of a. By the local Nullstellensatz h divides both f and g in $\mathcal{O}_{X,a}$.

In the same vein we obtain

<u>Corollary 2.5.</u> Let again f and g be non zero functions on X. Suppose that for some open semialgebraic subset U of X(R) the set $Z_t(f) \cap U$ is not empty and contained in $Z(g)$. Then the complex hypersurfaces N(f) and N(g) have a common irreducible component. In particular, if A is factorial then f and g have a non trivial common factor in A.

<u>Proposition 2.6.</u> Let f and g be non zero regular functions on X, and assume that the hypersurfaces N(f) and N(g) have no irreducible component in common. Then

$$Z_t(fg) = Z_t(f) \cup Z_t(g).$$

<u>Proof.</u> a) Let <u>a</u> be a point of X(R) which is not contained in $Z_t(f) \cup Z_t(g)$. Then there exists a neighbourhood U of <u>a</u> in X(R) such that both f and g are semidefinite on U (positive or negative). Then also the product fg is semidefinite on U, and <u>a</u> is not a transversal zero of fg. This proves that $Z_t(fg)$ is contained in $Z_t(f) \cup Z_t(g)$. (Our hypothesis, that N(f) and N(g) have no common component, is not yet needed for that.)

b) We show that the set $M := Z_t(f)$ is contained in $Z_t(fg)$, which will finish the proof. We may assume that M is not empty. By Proposition 2.3 M is pure of dimension n-1. On the other hand the set $N := Z_t(f) \cap Z_t(g)$ has dimension at most n-2, since N is contained in the intersection of the hypersurfaces N(f) and N(g) which have no common irreducible component. Thus the set $M \setminus N$ is dense in M (a trivial argument, cf. [DK, §13]). Since $Z_t(fg)$ is closed it suffices to verify that $M \setminus N$ is contained in $Z_t(fg)$.

Let x be a point of $M \setminus N$, which means that $x \in Z_t(f)$, $x \notin Z_t(g)$. We choose a neighbourhood U_o of x on which g is semidefinite. Now f is indefinite on every neighbourhood $U \subset U_o$ of x. Thus also fg is indefinite on every such U. This implies that $x \in Z_t(fg)$.

Q.e.d.

Corollary 2.7. Assume that A is factorial. Let f be a non zero element of A and let

$$f = u\, p_1^{e_1} \cdots p_t^{e_t}$$

be the decomposition of f into powers of pairwise non associated prime elements $p_1, \dots, p_t$, with u a unit of A. Then $Z_t(f)$ is the union of the sets $Z_t(p_i)$ with e_i odd.

Proof. Apply Proposition 2.6 and observe that $Z_t(p_i^{e_i})$ is empty if e_i is even and $Z_t(p_i^{e_i}) = Z_t(p_i)$ if e_i is odd.

In the same vein we obtain for the semialgebraic set germ $Z_t(f)_a$ of a non zero function $f \in A$ at any point $a \in X(R)$:

Corollary 2.8. Let

$$f = u\, p_1^{e_1} \cdots p_t^{e_t}$$

be the decomposition of f into prime elements in the factorial ring $\mathcal{O}_{X,a}$. Then $Z_t(f)_a$ is the union of the set germs $Z_t(p_i)_a$ with e_i odd.

§ 3 Purely indefinite divisors

We still assume that X is an irreducible n-dimensional variety over R and that the set X(R) is not empty and contains no singular points of X. But we no longer assume that X is affine. Our terminology from §2 then takes over from functions to effective divisors $D \geq 0$ on X, by which we always mean effective Weil divisors.

Definition 3.1. Let D be an effective divisor on X and let a be a point of X(R). Let f be the local equation of D on some affine Zariski open neighbourhood V of a. We call D indefinite at a, if f is indefinite on every neighbourhood of a in V(R). Similarly we call D semidefinite (resp. definite) at a, if f is positive or negative semidefinite (resp. definite) on some neighbourhood of a in V(R). The points of X(R) where D is indefinite are called the transversal points of D, and the set of these points is denoted by $|D|_t$. This set is a closed semialgebraic subset of the set of real points $|D|_R := |D| \cap X(R)$ of the support $|D|$ of D.

Let $D = e_1 D_1 + \dots + e_t D_t$ be the decomposition of D into irreducible components.

<u>Proposition 3.2.</u> $|D|_t$ is the union of all sets $|D_i|_t$ with e_i odd.

This is clear from Proposition 2.6 in §2, or its corollary 2.8.

<u>Definition 3.3.</u> We call an effective divisor D <u>indefinite</u>, if $|D|_t$ is not empty, i.e. if D is indefinite at some point of X(R). We call D <u>semidefinite</u>, if $|D|_t$ is empty, and we call D <u>definite</u> if $|D|_R$ is empty. Finally, we call D <u>purely indefinite</u>, if $D \neq 0$ and there does <u>not</u> exist a semidefinite effective divisor $E \neq 0$ with $E \leq D$. This means that D is non zero, has no multiple components, and that all irreducible components of D are indefinite.

It is clear from Proposition 2.3 in §2 that for every effective divisor D on X the set $|D|_t$ is either empty or pure of dimension n-1. This result can be improved.

<u>Theorem 3.4.</u> Assume that D has no multiple components. Then the semialgebraic set $|D|_t$ of transversal points of D coincides with the pure (n-1)-dimensional part $\Sigma_{n-1}(|D|_R)$ of the set $|D|_R$ of real points on $|D|$.

<u>Proof.</u> It remains to verify that D is indefinite at any given point a of $|D|_R$ with $\dim_a |D|_R = n-1$. We choose a local equation f of D on some affine Zariski open neighbourhood W of a in X. Let U be any semialgebraic open neighbourhood of a in W(R). The set $U \cap |D|_R$ has dimension n-1, but the set of points in $|D|_R$ which are singular on $|D|$ has dimension at most n-2. Thus $U \cap |D|_R$ contains some regular point b of $|D|$. There exists a regular system of parameters $f_1, f_2, \ldots, f_n$ of the regular local ring $\mathcal{O}_{X,b}$ such that f_1 defines the germ of the variety $|D|$ at b. The functions f_1 and f differ in $\mathcal{O}_{X,b}$ only by a unit, hence we may assume that $f = f_1$. By the implicit function theorem the system $(f_1, \ldots, f_n)$ yields a semialgebraic isomorphism of some open semialgebraic neighbourhood $U' \subset U$ of b in X(R) onto some open semialgebraic subset of R^n. Since $f_1(b) = 0$ certainly $f = f_1$ changes sign on U'. A fortiori f is indefinite on U.

Q.e.d.

We mention that the theorem now proved implies a generalization of the "Sign-Changing Criterion" of Dubois and Efroymson for extending an ordering P of a field k to a given function field over k ([DE, Th.2.7], cf. also [ELW, §4 bis]

Corollary 3.5. (Dubois - Efroymson for $V = \mathbb{A}^n_k$). Let k be an ordered field and R be a real closure of k with respect to the given ordering. Let V be an absolutely irreducible variety without singular points over k and D a prime divisor on V. Let V_R denote the variety over R obtained from V by base extension and let $\tilde{D}$ denote the effective divisor on V_R obtained from D by base extension. Then the ordering of k can be extended to the function field k(D) of D if and only if $\tilde{D}$ is indefinite.

Proof. The ordering of k extends to k(D) if and only if there exists a field composite of k(D) and R over k, which is formally real. These field composites are the function fields $R(D_1),\ldots,R(D_s)$ of the irreducible components $D_1,\ldots,D_s$ of the divisor $\tilde{D}$. The prime divisors D_i all occur with multiplicity one in $\tilde{D}$. Thus $\tilde{D}$ is indefinite if at least one D_i is indefinite. By Theorem 3.4 a given D_i is indefinite if and only if the set of real points $D_i(R)$ of D_i has dimension n-1 with $n := \dim V = \dim V_R$. But $\dim D_i(R) = n-1$ means that the variety D_i has nonsingular real points, cf. §1. Now it is a well known fact, due to Artin, that D_i has nonsingular real points if and only if the field $R(D_i)$ is formally real ([A, §4], cf. also [E]).

We return to our irreducible variety X over R.

Proposition 3.6. Let D be an effective divisor $\neq 0$ without multiple components. Then $|D|_R$ is Zariski dense in $|D|$ if and only if D is purely indefinite. In this case even $|D|_t$ is Zariski dense in $|D|$.

Proof. Let $D_1,\ldots,D_r$ denote the irreducible components of D. Clearly

$$|D|_R = D_1(R) \cup \ldots \cup D_r(R)$$

is Zariski dense in D if and only if every $D_i(R)$ is Zariski dense in D_i. This means that $D_i(R)$ has the semialgebraic dimension n-1, i.e. that $\Sigma_{n-1}(D_i(R))$ is not empty, and in that case of course already $\Sigma_{n-1}(D_i(R))$ is Zariski dense in D_i. The proposition now follows from the preceding Theorem 3.4.

This is perhaps the appropriate place to indicate a relation between our investigations and the real Nullstellensatz of Dubois-Risler-Stengle [S, Theorem 2]. Assume that X is an affine variety over R and that W is a closed subvariety of X. Let A denote the affine ring of X and $\mathfrak{a}$ the ideal of functions in A vanishing on W. Then the real Nullstellensatz says in particular that W(R) is Zariski dense in W if and only if the ideal $\mathfrak{a}$ is "real", i.e.

$$h_1^2 + \ldots + h_r^2 \in \mathfrak{a} \Rightarrow h_1 \in \mathfrak{a}, \ldots, h_r \in \mathfrak{a}$$

for arbitrary elements $h_1, \ldots, h_r$ of A. (This is essentially Risler's version of the real Nullstellensatz [Ri], [Ri_1].) Thus if X is irreducible and has no singular real points then the proposition we just proved says the following:

Corollary 3.7. Let X be affine and I(D) denote the ideal of functions in R[X] vanishing on $|D|$ for D an effective divisor $\neq 0$ without multiple components. Then I(D) is real if and only if D is purely indefinite.

If D is a prime divisor then clearly I(D) is real if and only if the function field R(D) is formally real, and we are back to the arguments which led to the Sign-Changing Criterion above (Corollary 3.5).

Definition 3.8. We call a semialgebraic subset M of X(R) pure and full of dimension k in X, if dim M = k (hence the Zariski closure Z of M in X has dimension k) and M is the pure part $\Sigma_k(Z(R))$ of Z(R).

In this terminology we can say according to Theorem 3.4 and Proposition 3.6 that for every non zero purely indefinite divisor on X the set $|D|_t$ is pure and full of dimension n-1 in X. We now prove a converse of this statement.

Theorem 3.9. Let M be a pure and full (n-1)-dimensional semialgebraic subset of X(R). Then there exists a unique purely indefinite divisor D on X such that M coincides with the set $|D|_t$ of transversal points of D. The variety $|D|$ is the Zariski closure of M in X.

Proof. Let Z denote the Zariski closure of M in X and let $Z_1, \ldots, Z_r$ denote the irreducible components of Z. The set M is the union of the closed semialgebraic subsets $M_i := M \cap Z_i(R)$, $i = 1, \ldots, r$. Denoting by Z_i' the Zariski closure of M_i in X we have $Z_i' \subset Z_i$ and

$$Z_1' \cup \ldots \cup Z_r' = Z_1 \cup \ldots \cup Z_r,$$

and we conclude that $Z_i' = Z_i$ for $i = 1, \ldots, r$. This means that every M_i is Zariski dense in Z_i. Since Z_i is not contained in the union of the Z_j with $j \neq i$, also M_i is not contained in the union of the M_j with $j \neq i$. Thus

$$M_i' := M \setminus \bigcup_{j \neq i} M_j$$

is a non empty open subset of M, which is therefore pure of dimension n-1. This implies dim M_i = n-1 and dim Z_i = n-1 for every i = 1,...,n. The set $Z_i(R)$ contains M_i, hence has again dimension n-1. We now conclude from Theorem 3.4 that for every i = 1,...,r the prime divisor Z_i is indefinite. We introduce the purely indefinite divisor

$$D := Z_1 + \dots + Z_r.$$

By construction $|D|$ is the Zariski closure Z of M. Since M is pure and full, M coincides with $\Sigma_{n-1}(|D|_R)$. By Theorem 3.4 this last set is $|D|_t$. It is now also clear that D is the only purely indefinite divisor with $|D|_t = M$, since by Proposition 3.6 for any such divisor D' the variety $|D'|$ is the Zariski closure of M in X.

Q.e.d.

A mild generalization of these results is possible. Assume only that X is an irreducible n-dimensional variety which is normal at every real point, and that X(R) has dimension n. Let X' denote the open subvariety of all regular points of X. Then $X(R) \setminus X'(R)$ has dimension at most n-2. In particular X'(R) is not empty. Let D be an effective divisor on X and let D' denote the restriction of D to X'.

Definition 3.10. We call D indefinite (resp. semidefinite, resp. purely indefinite) if D' is indefinite (resp. semidefinite, resp. purely indefinite). We denote by $|D|_t$ the closure of the semialgebraic set $|D'|_t$ in X(R).

It is evident that all the theorems, propositions and corollaries in this section, except Corollary 3.5, remain true word by word in the present more general situation. Corollary 3.5 remains true for a normal irreducible variety V over k instead of a regular variety.

§ 4 A remark on semidefinite prime divisors

As before let X be an irreducible n-dimensional variety over R such that X(R) is also n-dimensional and contains only normal points. We regard on X(R) beside the strong topology also the coarser Zariski topology. This is the topology on X(R) induced by the Zariski topology of X. Every Zariski closed subset M of X(R) is a finite union of irreducible closed subsets $M_1,\dots,M_r$ with $M_i \not\subset M_j$ for $i \neq j$. We call these subsets M_i the irreducible components of M. They are uniquely determined by M.

Every irreducible Zariski closed subset M of X(R) which has dimension n-1 is clearly the set of real points of an indefinite prime divisor D on X uniquely determined by M (cf. Theorem 3.9, which says much more than this.) We now prove a weak analogue of this statement for lower dimensional irreducible Zariski closed subsets of X(R). Uniqueness of the prime divisor D can no longer be expected. Thus the following theorem is less valuable than Theorem 3.9.

<u>Theorem 4.1.</u> Suppose that X is also quasiprojective, i.e. a locally closed subscheme of some projective space $\mathbb{P}^N_R$. Let M be an irreducible Zariski closed subset of X(R) of dimension at most n-2. Then there exists some semidefinite prime divisor D on X such that M = D(R).

For the proof we replace X by its normalization, which does not change anything for the space X(R). Now the zero divisor $\text{div}(f)_+$ and the pole divisor $\text{div}(f)_-$ of any non zero rational function f on X are honestly defined as Weil divisors.

The set X(R) is contained in the affine open subscheme V of $\mathbb{P}^N_R$ which is the complement of the hypersurface $x_0^2 + \ldots + x_N^2 = 0$. We introduce the Zariski closure X_1 of X ∩ V in V. Then $X(R) = X_1(R)$ and X_1 is an affine variety. Let W denote the Zariski closure of M in X_1. We choose regular functions $g_1, \ldots, g_r$ on X_1 such that W is the reduced subscheme $N_{X_1}(g_1) \cap \ldots \cap N_{X_1}(g_r)$ of all common zeros of $g_1, \ldots, g_r$ on X_1. For the regular function

$$g := g_1^2 + \ldots + g_r^2$$

on X_1 we have

$$M = \{x \in X_1(R) \mid g(x) = 0\}.$$

We now extend the regular function $g \mid X \cap V$ to a rational function f on X in the unique possible way. The domain of definition of f contains X ∩ V, hence X(R). Thus the pole divisor $E := \text{div}(f)_-$ has in its support no real points, i.e. E is definite. On the other hand we have for the zero divisor $D := \text{div}(f)_+$

$$|D|_R = \{x \in X(R) \mid f(x) = 0\} = \{x \in X_1(R) \mid g(x) = 0\} = M.$$

Let $D = e_1 D_1 + \ldots + e_s D_s$ be the decomposition of D into prime divisors. M is the union of the Zariski closed subsets $D_1(R), \ldots, D_s(R)$. Since M is irreducible, M coincides with one of these sets, say $M = D_1(R)$. The prime divisor D_1 is semidefinite according to Theorem 3.4, or already Proposition 2.3, and our theorem is proved.

§ 5 Extremal positive semidefinite forms and extremal squares

Let X be the (n-1)-dimensional projective space $\mathbb{P}_R^{n-1}$ $(n \geq 2)$. Every effective divisor D on X is the divisor div(F) of a form $F(x_1,\ldots,x_n)$ with coefficients in R uniquely determined by D up to a multiplicative constant. In this way the prime divisors correspond with the irreducible forms, the indefinite divisors correspond with the indefinite forms in the usual sense - notice that X(R) is connected -, and the semidefinite (resp. definite) divisors correspond with the positive semidefinite (resp. definite) forms, of course also with the negative semidefinite (resp. definite) forms.

We call a form $F \in R[x_1,\ldots,x_n]$ purely indefinite, if the divisor div(F) is purely indefinite. This means that F is not constant, all irreducible factors of F are indefinite, and no irreducible factors occur with multiplicity > 1.

For any integral number $r \geq 0$ we denote by $\mathcal{F}(r)$ the set of all non zero forms of degree r in $R[x_1,\ldots,x_n]$ and by $\mathcal{F}$ the union of all $\mathcal{F}(r)$. For any even number $d \geq 0$ we denote by $\mathcal{P}(d)$ the convex cone in $\mathcal{F}(d)$ consisting of all psd (= postive semidefinite) forms of degree d in $R[x_1,\ldots,x_n]$, and by $\mathcal{P}$ the union of all $\mathcal{P}(d)$. Similarly we denote by $\Sigma(d)$ the convex subcone of $\mathcal{P}(d)$ consisting of all finite sums of squares of non zero forms in $R[x_1,\ldots,x_n]$ of degree $\frac{d}{2}$, and by Σ the union of the sets $\Sigma(d)$.

The cones $\mathcal{P}(d) \cup \{0\}$ and $\Sigma(d) \cup \{0\}$ are well known to be closed semialgebraic subsets of the vector space $\mathcal{F}(d) \cup \{0\}$. Our theory in §2 has some applications to the theory of the sets $E(\mathcal{P}(d))$ and $E(\Sigma(d))$ of extremal points of the cones $\mathcal{P}(d)$ and $\Sigma(d)$. We refer the reader to the paper [CL] for the background, some results, and concrete examples in this theory. Let again $E(\mathcal{P})$ denote the union of sets $E(\mathcal{P}(d))$ and $E(\Sigma)$ the union of the sets $E(\Sigma(d))$.

If nothing else is said all forms in the sequel are understood to be forms in $x_1,\ldots,x_n$ over R. For any two such forms we mean by "$F \geq G$" that $F - G$ lies in $\mathcal{P} \cup \{0\}$. In particular then F and G must have the same degree. Similarly we mean by "$F \gg G$" that $F - G$ lies in $\Sigma \cup \{0\}$. Clearly an element F of $\mathcal{P}$ lies in $E(\mathcal{P})$ if and only if $F \geq G \geq 0$ implies $G = \lambda F$ with some constant λ. Similarly an element F of Σ lies in $E(\Sigma)$ if and only if $F \gg G \gg 0$ implies $G = \lambda F$ with some constant λ. Of course in both cases the constant λ lies in the interval $[0,1]$.

Theorem 5.1. i) Let F and G be psd forms. Assume that $F \in E(P)$ and G divides F. Then $G \in E(P)$.

ii) Assume that $F \in E(\Sigma)$ and $F = G \cdot H^2$ with some forms G and H. Then $G \in E(\Sigma)$.

iii) Let G be a psd form and H a purely indefinite form. Then G lies in $E(P)$ if and only if GH^2 lies in $E(P)$.

iv) Let again G be a psd form and H a purely indefinite form. Then G lies in $E(\Sigma)$ if and only if GH^2 lies in $E(\Sigma)$.

Proof. i) We have $F = GH$ with some psd form H. Suppose that $G \geq G' \geq 0$. We have to verify that $G' = \lambda G$ with some constant λ. Since $H \geq 0$ we have $GH \geq G'H \geq 0$. Since F is extremal this implies $G'H = \lambda GH$ with some constant λ and then $G' = \lambda G$.

ii) We may induct on the number of irreducible factors of H and thus assume that H is irreducible. Since F is an extremal sum of squares F is actually a square L^2. Now H divides L. We have $L = HS$ with some form S and then $F = H^2S^2$. From this we obtain $G = S^2$. In particular $G \in \Sigma$. We see now by the same argument as in i) that G is extremal in Σ.

iii) If GH^2 is extremal then also G is extremal as has been proved above. Assume now that G is extremal. It suffices to consider the case that H is indefinite and irreducible, since we then obtain the full result by iteration. Let L be a non zero form with $GH^2 \geq L \geq 0$. The set of real zeros $Z(H)$ is contained in $Z(L)$. By a mild application of Corol. 2.5 we see that H divides L. (Restrict H and L to the n-standard open affine subvarieties of $\mathbb{P}^{n-1}_{\mathbb{R}}$.) Since H is indefinite then also H^2 divides L, cf. Proposition 3.2. We have $L = H^2L'$ with some psd form L' and obtain from $GH^2 \geq L'H^2 \geq 0$ that $G \geq L' \geq 0$. Since G is extremal this implies $L' = \lambda G$ with some constant λ and then $L = \lambda GH^2$.

iv) We again retreat to the case that H is irreducible and indefinite. If GH^2 lies in $E(\Sigma)$ then by ii) also G lies in $E(\Sigma)$. Assume now that $G \in E(\Sigma)$. Suppose that $GH^2 \gg L \gg 0$. We have

$$L = M_1^2 + \ldots + M_r^2$$

with some forms $M_1, \ldots, M_r$ of same degree. The set $Z(H)$ is contained in every zero set $Z(M_i)$. Thus by Corollary 2.5 we have $M_i = HN_i$ with some forms N_i and $L = H^2L_1$, where

$$L_1 = N_1^2 + \ldots + N_r^2 \in \Sigma.$$

We can apply the same argument to the sum of squares $GH^2 - L$ and have $GH^2 - L = H^2 S_1$ with some $S_1 \in \Sigma$. We obtain $G = L_1 + S_1$. Since G is extremal in Σ this implies $L_1 = \lambda G$ with some constant $\lambda \in [0,1]$ and then $L = \lambda G H^2$. Thus GH^2 is indeed extremal in Σ. Theorem 5.1 is now completely proved.

We may ask for which forms F the square F^2 is extremal in Σ or even in P. By part iii) of Theorem 5.1 the latter is true for any product F of irreducible indefinite forms. We also know from parts i) and ii) of the theorem that

$$(F_1F_2)^2 \in E(\Sigma) \Rightarrow F_1^2 \in E(\Sigma),\ F_2^2 \in E(\Sigma);$$

$$(F_1F_2)^2 \in E(P) \Rightarrow F_1^2 \in E(P),\ F_2^2 \in E(P).$$

To pursue this question further we may omit in a given form F all irreducible indefinite factors, according to Theorem 5.1, and assume that F is psd. We have the following partial result.

<u>Theorem 5.2.</u> Let F be a form in $E(P)$. Then F^2 has the following property: If $F^2 = G^2 + H$ with some psd form H and some form G then $G^2 = \varepsilon F^2$ with some constant ε. (Of course ε lies in the interval $[0,1]$.) In particular $F^2 \in E(\Sigma)$.

<u>Proof.</u> We may assume that $F \neq \pm G$. We distinguish two cases.

<u>Case 1:</u> $F - G$ is semidefinite. If $F - G$ would be negative semidefinite then also $F + G$ would be negative semidefinite, since $F^2 - G^2 \geq 0$. Thus the sum 2F of $F - G$ and $F + G$ would be negative semidefinite, which is not true. Thus $F - G \geq 0$. Since $F^2 - G^2 = (F - G)(F + G)$ is psd, also $F + G \geq 0$. From the relation

$$F = (F + G)/2 + (F - G)/2$$

we obtain, since F is extremal,

$$(F - G)/2 = \lambda F,\ (F + G)/2 = \mu F$$

with constants $\lambda > 0$, $\mu > 0$ such that $\lambda + \mu = 1$. This implies $G = (\mu - \lambda)F$ and then $G^2 = (\mu - \lambda)^2 F^2$, as desired.

<u>Case 2.</u> $F - G$ is indefinite. According to Proposition 3.2 there exists an irreducible indefinite form P which divides $F - G$ with an odd multiplicity. Since $F^2 - G^2 \geq 0$ the form P occurs in $F^2 - G^2$ with even multiplicity, again by Propostion 3.2. Thus P divides also $F + G$, hence P divides both F and G. Since F is psd even P^2 divides F. We have $F = P^2 F_1$ with a form $F_1 \in E(\mathcal{P})$ by Theorem 5.1.i. We also have $G = P\,G'$ with some form G' and the equation

$$P^4 F_1^2 = P^2 G'^2 + H.$$

Thus $H = P^2 H'$ with a form $H' \in \mathcal{P}$, and

$$P^2 F_1^2 = G'^2 + H'.$$

The zero set $Z(P)$ is contained in $Z(G')$ and also in $Z(H')$. Thus by §2 the irreducible indefinite from P divides both G' and H', the latter one with an even multiplicity. We obtain $G' = P\,G_1$, $H' = P^2 H_1$ with $H_1 \in \mathcal{P}$, and

$$F_1^2 = G_1^2 + H_1.$$

The proof can now be completed by induction on the degree of F, since F_1 has smaller degree than F.

Q.e.d.

<u>Remark.</u> In all these considerations we could have replaced our projective space $\mathbb{P}_R^{n-1}$ by a product $\mathbb{P}_R^{n_1} \times \ldots \times \mathbb{P}_R^{n_r}$, i.e. work with multiforms instead of forms. Thus Theorems 5.1 and 5.2 <u>remain true for multiforms</u> instead of forms.

§ 6 Comparison of the sets $EP(n,d)$ and $E\Sigma(n,d)$.

Looking again for forms F such that F^2 is extremal in Σ or even in P it is natural to ask whether every $F^2 \in E(\Sigma)$ actually lies in $E(P)$. In case of a positive answer we would know from Theorems 5.1 and 5.2 for any psd form F that F^2 lies in $E(\Sigma)$ if and only if F lies in $E(P)$, and the relation between the sets $E(\Sigma)$ and $E(P)$ would be well understood.

Unfortunately things turn out to be not that simple. Let us write more precisely $P(n,d)$ instead of $P(d)$ and $\Sigma(n,d)$ instead of $\Sigma(d)$ to indicate the number n of variables of the forms under consideration. We ask for which pairs (n,d) with $n \geq 2$, $d \geq 2$ and even, the set $E\Sigma(n,d)$ of extremal points of the cone $\Sigma(n,d)$ is contained in the set $EP(n,d)$ of extremal points of the cone $P(n,d)$. The following theorem gives a complete answer to this question.

<u>Theorem 6.1.</u> Let $n \geq 2$ be a natural number and d be an even natural number. Then $E\Sigma(n,d) \subset EP(n,d)$ precisely in the following cases.
i) $n = 2$; ii) $d \leq 6$; iii) $(n,d) = (3,8)$; iv) $(n,d) = (3,10)$.

Thus the question, whether $E\Sigma(n,d)$ is contained in $EP(n,d)$ is answered by the following chart:

n \ d	2	4	6	8	10	12	14
2	✓	✓	✓	✓	✓	✓	✓
3	✓	✓	✓	✓	✓	x	x
4	✓	✓	✓	x	x	x	x
5	✓	✓	✓	x	x	x	x
6	✓	✓	✓	x	x	x	x

Legend: ✓ = positive answer
x = negative answer

The rest of the section is devoted to a proof of this theorem. If $n = 2$ or $d = 2$ then $\Sigma(n,d) = P(n,d)$ and there is nothing to be proved. Thus we assume henceforth that $n \geq 3$ and $d \geq 4$.

Consider now the case that $d = 4$ or $d = 6$. Let F be a form with

$F^2 \in E\Sigma(n,d)$. Suppose that F^2 does not lie in $EP(n,d)$. Cancelling out in F all indefinite irreducible factors we obtain a form with the same properties, as follows from Theorem 5.1. Thus we may assume that F has only psd factors. Then F cannot have degree 3. Thus F is a psd quadratic form. After a linear change of coordinates we have

$$F = x_1^2 + \ldots + x_r^2$$

with $1 < r \le n$. Now

$$F^2 = x_1^4 + 2\,x_1^2\,(x_2^2 + \ldots + x_r^2) + (x_2^2 + \ldots + x_r^2)^2.$$

We see that F^2 is not extremal in $\Sigma(n,4)$. This contradiction proves that $E\Sigma(n,d)$ is contained in $EP(n,d)$ for $d \le 6$.

Suppose now that F is a form of degree 4 in n variables such that F^2 lies in $E\Sigma$ but not in EP. If F would contain an indefinite irreducible factor then taking out this factor we would obtain a form G with $G^2 \in E\Sigma(n,d)$ but $G^2 \notin EP(n,d)$ for some $d \le 6$ (Theorem 5.1). This has been proved to be impossible. Thus F does not contain an indefinite factor and we may assume in particular that F is psd. If F would be reducible then $F = Q_1Q_2$ with psd quadratic forms Q_1 and Q_2. But then also the factors Q_1^2 and Q_2^2 of $Q_1^2Q_2^2$ would lie in $E\Sigma$ (Theorem 5.1), which means that Q_1 and Q_2 would be squares of linear forms. This contradicts the fact that F has no indefinite factors. Thus F must be an irreducible positive semidefinite quartic.

It is known since Hilbert that $P(3,4) = \Sigma(3,4)$, cf. [CL, §6] for an elementary proof in the case $R = \mathbb{R}$ *). Thus in the case n = 3 our form F has to be a sum of squares, but not a square, and we obtain as above a contradiction to the assumption that F^2 is extremal in $\Sigma(3,8)$. We have proved that $E\Sigma(3,8)$ is contained in $EP(3,8)$.

Assume now that F is a form in 3 variables of degree 5 such that F^2 is extremal in $\Sigma(3,10)$. F contains an irreducible factor H of odd degree, F = HG. By Theorem 5.1 the form G^2 is extremal in Σ. Since $\deg G^2 \le 8$ we know that G^2 is extremal in P. Thus, again by Theorem 5.1, the form F^2 is extremal in P. We have proved that $E\Sigma(3,10)$ is contained in $EP(3,10)$.

---

*) This proof works equally well over all real closed fields R, taking into account the rudiments of [DK, §9]. No appeal to Tarski's principle is necessary.

We now have verified all the affirmative answers in the chart above. To get all negative answers it suffices to check that $E\Sigma(3,12)$ is not contained in $EP(3,12)$ and $E\Sigma(4,8)$ is not contained in $EP(4,8)$. Indeed, regarding a form F in the variables $x_1,\ldots,x_n$ also as a form in the variables $x_1,\ldots,x_{n+1}$, it is an easy exercise to prove that

$$F^2 \in E\Sigma(n,d) \Rightarrow F^2 \in E\Sigma(n+1, d),$$

and it is trivial that

$$F^2 \notin EP(n,d) \Rightarrow F^2 \notin EP(n+1, d).$$

Furthermore choosing some linear form L in the variables $x_1,\ldots,x_n$, it is evident from Theorem 5.1 that

$$F^2 \in E\Sigma(n,d) \Rightarrow F^2L^2 \in E\Sigma(n, d+2)$$

and

$$F^2 \notin EP(n,d) \Rightarrow F^2L^2 \notin EP(n, d+2).$$

We shall now exhibit a form in $E\Sigma(3,12)$ which is not extremal in $P(3,12)$. Fortunately a counterexample for $(n,d) = (4,8)$ can be constructed by similar principles. Thus it will be sufficient to devote our main efforts to the case $(n,d) = (3,12)$.

We start with the ternary sextic

$$S(x,y,z) = x^4y^2 + y^4z^2 + z^4x^2 - 3x^2y^2z^2$$

in [CL]. This form has seven zeros: (1,0,0), (0,1,0), (0,0,1), (1,1,1), (-1,1,1), (1,-1,1) and (1,1,-1). We shall look at an auxiliary form

$$T(x,y,z) = (x^2y + y^2z - z^2x - xyz)^2$$

which is chosen in such a way that it vanishes on all zeros of S, except (-1,1,1).

<u>Theorem 6.2.</u> Let $f(x,y,z) = S(x,y,z) + T(x,y,z)$. Then $p := f^2$ lies in $E\Sigma(3,12)$ but not in $EP(3,12)$.

The fact that p is not extremal in $P(3,12)$ will be deduced from an easy lemma (Lemma 1), and follows by the way also from Theorem 5.1. i, while the fact that p is extremal in $\Sigma(3,12)$ will be deduced from a difficult lemma (Lemma 2).

Lemma 1. The forms S^2, ST, T^2 are linearly independent over R.

Proof. Suppose $\alpha S^2 + \beta ST + \gamma T^2 = 0$, where $\alpha,\beta,\gamma \in R$. Evaluating at $(-1,1,1) \in Z(S) \setminus Z(T)$, we get $\gamma = 0$. Dividing by S, we get $\alpha S + \beta T = 0$, so clearly $\alpha = \beta = 0$.

Q.e.d.

Since $p = f^2 = S^2 + 2ST + T^2$, this lemma clearly implies that p cannot be extremal in $P(3,12)$. It remains to be shown that p is extremal in $\Sigma(3,12)$.

Lemma 2. Let f be as in the theorem. If $f^2 = h_1^2 + \ldots + h_r^2$ in $R[x,y,z]$ then each h_i is an R-linear combination of S and T.

Using this lemma we can show that $p = f^2$ is extremal in $\Sigma(3,12)$ as follows. If $f^2 = h_1^2 + \ldots + h_r^2$, we write $h_i = a_i S + b_i T$ with $a_i, b_i \in R$. Then

$$f^2 = S^2 + 2ST + T^2 = (\sum_1^r a_i^2)\, S^2 + 2(\sum_1^r a_i b_i)\, ST + (\sum_1^r b_i)\, T^2,$$

so by Lemma 1,

$$\sum_1^r a_i^2 = \sum_1^r b_i^2 = \sum_1^r a_i b_i = 1.$$

This implies that $a_i = b_i$ for $1 \le i \le r$, so $h_i^2 = a_i^2 (S+T)^2 = a_i^2 p$, as desired.

Our job is now to prove Lemma 2. For this we need a third lemma which is true for arbitrary polynomials instead of just ternary forms.

Lemma 3. Suppose $f \in R[x_1,\ldots,x_n]$ is positive semidefinite and $f^2 = h_1^2 + \ldots + h_r^2$ with polynomials $h_i \in R[x_1,\ldots,x_n]$. Let $a \in R^n$ be a zero of f. Then a is also a zero of h_i and of every partial derivative $\partial h_i/\partial x_j$ $(1 \le i \le r,\ 1 \le j \le n)$.

Proof. Since f is psd clearly a is a zero of every $\partial f/\partial x_j$, $1 \le j \le n$. Computing the partial derivatives of f^2, we have

$$\frac{\partial}{\partial x_j} f^2 = 2f \frac{\partial f}{\partial x_j} ,$$

$$\frac{\partial^2}{\partial x_j \partial x_k} f^2 = 2f \frac{\partial^2 f}{\partial x_j \partial x_k} + 2 \frac{\partial f}{\partial x_j} \frac{\partial f}{\partial x_k} ,$$

so these partial derivatives all vanish at a. (In fact even the third order partial derivatives of f^2 vanish at a. We do not need this in the following.) From $0 = h_1(a)^2 + \ldots + h_r(a)^2$ we have of course $h_1(a) = \ldots = h_r(a) = 0$. Computing $(\partial^2/\partial x_j^2)(f^2)$ from the expression $f^2 = h_1^2 + \ldots + h_r^2$, we get

$$0 = \sum_{i=1}^{r} [2h_i(a)\frac{\partial^2 h_i}{\partial x_j}(a) + 2(\frac{\partial h_i}{\partial x_j}(a))^2] = 2 \sum_{i=1}^{r} \frac{\partial h_i}{\partial x_j}(a)^2,$$

so $\frac{\partial h_i}{\partial x_j}(a) = 0$ for all i,j.

Q.e.d.

We now enter the proof of Lemma 2. Thus f = S + T, and a decomposition $f^2 = h_1^2 + \ldots + h_r^2$ with forms $h_i \in R[x,y,z]$ of degree 6 is given. Let h be any of the forms h_i. The first step in the proof is to determine which are the sextic monomials which may occur in h. This can be done by inspection - but it is easier to invoke the general method of "cages", cf. [R].*) Denoting the cage of a form g by C(g) we have by the latter method

$$C(h) \subset \tfrac{1}{2}C(f^2) = C(f),$$

and C(f) contains the lattice points (4,2,0), (0,4,2), (2,0,4), (2,2,2), (3,2,1), (3,1,2), (2,3,1), (2,1,3), (1,2,3), (1,3,2). If we represent the points of C(f) by their first two coordinates, we have the following picture of a "projection" of C(f).

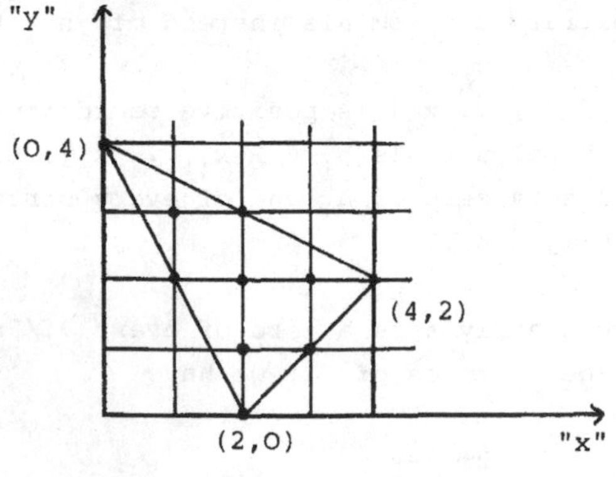

(Actually all lattice points of C(f) occur as monomials in f.) Thus we may express the sextic form h in the following way:

*) A more detailed account of this method will be given in [CLR].

$$h(x,y,z) = ax^4y^2 + by^4z^2 + cx^2z^4 + dx^2y^2z^2 + ex^3y^2z + gx^3yz^2 + ix^2y^3z + jx^2yz^3 + kxy^3z^2 + lxy^2z^3.$$

By Lemma 3 the partial derivatives $\partial h/\partial x$, $\partial h/\partial y$, $\partial h/\partial z$ must vanish at the points (1,1,1), (1,1,-1) and (1,-1,1) of $Z(f)$. This leads to the following system of nine linear homogeneous equations in the ten "unknowns" a,b,...,k,l.

$$\begin{cases}
(1) & 4a + 2c+2d+3e+3g+2i+2j+k+l=0 & (\frac{\partial}{\partial x} \text{ at } (1,1,1)) \\
(2) & 4a + 2c+2d-3e+3g-2i-2j+k-l=0 & (\ldots\ldots(1,1,-1)) \\
(3) & 4a + 2c+2d+3e-3g-2i-2j-k+l=0 & (\ldots\ldots(1,-1,1)) \\
(4) & 2a+4b+2d+2e+g+3i+j+3k+2l=0 & (\frac{\partial}{\partial y} \text{ at } (1,1,1)) \\
(5) & 2a+4b+2d-2e+g-3i-j+3k-2l=0 & (\ldots\ldots(1,1,-1)) \\
(6) & 2a+4b+2d+2e-g-3i-j-3k+2l=0 & (\ldots\ldots(1,-1,1)) \\
(7) & 2b+4c+2d+e+2g+i+3j+2k+3l=0 & (\frac{\partial}{\partial z} \text{ at } (1,1,1)) \\
(8) & 2b+4c+2d-e+2g-i-3j+2k-3l=0 & (\ldots\ldots(1,1,-1)) \\
(9) & 2b+4c+2d+e-2g-i-3j-2k+3l=0 & (\ldots\ldots(1,-1,1))
\end{cases}$$

By explicit computation we shall show that this linear system of equations has a solution space of dimension 2 (with a basis corresponding, of course, to S and T). We proceed as follows:

$$\begin{cases}
(1') = \frac{(1)-(2)}{2}: & 3e + 2i + 2j + l = 0 \\
(2') = \frac{(1)-(3)}{2}: & 3g + 2i + 2j + k = 0 \\
(3') = \frac{(1)+(2)}{2}: & 4a + 2c + 2d + 3g + k = 0 \\
(4') = \frac{(4)-(5)}{2}: & 2e + 3i + j + 2l = 0 \\
(5') = \frac{(4)-(6)}{2}: & g + 3i + j + 3k = 0 \\
(6') = \frac{(4)+(5)}{2}: & 2a + 4b + 2d + g + 3k = 0 \\
(7') = \frac{(7)-(8)}{2}: & e + i + 3j + 3l = 0 \\
(8') = \frac{(7)-(9)}{2}: & 2g + i + 3j + 2k = 0 \\
(9') = \frac{(7)+(8)}{2}: & 2b + 4c + 2d + 2g + 2k = 0
\end{cases}$$

Note that $\frac{(1')+(4')+(7')}{6}$ gives (1") $e + i + j + l = 0$

$\frac{(2')+(5')+(8')}{6}$ gives (2") $g + i + j + k = 0$

From (1"), (4') and (7'), we get $i = j = -e = -l$.
From (2"), (5') and (8'), we get $i = j = -g = -k$.
Eliminating g from (3'), (6') and (9') and dividing by 2, we get

$$\begin{aligned}
&(3'') \quad 2a + c + d + 2k = 0,\\
&(6'') \quad a + 2b + d + 2k = 0,\\
&(9'') \quad b + 2c + d + 2k = 0,
\end{aligned}$$

which leads easily to $a = b = c$ and $d = -3a - 2k$. Thus, a and k are the free parameters, and the solution space to our linear system of equations has dimension 2. Since S and T do give rise to independent solutions in the solution space, we can conclude that $h = \alpha S + \beta T$ ($\alpha, \beta \in R$). More explicitely, the general solution to the linear system is given by

$$\begin{aligned}
(a,b,c,d,e,g,i,j,k,l) &= (a,a,a,-3a-2k,k,k,-k,-k,k,k)\\
&= a(1,1,1,-3,0,\ldots,0) + k(0,0,0,-2,1,1,-1,-1,1,1)\\
&= \left(a+\frac{k}{2}\right)(1,1,1,-3,0,\ldots,0) - \frac{k}{2}(1,1,1,1,-2,-2,2,2,-2,-2)
\end{aligned}$$

So we are finished by noting that $(1,1,1,-3,0,\ldots,0)$ corresponds to S and $(1,1,1,1,-2,-2,2,2,-2,-2)$ corresponds to T. We now have proved Lemma 2 and Theorem 6.2.

The counterexample needed to show that $E\Sigma(4,8)$ is not contained in $E\mathcal{P}(4,8)$ is entirely analogous. We use $p := (Q+U)^2$ where

$$Q(w,x,y,z) = w^4 + x^2y^2 + y^2z^2 + z^2x^2 - 4xyzw,$$

$$U(w,x,y,z) = (w^2 + xy - yz - zx)^2.$$

The form Q has seven zeros: (0,1,0,0), (0,0,1,0), (0,0,0,1), (1,1,1,1), (1,1,-1,-1), (1,-1,1,-1), (1,-1,-1,1), all of which are zeros of U except the last one. By a cage consideration similar to the one used before we can see that, if $p = h_1^2 + \ldots + h_r^2$, then any of the h_i's has the form

$$\begin{aligned}
h(w,x,y,z) = aw^4 &+ bx^2y^2 + cy^2z^2 + dz^2x^2 + exyzw\\
&+ gw^2xy + iw^2yz + jw^2zx\\
&+ kz^2xy + lx^2yz + my^2zx,
\end{aligned}$$

with eleven possible terms. By Lemma 3 the four first partial derivatives of h must vanish on (1,1,1,1), (1,1,-1,-1) and (1,-1,1,-1). This gives us 12 linear homogeneous equations in the 11 unknowns a,b,...,l,m. A calculation similar to the one we did shows that the solution space has dimension 2, hence is spanned by the 11-tuples corresponding to Q and U.

There remains one problem open which fits naturally into the circle of ideas of this paper:

<u>Question.</u> For which (n,d) does there exist a form $F \in EP(n,d)$ such that $F^2 \notin EP(n,2d)$?

Notice that by Theorem 5.2 the form F^2 lies in $E\Sigma(n,2d)$. Thus the question is for a "stronger" counterexample to the inclusion $E\Sigma \subset EP$.

R e f e r e n c e s

[A] E. Artin, Über die Zerlegung definiter Funktionen in Quadrate, Abh. Math. Seminar, Universität Hamburg 5, 100-115 (1927).

[CL] M.D. Choi, T.Y. Lam, Extremal positive semidefinite forms, Math. Ann. 231, 1-18 (1977).

[CL_1] M.D. Choi, T.Y. Lam, An old question of Hilbert, Proceedings Quadratic Form Conference 1976 (ed. G. Orzech), Queen's Papers in Pure and Appl. Math. 46, 385-405.

[DK] H. Delfs, M. Knebusch, Semialgebraic topology over a real closed field II: Basic theory of semialgebraic spaces, Math. Z. 178, 175-213 (1981).

[DK_1] H. Delfs, M. Knebusch, On the homology of algebraic varieties over real closed fields, to appear, preprint Univ. Regensburg.

[DE] D.W. Dubois, G. Efroymson, Algebraic theory of real varieties I, Studies and Essays presented to Yu-Why Chen on his sixtieth birthday (1970), 107-135.

[E] G. Efroymson, Henselian fields and solid k-varieties II, Proc. Amer. Math. Soc. 35, 362-366 (1972).

[ELW] R. Elman, T.Y. Lam, A. Wadsworth, Orderings under field extensions, J. reine angew. Math. 306, 7-27 (1979).

[R] B. Reznick, Extremal psd forms with few terms, Duke Math. J. 45, 363-374 (1978).

[R_i] J.J. Risler, Une caractérisation des idéaux des variétés algébriques réelles, C.R. Acad. Sc. Paris 271, 1171-1173 (1970).

[R_{i_1}] J.J. Risler, Le théorème des zéros en géometrie algébrique et analytique réelles, Bull. Soc. math. France 104, 113-127 (1976).

[S] G. Stengle, A Nullstellensatz and a Positivstellensatz in semialgebraic geometry, Math. Ann. 207, 87-97 (1974).

[CLR] M.D. Choi, T.Y. Lam, B. Reznick, A combinatorial theory for sums of squares of polynomials, in preparation.

Man-Duen Choi
Department of Mathematics, University of Toronto,
Toronto, M5S 1A1, Canada.

Manfred Knebusch
Fakultät für Mathematik der Universität,
D-8400 Regensburg, Universitätsstr.31, F.R.G.

Tsit-Yuen Lam
Department of Mathematics, University of California,
Berkeley CA9 4720, U.S.A.

Bruce Reznick
Department of Mathematics, University of Illinois,
Urbana, Ill. 61801, U.S.A.

Zur Theorie der semialgebraischen Wege und Intervalle über einem reell abgeschlossenen Körper.

Hans Delfs und Manfred Knebusch (Regensburg)

Bei unserem Aufbau der semialgebraischen Topologie über einem reell abgeschlossenen Grundkörper R in den Arbeiten [DK_1] und [DK_2] benutzen wir bei der Diskussion der semialgebraischen Wege die von dem zweiten Autor in [K] entwickelte Theorie der Intervalle auf der Menge X(R) der reellen Punkte einer glatten projektiven Kurve X über R. Durch diesen Aufbau entsteht ein Mißverhältnis zwischen Methoden und erzielten Resultaten. In [K] werden Tatsachen über quadratische Formen {z.B. der Residuensatz für differentialwertige quadratische Formen} und Jacobi-Varietäten {insbesondere der Dualitätssatz von Geyer für abelsche Varietäten über R, [G]} ausgenutzt, die methodisch gesehen schon zu den höheren Teilen der algebraischen Geometrie gehören. Semialgebraische Topologie handelt aber nur von den einfachsten, nämlich den "topologischen" Eigenschaften des Raumes X(R) der reellen Punkte einer Varietät X über R.

Nun lassen sich aber die Abschnitte §6-§9 und §11 in [DK_2] unabhängig von den Ergebnissen in [K] verstehen. (Theorem 9.2 in [DK_2] kann man, wie schon dort bemerkt ist, mit Hilfe des Tarski-Prinzips durch Übertragen aus dem klassischen Fall gewinnen, der Hinweis auf Lemma 9.3 im Beweis von Proposition 11.1 ist unwesentlich). Ziel dieser Note ist es, ausgehend von dieser Beobachtung, einen neuen Zugang zur Theorie der semialgebraischen Wege zu geben, der mit semialgebraischen Standard-Methoden auskommt und somit nur sehr elementare algebraische Geometrie benutzt. Damit wird die semialgebraische Topologie von der Arbeit [K] unabhängig. Überdies erhalten wir neue Beweise für diejenigen Sätze aus [K], die rein semialgebraischer Natur sind, also nicht auf algebraische Funktionen Bezug nehmen. Alles dies wird in §1-§5 geleistet.

In einem letzten Abschnitt, §6, analysieren wir die Verhältnisse auf dem Einheitskreis

$$S(R) = \{(x,y) \in R^2 \mid x^2+y^2 = 1\}.$$

Wir zeigen, daß dort vieles "genauso ist" wie man es im Falle $R = \mathbb{R}$ gewohnt ist, obwohl vom Standpunkt der klassischen Topologie aus S(R) fast immer ein total unzusammenhängender Raum ist. Weil uns keine Exponentialfunktion zur Verfügung steht, muß dies jedoch mit gänzlich anderen Argumenten bewiesen werden, als man sie aus der Analysis kennt. Besonderes Interesse verdient unseres Erachtens die sich dabei ergebende Möglichkeit, für jede natürliche Zahl n in einem algebraisch abgeschlossenen Körper C der Charakteristik O nach Wahl eines reell abgeschlossenen Teilkörpers R mit [C:R] = 2 und Wahl einer Quadratwurzel aus -1 eine primitive n-te Einheitswurzel ζ_n auszuzeichnen.

§1 Wegekomponenten.

Wir setzen die in [DK_2, §6-§9] entwickelte allgemeine Theorie der semialgebraischen Räume über einem reell abgeschlossenen Grundkörper R voraus, und benutzen dieselben Bezeichnungen wie dort.

<u>Definition 1.</u> Ein semialgebraischer Weg in einem semialgebraischen Raum M ist eine semialgebraische Abbildung $\alpha : [0,1] \to M$. Dabei bedeutet [0,1] das abgeschlossene Einheitsintervall in R. $\alpha(0)$ heißt der <u>Anfangspunkt</u> P des Weges und $\alpha(1)$ der <u>Endpunkt</u> Q des Weges.

<u>Definition 2.</u> Zwei Punkte $P,Q \in M$ heißten <u>verbindbar</u>, wenn es einen Weg α in M mit Anfangspunkt P und Endpunkt Q gibt. Ersichtlich ist "verbindbar" eine Äquivalenzrelation auf M. Die Äquivalenzklassen heißen die <u>Wegekomponenten</u> von M. Liegt nur eine Äquivalenzklasse vor, so heißt M <u>wegezusammenhängend</u>.

Ausgehend von diesen beiden Definitionen wurden in [DK_2, §11] die folgenden beiden Sätze bewiesen, unter alleiniger Benutzung von [DK_2, §6-§9]und eines elementaren Satzes von Cohen [C, §1, Theorem B_n].

<u>Satz 1.1.</u> Ist M wegezusammenhängend, so gibt es <u>keine</u> Partition $M = U_1 \cup U_2$ von M in disjunkte offene nichtleere semialgebraische Teilmengen U_1, U_2 von M. (Das ist fast trivial.)

Theorem 1.2. Jeder semialgebraische Raum M hat nur endlich viele Wegekomponenten. Diese sind semialgebraische Teilmengen von M.

Beim Beweis dieses Theorems wurde eine Beschreibung einer vorgegebenen semialgebraischen Teilmenge M von $R^n \times R$ durch "Schichten" über einer geeigneten Partition der Koordinantenhyperebene R^n in endlich viele semialgebraische Mengen benutzt[*)] [DK_2, p.204], die auch in dieser Note eine Rolle spielen wird und überhaupt grundlegend für die semialgebraische Geometrie ist. Dabei ging der oben genannte Satz von Cohen ein.

Wir wollen noch sämtliche wegezusammenhängenden semialgebraischen Teilräume von R angeben. Aus der Tatsache, daß ein Polynom in einer Variablen mit Koeffizienten in R nur endlich viele Nullstellen hat und zwischen zwei reellen Nullstellen nicht das Vorzeichen wechseln kann, folgt, daß jede echte nicht leere semialgebraische Teilmenge M von R disjunkte Vereinigung von endlich vielen Mengen folgender Art ist:

a) $]a,b[:= \{x \in R \mid a < x < b\}$

$[a,b[:=]a,b[\cup \{a\}$

$]a,b] :=]a,b[\cup \{b\}$

$[a,b] :=]a,b[\cup \{a\} \cup \{b\}$

mit Elementen $a < b$ von R. Das sind die "Intervalle" auf R (offen, halboffen, abgeschlossen).

b) $]-\infty,a[:= \{x \in R \mid x < a\}$

$]-\infty,a] :=]-\infty,a[\cup \{a\}$

$]a,\infty[:= \{x \in R \mid x > a\}$

$[a,\infty[:=]a,\infty[\cup \{a\}$

Das sind die "Halbgeraden" auf R ($a \in R$).

c) Einpunktige Mengen.

Jetzt ist evident

Satz 1.3. Die wegezusammenhängenden semialgebraischen Teilmengen von R sind genau die Mengen $R,\emptyset$, und die in a),b),c) aufgeführten Mengen.

*) siehe auch [Cs, §2].

§ 2 Vervollständigung von Wegen.

Theorem 2.1. Sei $J \subset R$ ein Intervall und $\gamma : J \to M$ eine semialgebraische Abbildung in einen vollständigen semialgebraischen Raum. Dann läßt sich γ zu einer semialgebraischen Abbildung $\bar{\gamma} : \bar{J} \to M$ auf den Abschluß $\bar{J}$ von J in R fortsetzen.

Beweis. Ersichtlich genügt es, den Fall $J =]0,1]$ zu behandeln. Sei $\Gamma \subset]0,1] \times M$ der Graph von γ. Wir haben die folgende kanonische Faktorisierung von γ.

$$\begin{array}{ccc}]0,1] & \xrightarrow[\alpha]{\sim} & \Gamma \\ & \searrow^{\gamma} & \downarrow p \\ & & M \end{array}$$

mit dem Isomorphismus α, definiert durch $\alpha(t) = (t,\gamma(t))$, und der kanonischen Projektion p von Γ nach M, $p(t,x) = x$. Wir bezeichnen mit $\bar{\Gamma}$ den Abschluß von Γ in dem vollständigen Raum $N := [0,1] \times M$. $\bar{\Gamma}$ ist vollständig, aber $\Gamma \cong]0,1]$ ist nicht vollständig. Somit ist $\bar{\Gamma} \neq \Gamma$. Weiter ist [DK_2, Cor. 8.11]

$$\dim(\bar{\Gamma} \setminus \Gamma) < \dim \Gamma = 1.$$

Also ist $\bar{\Gamma} \setminus \Gamma$ eine endliche Menge $\{P_1,\ldots,P_r\}$ mit paarweise verschiedenen Punkten P_i. Wir wollen zeigen, daß $r = 1$ ist.

Angenommen, $r > 1$. Wir wählen offene semialgebraische Umgebungen U_1, U_2 von P_1, P_2 in N mit $U_1 \cap U_2 = \emptyset$. Der Punkt P_1 liegt schon im Abschluß der semialgebraischen Menge

$$U_1 \cap \Gamma = \alpha(\alpha^{-1}(U_1)).$$

Aufgrund unserer genauen Kenntnis der semialgebraischen Teilmengen von $]0,1]$ (s. Ende von §1) wissen wir:

$$\alpha^{-1}(U_1) = J_1 \cup \ldots \cup J_r$$

mit paarweise disjunkten in $]0,1]$ offenen Intervallen J_i und $t < u$ für $t \in J_i$, $u \in J_{i+1}$ $(1 \leq i \leq r-1)$. Für $i \geq 2$ ist der Abschluß $\bar{J}_i$ von J_i in R in $]0,1]$ enthalten, also wegen der Vollständigkeit von $\bar{J}_i$

$$P_1 \notin \alpha(\overline{J}_i) = \overline{\alpha(J_i)}.$$

Somit ist $P_1 \in \overline{\alpha(J_1)}$. Wäre der Anfangspunkt des Intervalls J_1 von Null verschieden, so wäre auch $\overline{J_1} \subset]0,1]$ und $P_1 \notin \overline{\alpha(J_1)}$. Somit hat J_1 die Gestalt $]0,c_1[$ mit $c_1 \in]0,1]$. Wir haben damit ein $c_1 \in]0,1]$ gefunden so daß

$$\alpha(]0,c_1[) \subset U_1$$

ist. Ebenso findet man ein $c_2 \in]0,1]$ mit

$$\alpha(]0,c_2[) \subset U_2.$$

Es folgt $U_1 \cap U_2 \neq \emptyset$, und das ist der gesuchte Widerspruch. Somit ist $\overline{\Gamma} \setminus \Gamma$ eine einpunktige Menge $\{P\}$. Es ist $P = (\tau, Q)$ mit $\tau \in [0,1]$, $Q \in M$. Da $\alpha^{-1}(\Gamma) =]0,1]$ und das Bild von $\overline{\Gamma}$ unter der Projektion auf $[0,1]$ vollständig, also gleich $[0,1]$ ist, muß $\tau = 0$ sein. $\overline{\Gamma}$ ist der Graph der Abbildung $\overline{\gamma} : [0,1] \to M$, definiert durch

$$\overline{\gamma}(t) = \begin{cases} \gamma(t) & 0 < t \leq 1 \\ Q & t = 0 \, . \end{cases}$$

Weil $\overline{\Gamma}$ abgeschlossen in $[0,1] \times M$ und M vollständig ist, ist $\overline{\gamma}$ stetig, also eine semialgebraische Abbildung, vgl. [DK_2, Th. 9.9]. Damit ist das Theorem bewiesen.

Schon ein Spezialfall dieses Theorems, nämlich das Lemma 12.2 in [DK_2], reicht aus, um den Kurvenauswahlsatz zu beweisen, wie in [DK_2, §12] näher ausgeführt ist.

Theorem 2.2 (Kurvenauswahlsatz). Sei M semialgebraischer Teilraum eines semialgebraischen Raumes L und P ein Punkt in dem Abschluß $\overline{M}$ von M in L. Dann gibt es einen Weg $\gamma : [0,1] \to L$ mit $\gamma(0) = P$ und $\gamma(]0,1]) \subset M$.

Aufgrund dieses Satzes ist klar, daß die Wegekomponenten eines semialgebraischen Raumes M sämtlich abgeschlossen, also auch offen in M sind. Sie sind somit in jedem vernünftigen Sinne als die "Zusammenhangskomponenten" von M anzusehen. (Man erinnere sich an Satz 1.1.). Wir nennen deshalb ab jetzt die Wegekomponenten suggestiver "Zusammenhangskomponenten" und nennen insbesondere einen Raum M kurz "zusammenhängend", wenn er wegezusammenhängend ist.

Ohne weitere Arbeit erhält man jetzt als Korollar zu Theorem 2.1 einen Satz über die Liftung von Wegen. Ein Spezialfall dieses Satzes spielte in [DK_1, §3] eine wesentliche Rolle [loc.cit, Th.3.3].

Korollar 2.3. Sei $\pi : M \to N$ eine eigentliche semialgebraische Abbildung. Sei J ein Intervall in R und $\alpha : \bar{J} \to N$ eine semialgebraische Abbildung. Weiter sei eine semialgebraische Liftung $\beta : J \to M$ von $\alpha|J$ vorgegeben, d.h. β sei eine semialgebraische Abbildung mit $\pi \circ \beta = \alpha|J$. Dann läßt sich β fortsetzen zu einer semialgebraischen Liftung $\bar{\beta}$ von α.

Beweis. $\alpha(\bar{J}) = L$ ist ein vollständiger semialgebraischer Teilraum von N. Wir dürfen N durch L und M durch den vollständigen Raum $\pi^{-1}(L)$ ersetzen, somit ohne Einschränkung der Allgemeinheit M als vollständig voraussetzen. Nach Theorem 2.1 setzt sich β zu einer semialgebraischen Abbildung $\bar{\beta} : \bar{J} \to M$ fort. Die semialgebraischen Abbildungen $\pi \circ \bar{\beta}$ und α stimmen auf J überein, wegen ihrer Stetigkeit also auch auf $\bar{J}$.

q.e.d.

§ 3 Genauere Analyse eines Weges.

Wir formulieren zunächst das semialgebraische Analogon zum Zwischenwertsatz und zum Satz von der Existenz von Maximum und Minimum für stetige Funktionen einer reellen Veränderlichen.

Satz 3.1. Sei $f : [a,b] \to R$ eine semialgebraische nicht konstante Funktion auf einem abgeschlossenen Intervall $[a,b] \subset R$. Dann ist $f([a,b])$ ein abgeschlossenes Intervall $[c,d]$.

Beweis. Das Bild $N := f([a,b])$ von f ist vollständig, weil $[a,b]$ vollständig ist (vgl. [DK_2, §9]). Somit ist N eine abgeschlossene beschränkte semialgebraische Teilmenge von R. Weil $[a,b]$ wegezusammenhängend ist, gilt gleiches für N. Aus der am Ende von §1 angegebenen Liste aller wegezusammenhängenden Teilmengen von R ersieht man, daß N ein abgeschlossenes Intervall ist.

Ziel dieses Abschnittes ist nun der Beweis des folgenden Satzes.

Theorem 3.2. Sei $\gamma : [a,b] \to M$ eine semialgebraische Abbildung von einem abgeschlossenen Intervall $[a,b] \subset R$ in einen semialgebraischen Raum M. Dann gibt es eine Unterteilung

$$t_o = a < t_1 < \ldots < t_r = b$$

so daß jede Einschränkung $\gamma|[t_{k-1},t_k]$ $(1 \le k \le r)$ entweder eine konstante Abbildung oder eine Einbettung (= Isomorphismus aufs Bild) ist.

Zum Beweis wählen wir eine endliche Überdeckung $(M_i | i \in I)$ von M durch offene affine semialgebraische Teilräume M_i. Dann ist $(\gamma^{-1}(M_i)|i \in I)$ eine Überdeckung von $[a,b]$ durch endlich viele offene semialgebraische Teilmengen. Jedes $\gamma^{-1}(M_i)$ ist disjunkte Vereinigung endlich vieler in $[a,b]$ offener Intervalle. Man findet somit leicht eine Unterteilung

$$a_o = a < a_1 < \ldots < a_m = b,$$

so daß γ jedes abgeschlossene Teilintervall $[a_{j-1},a_j]$ in eine der Mengen M_i abbildet. Daher können wir uns von vornherein auf den Fall zurückziehen, daß M affin ist. M läßt sich dann in einen Raum R^n einbetten und wir dürfen somit sogar $M = R^n$ voraussetzen. Seien $\gamma_1,\ldots,\gamma_n$ die Komponenten der Abbildung γ von $[a,b]$ nach R^n. Es genügt, die Behauptung für jede Komponente $\gamma_i : M \to R$ separat einzusehen. Dann ist die Behauptung auch für γ evident.

Damit haben wir uns auf den Fall $M = R$ zurückgezogen. Wir benötigen jetzt das folgende Lemma.

Lemma 3.3. Sei $\gamma : [a,b] \to R$ eine semialgebraische Abbildung. Dann gibt es zu jedem $x \in [a,b[$ ein $\varepsilon > 0$ in R, so daß entweder $\gamma(y) > \gamma(x)$ oder $\gamma(y) = \gamma(x)$ oder $\gamma(y) < \gamma(x)$ jeweils für alle $y \in [a,b]$ mit $x < y < x+\varepsilon$ gilt.

Beweis. Die Mengen

$$A := \{y \in [a,b] \mid \gamma(y) > \gamma(x)\}$$

$$B := \{y \in [a,b] \mid \gamma(y) < \gamma(x)\}$$

$$C := \{y \in [a,b] \mid \gamma(y) = \gamma(x)\}$$

sind semialgebraisch nach Tarski, also disjunkte Vereinigungen endlich vieler Intervalle und endlich vieler Punkte (vgl. Satz 1.3). Damit ist schon klar, daß $]x,x+\varepsilon[$ für ein $\varepsilon > 0$ ganz in einer der Mengen A,B,C liegt.

q.e.d.

Mit diesem Lemma läßt sich Theorem 3.2 im Falle $M = R$ leicht verifizieren. Wir betrachten die Teilmengen F,G,H von $[a,b[$, die aus allen Punkten $x \in [a,b[$ bestehen, zu denen es ein $\varepsilon > 0$ gibt, so daß $\gamma(y) > \gamma(x)$ bzw. $\gamma(y) < \gamma(x)$ bzw. $\gamma(y) = \gamma(x)$ jeweils für alle $y \in [a,b]$ mit $x < y < x+\varepsilon$ gilt. Nach Tarski sind diese Mengen sicherlich semialgebraisch, also Vereinigungen von endlich vielen Intervallen und endlich vielen Punkten. Nach Lemma 3.3 ist $[a,b[$ die Vereinigung dieser drei Mengen F,G,H. Daher gibt es eine Unterteilung

$$a = t_o < t_1 < \ldots < t_r = b,$$

so daß jedes Intervall $]t_{k-1},t_k[$, $1 \leq k \leq r$, ganz in einer der Mengen F,G,H enthalten ist. Ist $]t_{k-1},t_k[$ in H enthalten, so ist γ auf $[t_{k-1},t_k]$ konstant. Wir betrachten jetzt den Fall, daß ein vorgegebenes Intervall $I =]t_{k-1},t_k[$ ganz in F enthalten ist. Auf jedem abgeschlossenen Teilintervall $[c,d]$ von I muß die Funktion γ nach Satz 3.1 ihr Maximum annehmen. Aufgrund der Definition von F kann das Maximum nur in dem Punkt d angenommen werden. Daher ist γ auf I streng monoton wachsend. Weil γ stetig ist, folgt, daß γ auch auf dem Abschluß $\overline{I} = [t_{k-1},t_k]$ von I streng monoton wächst. Insbesondere ist $\gamma|\overline{I}$ injektiv, somit wegen der Vollständigkeit von $\overline{I}$ eine Einbettung [DK_2,9.8]. Ist schließlich I in der Menge G enthalten, so ist $\gamma|\overline{I}$ streng monoton fallend und wieder eine Einbettung. Damit ist Theorem 3.2 völlig bewiesen.

§ 4 Die projektive Gerade $\mathbb{P}(R)$.

Wie üblich fassen wir die projektive Gerade

$$\mathbb{P}(R) = \{(x_o : x_1) \mid x_o, x_1 \in R,\ x_o \neq 0 \text{ oder } x_1 \neq 0\}$$

als Vereinigung der affinen Geraden R mit einem weiteren Punkt ∞ auf. Genauer identifizieren wir a mit (1:a) für $a \in R$ und setzen $\infty = (0:1)$. R ist dann mit seiner Standardstruktur als semialgebraischer Raum ein offener dichter Teilraum des semialgebraischen Raumes $\mathbb{P}(R)$. Zu je zwei verschiedenen Punkten a,b auf $\mathbb{P}(R)$ definieren wir ein "offenes Intervall"]a,b[wie folgt:

$a,b \in R,\ a < b:\quad]a,b[= \{x \in R \mid a < x < b\};$

$a,b \in R,\ a > b:\quad]a,b[= \{x \in R \mid x > a \text{ oder } x < b\} \cup \{\infty\};$

$]\infty,a[= \{x \in R \mid x < a\};$

$]a,\infty[= \{x \in R \mid x > a\}.$

Weiter definieren wir "halboffene Intervalle"

$[a,b[:=]a,b[\cup \{a\}$

$]a,b] :=]a,b[\cup \{b\}$

und "abgeschlossene Intervalle"

$[a,b] :=]a,b[\cup \{a\} \cup \{b\}.$

Um die so definierten semialgebraischen Teilmengen von $\mathbb{P}(R)$ besser zu verstehen, betrachten wir die folgende semialgebraische Einbettung des Intervalls $]-1,1[\subset R$ in $\mathbb{P}(R)$.

$$]-1,1[\xrightarrow{\sim} R \hookrightarrow \mathbb{P}(R),$$
$$x \mapsto x/1-|x|.$$

Nach §2 läßt sich diese Einbettung zu einer semialgebraischen Abbildung

$$\pi : [-1,1] \to \mathbb{P}(R)$$

fortsetzen, natürlich auf genau eine Weise. Diese Abbildung π ist in keiner Umgebung von -1 konstant, muß also nach Theorem 3.2 in einer Umgebung von -1 injektiv sein. Insbesondere ist $\pi(-1) = \infty$. Ebenso ist $\pi(1) = \infty$. Man prüft natürlich auch leicht elementar nach, daß die durch $\pi(x) = x/1-|x|$ für $x \in]-1,1[$, $\pi(-1) = \pi(1) = \infty$ definierte Abbildung von $[-1,1]$ nach $\mathbb{P}(R)$ semialgebraisch ist.

π ist eigentlich und surjektiv, somit gilt:

1) Eine Teilmenge M von $\mathbb{P}(R)$ ist genau dann semialgebraisch, wenn $\pi^{-1}(M)$ semialgebraisch ist. Eine semialgebraische Teilmenge M von $\mathbb{P}(R)$ ist genau dann abgeschlossen bzw. offen in $\mathbb{P}(R)$, wenn $\pi^{-1}(M)$ abgeschlossen bzw. offen in $[-1,1]$ ist.

2) Eine Abbildung $f : \mathbb{P}(R) \to X$ in einen semialgebraischen Raum X ist genau dann semialgebraisch, wenn die Komposition $f \circ \pi : [-1,1] \to X$ semialgebraisch ist. π ist also "identifizierend".

Mit Hilfe der Abbildung π verifiziert man nun leicht

<u>Satz 4.1.</u> Seien a,b verschiedene Punkte auf $\mathbb{P}(R)$.

1) $]a,b[$ ist eine offene und zusammenhängende semialgebraische Teilmenge von $\mathbb{P}(R)$.

2) $[a,b]$ ist der Abschluß von $]a,b[$ in $\mathbb{P}(R)$.

3) Es gibt einen - explizit angebbaren - semialgebraischen Isomorphismus λ von $[0,1]$ auf $[a,b]$ mit $\lambda(0) = a$, $\lambda(1) = b$.

Weiter dient uns die Abbildung π dazu, für jeden Punkt $p \in R$ einen Automorphismus α_p von $\mathbb{P}(R)$ zu konstruieren, der p auf ∞ abbildet. Sei c das Urbild von p unter π. Zunächst definieren wir eine semialgebraische Abbildung $\tilde{\alpha}_p : [-1,1] \to \mathbb{P}(R)$ vermöge

$$\tilde{\alpha}_p(x) = \begin{cases} \pi(x+1-c) & -1 \le x \le c \\ \pi(x-1-c) & c \le x \le 1. \end{cases}$$

Wir stellen fest, daß

$$\tilde{\alpha}_p(1) = \tilde{\alpha}_p(-1) = \pi(-c)$$

ist, und erhalten somit eine semialgebraische Abbildung $\alpha_p : \mathbb{P}(R) \to \mathbb{P}(R)$ mit $\alpha_p \circ \pi = \tilde{\alpha}_p$. Man prüft leicht, daß α_p bijektiv ist. Da $\mathbb{P}(R)$ vollständig ist [DK_2, §9], folgt, daß α_p ein Automorphismus von $\mathbb{P}(R)$ ist. Unter erneuter Benutzung von π verifiziert man mit Geduld, aber ohne Mühe

Lemma 4.2. α_p bildet p auf ∞ ab. Für je zwei verschiedene Punkte a,b auf $\mathbb{P}(R)$ ist

$$\alpha_p([a,b]) = [\alpha_p(a), \alpha_p(b)].$$

Der folgende Satz ist evident für $p=\infty$, aufgrund dieses Lemmas somit richtig für jeden Punkt p auf $\mathbb{P}(R)$.

Satz 4.3. i) Für jeden Punkt $p \in \mathbb{P}(R)$ ist die Menge $\mathbb{P}(R)\setminus\{p\}$ total geordnet vermöge der folgenden Relation:

$$a < b \iff a \neq b \text{ und } p \notin]a,b[.$$

ii) Mit dieser Anordnung gilt für zwei verschiedene Punkte a,b auf $\mathbb{P}(R)\setminus\{p\}$: Ist $a < b$, so ist

$$]a,b[= \{x \in \mathbb{P}(R)\setminus\{p\} \mid a < x < b\}.$$

Ist $a > b$, so ist

$$]a,b[= \{x \in \mathbb{P}(R)\setminus\{p\} \mid x > a \text{ oder } x < b\} \cup \{p\}.$$

iii) Sei $[a,b]$ ein abgeschlossenes in $\mathbb{P}(R)\setminus\{p\}$ enthaltenes Intervall. Dann gilt für Punkte $c,d \in]a,b[$: $a < c < b$. Genau dann ist $c \leq d$, wenn das Intervall $[a,c]$ in $[a,d]$ enthalten ist. Also auch: Genau dann ist $c \leq d$, wenn das Intervall $[c,b]$ das Intervall $[d,b]$ umfaßt. Insbesondere ist die von $\mathbb{P}(R)\setminus\{p\}$ auf $[a,b]$ induzierte totale Anordnung unabhängig von der Wahl des Punktes p außerhalb von $[a,b]$.

iv) $\mathbb{P}(R)\setminus\{p\}$ ist zusammenhängend. Sind $p_0 = p, p_1, \ldots, p_r$ verschiedene Punkte auf $\mathbb{P}(R)$, $r \geq 1$, und ist in der Anordnung von $\mathbb{P}(R)\setminus\{p\}$

$$p_1 < p_2 < \ldots < p_r,$$

so hat $\mathbb{P}(R)\setminus\{p_0, p_1, \ldots, p_r\}$ die folgenden Zusammenhangskomponenten: $]p_0,p_1[$, $]p_1,p_2[$, ..., $]p_r,p_0[$.

v) Die zusammenhängenden nicht leeren echten semialgebraischen Teilmengen von $\mathbb{P}(R)\setminus\{p\}$ sind genau die in $\mathbb{P}(R)\setminus\{p\}$ enthaltenen Intervalle und die einpunktigen Teilmengen.

Bemerkung. Auch $\mathbb{P}(R)$ ist als Bild von $[-1,1]$ unter π zusammenhängend. Damit haben wir alle zusammenhängenden semialgebraischen Teilmengen von $\mathbb{P}(R)$ gefunden.

Satz 4.4. Für jeden semialgebraischen Automorphismus α von $\mathbb{P}(R)$ gilt:
Entweder ist für je zwei verschiedene Punkte a,b auf $\mathbb{P}(R)$

$$\alpha([a,b]) = [\alpha(a), \alpha(b)],$$

oder es ist für je zwei verschiedene Punkte a,b auf $\mathbb{P}(R)$

$$\alpha([a,b]) = [\alpha(b), \alpha(a)].$$

Beweis. Sei $p := \alpha(\infty)$. Weil der oben konstruierte Automorphismus α_p die erste Alternative aus dem Satz erfüllt, genügt es, die Behauptung für $\beta := \alpha_p \circ \alpha$ anstelle von α zu verifizieren. Es ist $\beta(\infty) = \infty$. Die Einschränkung $f : R \xrightarrow{\sim} R$ von β muß aufgrund des Zwischenwertsatzes (s. Satz 3.1) entweder streng monoton wachsend oder streng monoton fallend sein. Man sieht nun leicht, daß dementsprechend β eine der beiden Alternativen in dem Satz erfüllt.

Genügt ein Automorphismus α von $\mathbb{P}(R)$ der ersten Alternative in Satz 4.4, so nennen wir α "wachsend", anderenfalls nennen wir α "fallend". Mit $\mathrm{Aut}(\mathbb{P}(R))$ bezeichnen wir die Gruppe aller semialgebraischen Automorphismen von $\mathbb{P}(R)$, mit $\mathrm{Aut}^+(\mathbb{P}(R))$ die Untergruppe der wachsenden Automorphismen, mit $\mathrm{Aut}^-(\mathbb{P}(R))$ die Menge der fallenden Automorphismen.

Beispiel. Die Abbildung $i : \mathbb{P}(R) \to \mathbb{P}(R)$, definiert durch $i(x) = -x$ für $x \in R$, $i(\infty) = \infty$, ist ein fallender Automorphismus von $\mathbb{P}(R)$. Überdies ist i^2 die Identität.

Somit ist $\mathrm{Aut}^-(\mathbb{P}(R))$ sicherlich nicht leer. $\mathrm{Aut}^+(\mathbb{P}(R))$ ist also ein Normalteiler vom Index 2 in $\mathrm{Aut}(\mathbb{P}(R))$.

Epilog. Vermöge einer stereographischen Projektion läßt sich $\mathbb{P}(R)$ mit dem Einheitskreis

$$S(R) := \{(x,y) \in R^2 \mid x^2 + y^2 = 1\}$$

identifizieren. Es scheint auf den ersten Blick natürlicher als unser Vorgehen zu sein, auf $S(R)$ eine "zirkulare Anordnung" einzuführen, und diese mit der stereographischen Projektion auf $\mathbb{P}(R)$ zu übertragen. Jedoch haben wir den Eindruck, daß dies schwieriger ist, als mit obiger Abbildung $\pi : [-1,1] \twoheadrightarrow \mathbb{P}(R)$ zu arbeiten, weil keine trigonometrischen Funktionen zur Verfügung stehen. Wir werden umgekehrt in §6 die jetzt konstruierte zirkulare Anordnung von $\mathbb{P}(R)$ auf $S(R)$ übertragen.

§ 5 Intervalle auf glatten vollständigen Kurven.

Ist S ein zu $[0,1]$ isomorpher semialgebraischer Raum, so bezeichnen wir als Rand ∂S von S die Menge aller $x \in S$ für die $S \setminus \{x\}$ zusammenhängend ist. ∂S besteht aus genau zwei Punkten.

Theorem 5.1. Sei M ein vollständiger eindimensionaler affiner semialgebraischer Raum, der keine isolierten Punkte besitzt (d.h. "rein" von der Dimension 1 ist, vgl. [DK_2, §13]). Dann gibt es eine endliche Familie $(S_i | i \in I)$ von zum Einheitsintervall $[0,1]$ isomorphen Teilräumen S_i von M, so daß gilt:

1) $M = \bigcup_{i \in I} S_i$

2) Für $i \neq j$ ist $S_i \cap S_j$ entweder leer oder besteht aus genau einem Punkt, und dieser liegt in ∂S_i und in ∂S_j.

Wir nennen eine solche Familie $(S_i | i \in I)$ eine Triangulierung von M, weiter die Mengen S_i die 1-Simplizes der Triangulierung und die sämtlichen Randpunkte sämtlicher S_i die Ecken der Triangulierung. (Das sind Begriffe ad hoc für den eindimensionalen Fall. Eine systematischere Terminologie findet man in [DK_3, §2].)

Theorem 5.1 ist ein Spezialfall des in [DK_3, §2] bewiesenen Triangulierungssatzes für beliebige affine semialgebraische Räume [loc. cit, Th.2.1]. Jedoch läßt sich dieser Spezialfall wesentlich leichter als der allgemeine Satz herleiten. Wir wollen das kurz skizzieren.

Wir betten M in einen R^n ein und machen Induktion nach n. Für $n = 1$ ist M eine disjunkte Vereinigung von abgeschlossenen Intervallen, und die Behauptung ist evident. Sei jetzt $n = 2$ und ohne wesentliche Einschränkung der Allgemeinheit M zusammenhängend. $p : R^2 \to R$ bezeichne die Projektion $(x,y) \mapsto x$. Die Menge $p(M)$ ist semialgebraisch, zusammenhängend und vollständig in R, also ein abgeschlossenes Intervall [a,b]. Wir haben nun bezüglich p eine Zerlegung von M in "Schichten", wie sie allgemein in [DK_2, p.204] und auch in §2 des Übersichtsartikels [Cs] von Coste angegeben wurde. Da M Dimension 1 hat, und da jede semialgebraische Teilmenge von [a,b] Vereinigung von endlich vielen Intervallen und Punkten ist, sieht diese Schichtenzerlegung wie folgt aus: Es gibt eine Unterteilung

$$a = a_o < a_1 < \ldots < a_r = b$$

von [a,b] und über jedem offenen Intervall $]a_{i-1},a_i[$, $1 \le i \le r$, semialgebraische Funktionen $\xi_1^i,\ldots,\xi_{m(i)}^i$ $(1 \le i \le r,\ m(i) \ge 1)$, so daß gilt:

1) Für jedes $x \in\ ]a_{i-1},a_i[$ ist
$$\xi_1^i(x) < \ldots < \xi_{m(i)}^i(x).$$

2) $M \cap (]a_{i-1},a_i[\times R)$ ist die disjunkte Vereinigung der Graphen
$$\Gamma_j^i := \{(x,\xi_j^i(x)) \mid a_{i-1} < x < a_i\}$$
mit $1 \le j \le m(i)$.

3) $M \cap (\{a_i\} \times R)$ ist disjunkte Vereinigung von einpunktigen Mengen und abgeschlossenen Intervallen auf der Geraden $\{a_i\} \times R$.

Nach §2 setzen sich die Funktionen ξ_j^i stetig fort zu semialgebraischen Abbildungen $\eta_j^i : [a_{i-1},a_i] \to M$. Der Graph Δ_j^i von η_j^i ist vollständig und somit ersichtlich der Abschluß von Γ_j^i in M. Überdies ist Δ_j^i zu $[a_{i-1},a_i]$ und damit zu [0,1] isomorph. Da M keine isolierten Punkte hat, ist jetzt klar, daß M die Vereinigung aller Mengen Δ_j^i, $(1 \le i \le r,\ 1 \le j \le m(i))$ und endlich vieler abgeschlossener Intervalle auf den Geraden $\{a_i\} \times R$ ist, und man kann die gesuchte Triangulierung von M "sehen".

Sei jetzt $n > 2$ und p die kanonische Projektion von $R^n = R^{n-1} \times R$ auf R^{n-1}. Wir dürfen M wieder als zusammenhängend voraussetzen. Ist p(M) ein Punkt x_o, so ist M in der Geraden $\{x_o\} \times R$ enthalten, und wir sind im eindimensionalen Fall, der erledigt ist. Anderenfalls haben wir nach Induktionsvoraussetzung eine Triangulierung $(T_j \mid j \in J)$ von p(M). Es genügt, jeden abgeschlossenen Unterraum $p^{-1}(T_j) \cap M$ von M zu triangulieren. Sammelt man dann die 1-Simplizes aller dieser Triangulierungen, so erhält man eine Triangulierung von M, wenn man noch die 1-Simplizes, die ganz über den Randpunkten der T_j liegen, genügend unterteilt. Damit können wir uns auf den Fall zurückziehen, daß es einen Isomorphismus $\varphi : p(M) \xrightarrow{\sim} [0,1]$ gibt. Unter dem Isomorphismus

$$\varphi \times id : p(M) \times R \to [0,1] \times R$$

wird M isomorph auf einen vollständigen eindimensionalen Teilraum M' von $[0,1] \times R$ abgebildet. M' läßt sich aufgrund des erledigten Falles $n = 2$ triangulieren. Damit erhalten wir auch eine Triangulierung von M.

Bemerkung. Ausgehend von Theorem 5.1 läßt sich ziemlich leicht zeigen, daß jeder vollständige eindimensionale (natürlich separierte) semialgebraische Raum affin ist. Nachträglich läßt sich also in Theorem 5.1 das Wort "affin" streichen.

Theorem 5.2. Sei X eine glatte Kurve über R und M eine Zusammenhangskomponente des semialgebraischen Raumes X(R). Angenommen, M ist vollständig. Dann ist M zu der reellen projektiven Geraden $\mathbb{P}(R)$ isomorph.

Beispiel. Ist X überdies vollständig, also projektiv, so ist jede Zusammenhangskomponente von X(R) vollständig, vgl. [DK_2, §9].

Beweis des Theorems. Aufgrund des Satzes über implizite Funktionen ist M eine eindimensionale semialgebraische Mannigfaltigkeit, d.h. jeder Punkt von M hat eine zu $]0,1[$ isomorphe offene semialgebraische Umgebung (vgl. [DK_2], Beweis von Prop. 8.6). Überdies ist M affin, da X quasiprojektiv ist. Wir wählen nun eine Triangulierung $(S_i | i \in I)$ von M, wie in Theorem 5.1 angegeben. Sei x_o eine Ecke der Triangulierung, und sei $\{S_i | i \in J\}$ die Menge der 1-Simplizes, die x_o enthalten und somit x_o als Randpunkt haben. J enthalte r Elemente. Dann hat für jede Umgebung U von x_o, welche die Simplizes S_i mit $i \notin J$ vermeidet, die Menge $U \setminus \{x_o\}$ mindestens r Zusammenhangskomponenten, denn $U \setminus \{x_o\}$ ist disjunkte Vereinigung der relativ abgeschlossenen Teilmengen $(U \cap S_i) \setminus \{x_o\}$ mit $i \in J$. Weiter gibt es ein Fundamentalsystem von Umgebungen U, bei denen $U \setminus \{x_o\}$ aus genau r Zusammenhangskomponenten besteht, weil x_o in jedem Raum S_i ein Fundamentalsystem von Umgebungen U_i besitzt mit $U_i \setminus \{x_o\}$ zusammenhängend. Da M eine eindimensionale semialgebraische Mannigfaltigkeit ist, muß $r = 2$ sein. In jeder Ecke der Triangulierung treffen also genau zwei Simplizes zusammen.

Wir wählen jetzt ein 1-Simplex aus und nennen es S_0. Die Randpunkte von S_0 bezeichnen wir in beliebig gewählter Reihenfolge mit P_0, P_1. Es gibt dann genau ein weiteres Simplex, genannt S_1, mit Randpunkt P_1. Entweder der andere Randpunkt von S_1 ist P_0 oder er ist ein von P_0, P_1 verschiedener Punkt P_2. Dann gibt es genau ein weiteres Simplex, genannt S_2, mit Randpunkt P_2. Falls der andere Randpunkt von S_2 nicht P_0 ist, muß er ein von P_0, P_1, P_2 verschiedener Punkt P_3 sein, und wir setzen das Verfahren fort. Schließlich erhalten wir eine Folge paarweise verschiedener 1-Simplizes $S_0, S_1, \ldots, S_r$ mit Randpunkten $P_0, P_1; P_1, P_2; \ldots; P_r, P_0$. Kein anderes 1-Simplex der Triangulierung kann die Menge $S_0 \cup S_1 \cup \ldots \cup S_r$ treffen, weil jede Ecke in dieser Menge schon Randpunkt zweier in der Menge enthaltener 1-Simplizes ist. Da M nicht disjunkte Vereinigung zweier nicht leerer semialgebraischer abgeschlossener Teilmengen sein kann, ist also

$$M = S_0 \cup S_1 \cup \ldots \cup S_r .$$

Wir wählen jetzt irgendwie r Punkte $a_1 < a_2 < \ldots < a_r$ auf der affinen Geraden R. Dann ist $\mathbb{P}(R)$ die Vereinigung der abgeschlossenen Intervalle $[a_i, a_{i+1}]$, $0 \le i \le r$, mit $a_0 = a_{r+1} = \infty$. Für jedes dieser Intervalle wählen wir einen semialgebraischen Isomorphismus $\varphi_i : [a_i, a_{i+1}] \xrightarrow{\sim} S_i$, der a_i auf P_i und a_{i+1} auf P_{i+1} abbildet ($P_{r+1} := P_0$). Diese Isomorphismen fügen sich zusammen zu einer semialgebraischen Abbildung $\varphi : \mathbb{P}(R) \to M$. Aufgrund unserer Kenntnis aller Durchschnitte $S_i \cap S_j$ sehen wir, daß φ bijektiv ist. Weil $\mathbb{P}(R)$ vollständig ist, ist φ ein Isomorphismus. Damit ist Theorem 5.2 bewiesen. Der Beweis wurde nur deshalb in den Einzelheiten ausgeführt, damit keine Bedenken entstehen, daß in unsere Theorie unzulässige intuitive Vorstellungen aus dem klassischen Fall $R = \mathbb{R}$ einfließen.

Wir wollen jetzt auf einer vorgegebenen vollständigen Zusammenhangskomponente M von X(R) zu unserer glatten Kurve X "Intervalle" einführen, die den in §4 im Spezialfall der projektiven Geraden $X = \mathbb{P}_R$ definierten Intervallen entsprechen. Dazu müssen wir M zunächst "orientieren".

Wir führen auf der Menge $\mathrm{Iso}(\mathbb{P}(R), M)$ aller Isomorphismen von

$\mathbb{P}(R)$ auf M die folgende Äquivalenzrelation ein:

$$\varphi \sim \psi :\Longleftrightarrow \psi^{-1} \circ \varphi \in \mathrm{Aut}^+(\mathbb{P}(R)).$$

Weil $\mathrm{Aut}^+(\mathbb{P}(R))$ Untergruppe vom Index 2 in $\mathrm{Aut}(\mathbb{P}(R))$ ist, zerfällt $\mathrm{Iso}(\mathbb{P}(R), M)$ in zwei Äquivalenzklassen.

Definition 1. Eine Orientierung von M ist die Auszeichnung einer dieser beiden Äquivalenzklassen. Sie wird dann mit $\mathrm{Iso}^+(\mathbb{P}(R), M)$ bezeichnet und die andere Äquivalenzklasse mit $\mathrm{Iso}^-(\mathbb{P}(R), M)$.

In dem Spezialfall $X = \mathbb{P}_R$ orientieren wir $X(R)$ immer durch

$$\mathrm{Iso}^+(\mathbb{P}(R), \mathbb{P}(R)) := \mathrm{Aut}^+(\mathbb{P}(R)).$$

(Standardorientierung). Sind allgemein orientierte vollständige Zusammenhangskomponenten M,N von X(R), Y(R) zu glatten Kurven X,Y über R vorgegeben, so heiße ein Isomorphismus $f : M \xrightarrow{\sim} N$ orientierungserhaltend, (oder orientierungstreu), wenn für ein - und damit jedes - Element $\varphi \in \mathrm{Iso}^+(\mathbb{P}(R)), M)$ die Komposition $f \circ \varphi$ Element von $\mathrm{Iso}^+(\mathbb{P}(R), N)$ ist. Anderenfalls nennen wir f orientierungsumkehrend.

Definition 2. Sei M eine orientierte vollständige Zusammenhangskomponente von X(R) zu einer glatten Kurve X über R. Seien P,Q verschiedene Punkte auf M. Wir wählen einen orientierungstreuen Isomorphismus φ von $\mathbb{P}(R)$ auf M und definieren das offene Intervall $]P,Q[$ wie folgt:

$$]P,Q[:= \varphi(]\varphi^{-1}(P), \varphi^{-1}(Q)[).$$

Weiter definieren wir die halboffenen Intervalle $[P,Q[$, $]P,Q]$ und das abgeschlossene Intervall $[P,Q]$ in völlig analoger Weise.

Aufgrund von §4 ist evident, daß die so definierten Intervalle nicht von der Wahl von φ abhängen, und daß gilt:

$$[P,Q[=]P,Q[\cup \{P\},]P,Q] =]P,Q[\cup \{Q\},$$

$$[P,Q] =]P,Q[\cup \{P\} \cup \{Q\}.$$

Weiter überträgt sich alles, was in §4 über Intervalle und zusammenhängende semialgebraische Teilmengen von $\mathbb{P}(R)$ gesagt wurde, auf M. Zum Beispiel haben wir vermöge der Orientierung jede Menge $M \setminus \{p\}$ $(p \in M)$ total geordnet.

Insbesondere erhalten wir eine Beschreibung der Zusammenhangskomponenten von Z(R) für eine beliebige glatte Kurve Z über R. Die Varietät Z besitzt eine kanonische Einbettung $Z \subset X$ in eine vollständige glatte Kurve X über R, und $S := X(R) \setminus Z(R)$ ist eine endliche Menge. Seien $M_1, \ldots, M_r$ die Zusammenhangskomponenten von X(R). Wir orientieren jedes M_i. Jede Zusammenhangskomponente von Z(R) ist in einer Menge M_i enthalten. Aus Satz 4.3.iv folgt

Theorem 5.3. Sei $S_1 := S \cap M_1$.

a) Ist S_1 leer oder einpunktig, so ist $M_1 \setminus S_1$ eine Zusammenhangskomponente von Z(R).

b) Besteht S_1 aus r+1 Punkte $P_o, P_1, \ldots, P_r$ $(r \geq 1)$, und ist in der totalen Anordnung von $M_1 \setminus \{P_o\}$

$$P_1 < P_2 < \ldots < P_r,$$

so sind die in M_1 enthaltenen Zusammenhangskomponenten von Z(R) die r+1 Intervalle

$$]P_o, P_1[,\ ]P_1, P_2[,\ \ldots,\ ]P_r, P_o[.$$

§ 6 Der Einheitskreis.

Wir betrachten jetzt den Raum S(R) der reellen Punkte der affinen Varietät $S = S^1_R$ im zweidimensionalen affinen Standardraum $\mathbb{A}^2$, die durch die Gleichung $x^2 + y^2 = 1$ definiert ist. Es ist also

$$S(R) = \{(x,y) \in R^2 \mid x^2 + y^2 = 1\}.$$

Die stereographische Projektion $p : S(R) \to \mathbb{P}(R)$ mit Zentrum (O,1), definiert durch

$$(6.1) \qquad p(x,y) = \begin{cases} (x : 1+y) & y \neq -1 \\ (1-y : x) & y \neq 1 \end{cases}$$

bildet den semialgebraischen Raum S(R) isomorph auf die reelle projektive Gerade $\mathbb{P}(R)$ ab. Insbesondere ist S(R) zusammenhängend und vollständig, und die in §5 gewonnenen Erkenntnisse lassen sich auf S(R) anwenden. Wir orientieren S(R) durch die Festsetzung, daß die stereographische Projektion p orientierungserhaltend ist. Damit sind für je zwei verschiedene Punkte P,Q auf S(R) die Intervalle]P,Q[, [P,Q[, etc. definiert, und für jeden Punkt $P \in S(R)$ ist die Menge $S(R) \setminus \{P\}$ total geordnet.

Im folgenden wählen wir in dem algebraischen Abschluß C von R eine Quadratwurzel $i = \sqrt{-1}$ fest aus und identifizieren den Standardraum R^2 mit C vermöge

$$(x,y) = x + i\,y .$$

Wir bezeichnen die Punkte von S(R) meist durch die zugehörigen Elemente von C. Der Einheitskreis S(R) ist eine Untergruppe der multiplikativen Gruppe $C^* = C \setminus \{O\}$ von C. Die Abbildung aufs Inverse

$$z = x + i\,y \mapsto \bar{z} := x - i\,y$$

ist ein Automorphismus des semialgebraischen Raumes S(R). Ebenso ist die Multiplikation $(z_1,z_2) \mapsto z_1 z_2$ eine semialgebraische Abbildung von $S(R) \times S(R)$ auf S(R). Somit ist S(R) eine "semialgebraische Gruppe' über R, in Wahrheit sogar eine "reell-algebraische Gruppe". Insbesondere ist für jeden Punkt a von S(R) die Abbildung

$$L(a) : z \mapsto a\,z$$

von S(R) nach S(R) ein semialgebraischer Automorphismus von S(R).

Es treten nun durchaus nichttriviale Fragen über den Zusammenhang zwischen der zirkularen Anordnung, d.h. der "Intervallstruktur" von S(R), und der Gruppenstruktur von S(R) auf. Zum Beispiel: Sind die Automorphismen L(a) alle orientierungstreu?

Um diese und andere Fragen zu lösen, müssen wir uns die zirkulare Anordnung genauer ansehen. Wenn es die Deutlichkeit erfordert, bezeichnen wir Intervalle auf S(R), $\mathbb{P}(R)$, R mit einem tiefgestellten Buchstaben S, bzw. $\mathbb{P}$, bzw. R. Wir führen weiter folgende Teilmengen von S(R) ein:

$$S_+ := \{(x,y) \in S(R) \mid y \geq O\},$$
$$S_- := \{(x,y) \in S(R) \mid y \leq O\},$$
$$S_l := \{(x,y) \in S(R) \mid x \leq O\},$$
$$S_r := \{(x,y) \in S(R) \mid x \geq O\}.$$

(l = "links", r = "rechts"), und bezeichnen der Reihe nach $S_+ \cap S_r$, $S_+ \cap S_l$, $S_- \cap S_l$, $S_- \cap S_r$ als ersten, zweiten, dritten und vierten Quadranten des Einheitskreises S(R). Schließlich führen wir in Anlehnung an den nicht vorhandenen Cosinus und Sinus auf S(R) zwei semialgebraische Funktionen ein:

$$c : S(R) \to [-1,1]_R, \ (x,y) \mapsto x.$$

$$s : S(R) \to [-1,1]_R, \ (x,y) \mapsto y.$$

Die Einschränkung von s auf S_l hat die Umkehrabbildung

$$y \mapsto (-\sqrt{1-y^2}, \ y)$$

und ist somit ein semialgebraischer Isomorphismus von S_l auf $[-1,1]_R$. Ebenso ist die Einschränkung von s auf S_r ein semialgebraischer Isomorphismus von S_r auf $[-1,1]_R$, und die Einschränkungen von c auf S_+ und S_- sind auch semialgebraische Isomorphismen von S_+ bzw. S_- auf $[-1,1]_R$. Insbesondere sind die vier semialgebraischen Teilmengen S_l, S_r, S_+, S_- von S(R) vollständig und zusammenhängend, also abgeschlossene Intervalle von S(R).

Wir wollen für jedes dieser Intervalle Anfangspunkt und Endpunkt bestimmen.

Man liest aus den Formeln (6.1) für die stereographische Projektion ab:

$$p(1) = 1, \ p(i) = \infty, \ p(-1) = -1, \ p(-i) = 0;$$

$$p(S_l) \subset [\infty,0]_{\mathbb{P}}, \ p(S_-) \subset [-1,1]_{\mathbb{P}}.$$

Weil p orientierungserhaltend ist, gilt:

$$p([i,-i]_S) = [\infty,0]_{\mathbb{P}}, \ p([-1,1]_S) = [-1,1]_{\mathbb{P}}.$$

Daher ist

$$S_l \subset [i,-i]_S, \ S_- \subset [-1,1]_S.$$

Andererseits bildet s - wie jeder semialgebraische Isomorphismus - die Randpunkte von S_l auf die Randpunkte -1,1 von $[-1,1]_R$ ab. Somit hat S_l die Randpunkte i,-i. Ebenso sieht man mit Hilfe der Funktion c, daß S_- die Randpunkte 1,-1 hat. Es ist also

(6.2a) $S_l = [i,-i]_S, \ S_- = [-1,1]_S,$

und daher auch

(6.2b) $S_r = [-i,i]_S$, $S_+ = [1,-1]_S$.

Der Isomorphismus $s : S_l \xrightarrow{\sim} [-1,1]_R$ muß - letztlich aufgrund des Zwischenwertsatzes 3.1 - als Abbildung zwischen den total geordneten Mengen S_l und $[-1,1]_R$ streng isoton oder streng antiton sein. Weil $s(i) = 1$, $s(-i) = -1$ ist, muß s auf S_l also streng antiton sein. Ebenso sieht man, daß s auf S_r streng isoton ist, daß c auf S_- streng isoton ist, und daß c auf S_+ streng antiton ist. Damit haben wir ein gutes Verständnis der totalen Anordnungen auf den vier Mengen S_l, S_r, S_-, S_+ erzielt. Insbesondere sieht man sofort, daß die 4 Quadranten der Reihe nach die folgenden Intervalle sind:

(6.3) $S_+ \cap S_r = [1,i]_S$, $S_+ \cap S_l = [i,-1]_S$,

$S_- \cap S_l = [-1,-i]_S$, $S_- \cap S_r = [-i,1]_S$.

Die totale Anordnung auf jeder dieser 4 Mengen wird durch die folgende Tabelle beschrieben.

(6.4)

Quadrant	P < Q wenn	P < Q wenn
$S_+ \cap S_r$	$c(P) > c(Q)$	$s(P) < s(Q)$
$S_+ \cap S_l$	$c(P) > c(Q)$	$s(P) > s(Q)$
$S_- \cap S_l$	$c(P) < c(Q)$	$s(P) > s(Q)$
$S_- \cap S_r$	$c(P) < c(Q)$	$s(P) < s(Q)$

Jetzt können wir die oben aufgeworfene Frage über die Automorphismen $L(a) : z \mapsto az$ von $S(R)$ beantworten.

Satz 6.5. Für jedes $a \in S(R)$ ist $L(a)$ orientierungstreu.

Beweis. Für zwei Punkte $z = x + iy$, $w = u + iv$ aus dem 1. Quadranten hat zw den Imaginärteil $xv + yu \geq 0$, somit liegt zw in S_+. Es ist also

$$[1,i]\cdot[1,i] \subset [1,-1]_S.$$

Jetzt läßt sich die Orientierungstreue von $L(a)$ für einen Punkt $a = \alpha + i\beta$ aus dem ersten Quadranten ($\alpha \geq 0$, $\beta \geq 0$) wie folgt einsehen. A priori ist $a[1,i] = [a,ia]$ oder $= [ia,a]$ und $a[1,i] \subset [1,-1]_S$. Der Punkt $ia = -\beta + i\alpha$ liegt aber im zweiten Quadranten und somit gilt in der Totalordnung von $[1,-1]_S$ die Relation $a < ia$. Das Intervall $[ia,a]$ ist also nicht in $[1,-1]_S$ enthalten, und es muß $a\cdot[1,i] = [a,ia]$ sein. Weil $L(a)$ Anfangspunkt auf Anfangspunkt und Endpunkt auf Endpunkt abbildet, ist $L(a)$ orientierungstreu. (N.B. Man braucht zum Nachweis der Orientierungstreue nur das Bild eines Intervalles zu betrachten!)

Insbesondere $(a = i)$ ist $i\cdot[1,i] = [i,-1]$. Jeder Punkt a aus dem zweiten Quadranten hat also die Gestalt $a = ib$ mit b aus dem ersten Quadranten (wie man auch leicht ab ovo sieht), und somit ist auch $L(a) = L(i)L(b)$ für jedes a aus dem zweiten Quadranten orientierungstreu. Schließlich liegt für jedes $a \in S_-$ das Inverse $\overline{a}$ in S_+ und somit ist $L(a) = L(\overline{a})^{-1}$ auch für diese Punkte a orientierungstreu.

q.e.d.

Andererseits bildet der Automorphismus $z \mapsto \overline{z}$ das Intervall $[1,-1]_S = S_+$ auf das Intervall $[-1,1]_S = S_-$, 1 auf 1 und -1 auf -1 ab.

Zusatz 6.6. Der Automorphismus $z \mapsto \overline{z}$ von $S(R)$ ist orientierungsumkehrend.

Wir wollen jetzt für jede natürliche Zahl $n > 1$ die Untergruppe

$$\mu_n(C) = \{z \in C \mid z^n = 1\}$$

der n-ten Einheitswurzeln von $S(R)$ studieren. Bekanntlich besteht $\mu_n(C)$ aus n Elementen. Wir numerieren die von 1 verschiedenen n-ten Einheitswurzeln gemäß der totalen Anordnung von $S(R) \setminus \{1\}$,

$$\varepsilon_1 < \varepsilon_2 < \ldots < \varepsilon_{n-1},$$

und setzen $\varepsilon_o := 1$. Weiter setzen wir für jede ganze Zahl r fest: $\varepsilon_r := \varepsilon_j$, mit $0 \le j < n$, $j \equiv r \bmod n$.

Satz 6.7. Für jedes $r \in \mathbb{Z}$ ist $\varepsilon_r = \varepsilon_1^r$.

Beweis. Nach Theorem 5.3 zerfällt $S(R) \setminus \mu_n(C)$ in die Zusammenhangskomponenten

$$]\varepsilon_o,\varepsilon_1[, \ \ldots, \ ]\varepsilon_{n-1},\varepsilon_o[.$$

Somit bildet $L(\varepsilon_1)$ jedes Intervall $]\varepsilon_j,\varepsilon_{j+1}[$ auf ein Intervall $]\varepsilon_k,\varepsilon_{k+1}[$ ab, also auch $[\varepsilon_j,\varepsilon_{j+1}]$ auf $[\varepsilon_k,\varepsilon_{k+1}]$. Nach Satz 6.5 ist diese Abbildung streng isoton. Wir sehen also: Ist $\varepsilon_1\varepsilon_j = \varepsilon_k$, so ist $\varepsilon_1\varepsilon_{j+1} = \varepsilon_{k+1}$. Weil $\varepsilon_1\varepsilon_0 = \varepsilon_1$ ist, folgt nun der Reihe nach:

$$\varepsilon_1\varepsilon_1 = \varepsilon_2,\ \varepsilon_1\varepsilon_2 = \varepsilon_3,\ \ldots,\ \varepsilon_1\varepsilon_{n-1} = \varepsilon_0,$$

und damit der Satz.

q.e.d.

Bemerkung. Wir haben im Falle der Charakteristik Null zugleich einen neuen Beweis der wohlbekannten Tatsache gefunden, daß die Einheitswurzelgruppe $\mu_n(C)$ eines jeden (algebraisch abgeschlossenen) Körpers C zyklisch ist.

Wir bezeichnen ab jetzt die obige, durch die Wahl des Körpers R in C und die Wahl von $\sqrt{-1}$ ausgezeichnete n-te Einheitswurzel ε_1 mit ζ_n.

Zusatz 6.8. ζ_n ist dasjenige Element von $\mu_n(C) \setminus \{1\}$, welches in S_+ liegt und von allen in S_+ gelegenen Einheitswurzeln $\neq 1$ den kleinsten euklidischen Abstand zu 1 hat.

Beweis. Im Falle n = 2 ist nichts zu zeigen. Sei jetzt $n \geq 3$. Dann ist $\zeta_n = \varepsilon_1$ von $\zeta_n^{-1} = \varepsilon_{n-1}$ verschieden. Wäre $\varepsilon_1 \in S_-$, so wäre $\varepsilon_{n-1} \in S_+$, also würde das Intervall $]\varepsilon_{n-1},\varepsilon_1[$ den Punkt 1 nicht enthalten, wie aus unserer obigen expliziten Beschreibung der zirkularen Anordnung auf S(R) sofort folgt. Es ist aber $1 = \varepsilon_0 \in\,]\varepsilon_{n-1},\varepsilon_1[$. Daher ist $\zeta_n \in S_+$.

Ein Punkt $z = x + i\,y \in S_+$ hat von 1 das euklidische Abstandsquadrat $(1-x)^2 + y^2 = 2(1-x)$. Die Funktion 2(1-c) ist auf $S_+ = [1,-1]_S$ streng isoton, weil c streng antiton ist. Damit ist der Zusatz 6.8 evident.

Satz 6.9. Für beliebige natürliche Zahlen $r > 1$, $n > 1$ gilt

$$\zeta_{n\,r}^r = \zeta_n.$$

Beweis. In der total geordneten Menge $S(R) \setminus \{1\}$ gilt:

$$\zeta_{nr} < \zeta_{nr}^2 < \ldots < \zeta_{nr}^{nr-1},$$

insbesondere also

$$\zeta_{nr}^r < \zeta_{nr}^{2r} < \ldots < \zeta_{nr}^{(n-1)r}.$$

Das sind n-1 von 1 verschiedene n-te Einheitswurzeln, also sämtliche von 1 verschiedenen n-ten Einheitswurzeln. Weil ζ_{nr}^r die kleinste unter ihnen ist, muß sie mit ζ_n übereinstimmen.

q.e.d.

Jetzt können wir zeigen, daß für jedes natürliche n der Gruppenhomomorphismus $z \mapsto z^n$ von S(R) in sich "streng monoton wachsend im Sinne zirkularer Anordnung" ist. Genauer gilt

Satz 6.10. Für jeden Punkt $a \in S(R)$ nimmt die Funktion $z \mapsto z^n$ auf $]a, a\zeta_n[$ nicht den Wert a^n an. Die somit definierte Abbildung

$$]a, a\zeta_n[\to S(R) \setminus \{a^n\}, \quad z \mapsto z^n,$$

ist ein isotoner semialgebraischer Isomorphismus.

Beweis. Indem wir anstelle von $\varphi_n :]a, a\zeta_n[\to S(R), z \mapsto z^n$ die Abbildung

$$L(a^{-n}) \circ \varphi_n \circ L(a) :]1, \zeta_n[\to S(R), \quad z \mapsto z^n$$

betrachten, ziehen wir uns auf den Fall $a = 1$ zurück. Weil das Intervall $]1, \zeta_n[$ keine n-ten Einheitswurzeln enthält, haben wir also eine wohldefinierte semialgebraische Abbildung

$$\psi_n :]1, \zeta_n[\to S(R) \setminus \{1\}, \quad z \mapsto z^n.$$

Weil der Körper C algebraisch abgeschlossen ist, ist die Abbildung $z \mapsto z^n$ von S(R) nach S(R) surjektiv. Also ist auch ψ_n surjektiv. Da jede Translation $L(\zeta_n^k)$ zu einer Einheitswurzel $\zeta_n^k \neq 1$ das Intervall $]1, \zeta_n[$ in ein zu $]1, \zeta_n[$ fremdes Intervall $]\zeta_n^k, \zeta_n^{k+1}[$ überführt, ist ψ_n auch injektiv. Die Fortsetzung

$$[1, \zeta_n] \twoheadrightarrow S(R), \quad z \mapsto z^n$$

von ψ_n ist eigentlich, da $[1, \zeta_n]$ vollständig ist. Also ist auch ψ_n eigentlich, und somit ein semialgebraischer Isomorphismus. ψ_n muß streng isoton oder streng antiton sein.

Wir betrachten die Punkte ζ_{4n}, ζ_{4n}^2. Nach dem vorigen Satz 6.9 ist $\zeta_n = \zeta_{4n}^4$. Somit liegen ζ_{4n}, ζ_{4n}^2 in $]1,\zeta_n[$ und es ist $\zeta_{4n} < \zeta_{4n}^2$. Unter ψ_n haben diese Punkte die Bilder $\zeta_4 = i$ und $\zeta_4^2 = -1$ und in $S(R) \setminus \{1\}$ ist $i < -1$. Also ist ψ_n streng isoton.

q.e.d.

Literatur

[C] P.J. Cohen, Decision procedures for real and p-adic fields, Comm. Pure Appl. Math. 22, 131-151 (1969).

[Cs] M. Coste, Ensembles semi-algebriques, this volume.

[DK_1] H. Delfs, M. Knebusch, Semialgebraic topology over a real closed field I: Paths and components in the set of rational points of an algebraic variety, Math. Z. 177, 107-129 (1981).

[DK_2] H. Delfs, M. Knebusch, Semialgebraic topology over a real closed field II: Basic theory of semialgebraic spaces, Math. Z. 178, 175-213 (1981).

[DK_3] H. Delfs, M. Knebusch, On the homology of algebraic varieties over real closed fields, erscheint demnächst, preprint Univ. Regensburg 1981.

[G] W.D. Geyer, Dualität bei abelschen Varietäten über reell abgeschlossenen Körpern, J. reine angew. Math. 293/294, 62-66 (1977).

[K] M. Knebusch, On algebraic curves over real closed fields I, II. Math.Z. 150, 49-70; 151, 189-205 (1976).

Adresse: Hans Delfs, Manfred Knebusch
Fakultät für Mathematik
Universität
D - 8400 Regensburg

SUBORDINATE STRUCTURE SHEAVES

D. W. Dubois and Tomás Recio

For an arbitrary field $F|k$ over fixed orderd ground field k, we construct a ringed space $(S(F|k),\mathcal{F})$ where $S(F|k)$ is the suitably topologized set of all <u>suborders</u> of $F|k$ (a suborder is an order of a subfield) and $\mathcal{F}$ is a sheaf (called the "subordinate structure sheaf") of Boolean rings which reflects the dichotomy of order in a manner analogous to the generalized signature of a quadratic form. The construction takes on heightened relevance with the appearance of the brilliant work* of Coste and Coste-Roy on the "real spectrum." For there is an obvious way of associating to each point (p,α) of the spectrum of a real coordinate ring $k[x_1,\ldots,x_n]$ a set of suborders in $F|k$, if F is taken to be any large enough extension, such as a universal domain as defined below (paragraph 2). The relations of the subordinate structure sheaf to the real spectrum will be investigated further in subsequent papers.

The ringed space $(S(F|k),\mathcal{F})$ is shown to define a covariant functor with the expected features (including change of ground field) from the category of all grounded fields (with ordered ground field) and their morphisms to the category of all ringed spaces of Boolean rings.

---

* M. Coste and M.-F. Roy, La topologie du spectre réel, in Ordered Fields and Real Algebraic Geometry, Contemporary Mathematics. American Math. Soc., to appear.

§ 1. The Structure Sheaves

A. The space $S(F|k)$.

Let k be an ordered subfield of the field F, and let k^+ be the set of all non-negative elements of k; i.e., k^+ is a (fixed) order of k. Let S be the set of all suborders of $F|k$; a suborder $F|k$ is a set P, with $k^+ \subset P \subset F$ such that P is an order of $k(P)$. Whenever we speak below of a suborder of F or of just a suborder, we mean a suborder of $F|k$. For any suborder P_1 let

$$J(P_1) = \{P\ ;\ P \text{ is a suborder containing } P_1\}$$

The set of all unions of sets of the form $J(P_1)$, where P_1 ranges over the finitely generated suborders of $F|k$ ("P_1 is finitely generated" means that $k(P_1)$ is a finitely generated extension of k) is an open set topology for S; see 1 below. The idea comes from Zariski's Riemann surface of a field; see Zariski-Samuel, Commutative Algebra, pp. 110-115. The sets $J(P_1)$, for finitely generated P_1, are called distinuished open sets; also $J(P_1)$ is a distinguished neighborhood of any suborder contained in it.

1. Let P_1 and P_2 be finitely generated suborders of $F|k$. Since 1 belongs to every suborder, $k(P_1 \cup P_2) = k(P_1P_2)$, where P_1P_2 is the set of all sums of products x_1x_2 with $x_i \in P_i$. The intersection of $J(P_1)$ and $J(P_2)$ is the union of all $J(P_3)$ such that P_3 is an order of $k(P_1P_2)$ containing $P_1 \cup P_2$, and it is empty if no such P_3 exists. For if P belongs to $J(P_1) \cap J(P_2)$ then P contains $P_1 \cup P_2$ and intersects $k(P_1P_2)$ in an order P_3 of $k(P_1P_2)$; P_3 is finitely generated, it contains $P_1 \cup P_2$ and P belongs to $J(P_3)$. Conversely if P_3 is an order of $k(P_1P_2)$ containing $P_1 \cup P_2$ then every suborder P containing P_3 also contains $P_1 \cup P_2$. Hence $J(P_3)$ is contained in $J(P_1) \cap J(P_2)$.

The distinguished open sets form a base for the topology of S .

2. The closure of any singleton $\{P\}$, for P in S , is the set of all $\underset{\sim}{Q}$ contained in P . For if $\underset{\sim}{Q} \subset P$ then for every distinguished neighborhood $J(P_1)$ of $\underset{\sim}{Q}$, we have $P_1 \subset \underset{\sim}{Q} \subset P$, so P belongs to $J(P_1)$. Hence $\underset{\sim}{Q}$ belongs to $\overline{\{P\}}$. If $\underset{\sim}{Q}$ is not contained in P , let u belong to $\underset{\sim}{Q}$ but not to P . Then $\underset{\sim}{Q}_1 = k(u) \cap \underset{\sim}{Q}$ which is a finitely generated suborder, is an order of $k(u) \cap \underset{\sim}{Q}$ and it contains u . Hence $\underset{\sim}{Q}$ belongs to $J(\underset{\sim}{Q}_1)$ while P does not (since $u \notin P$). Thus $\underset{\sim}{Q}$ is not in the closure of $\{P\}$.

3. The set of all maximal suborders (maximal by set inclusion) is dense in S . A maximal suborder is topologically characterized as a point not in the closure of any other point.

4. Every open set is a dual ideal in the partially ordered set of all suborders of $F|k$: If U is open, if $P \in U$, and if $P \subset \underset{\sim}{Q}$ then $\underset{\sim}{Q}$ belongs to U .

5. S is compact. For the only open set that contains k^+ is S .

6. If P is a finitely generated maximal suborder then $\{P\}$ is open.

7. If F is finitely generated over k then the subspace of all maximal suborders is discrete. In general the subspace of all maximal suborders is a T_1-space.

8. DEFINITIONS. A <u>grounded field</u> is a pair (F,k) where k is an ordered subfield, with order k^+ , of F . A <u>morphism</u> of grounded fields $(F,k) \to (F',k')$ is a field embedding $\phi : F \to F'$ such that the restriction of ϕ to k is an order-isomorphism of k onto k' . The space of all

suborders of $F|k$ defined above is dentoed $S(F|k)$, or $S(F)$.

9. Let (F,k) and (F',k') be grounded fields, let ϕ be a morphism $(F,k) \to (F',k')$, and let $\phi_1 : S' \to S$ (where S and S' are the spaces of suborders) be defined, as for the prime spectrums of rings, by

$$\phi_1(P') = \phi^{-1}(P') \qquad (P' \in S') .$$

Then ϕ_1 is a continuous map of S' onto S. For if P' is a suborder of $F'|k'$ then $P' \cap \phi(F)$ is a suborder of $\phi(F)$ containing k' and its inverse image is an order of $\phi^{-1}(k'(P') \cap \phi(F))$. Hence ϕ_1 maps into S. Let $P = \phi_1(P')$ and let P_1 be a finitely generated suborder of $F|k$ contained in P. Then $\phi(P_1)$ is a finitely generated suborder of $F'|k'$ contained in P'. Every suborder P'' in $J(\phi(P_1))$ maps by ϕ_1 into $J(P_1)$. Hence ϕ_1 is continuous. Surjectivity is obvious.

10. NOTE. ϕ_1 need not map maximal suborders of S' to maximal suborders of S. Example. Let $k = k' = \mathbb{Q}^+$, $F = \mathbb{Q}(\omega \sqrt[3]{2})$, $F' = C$, where C is the field of all complex numbers, $\mathbb{Q}$ is the field of all rationals. The set R^+ of all nonnegative reals is a maximal suborder of C, but its intersection in F is $\mathbb{Q}^+$, which is not a maximal suborder of F. Here the morphism is just the inclusion map and ω is a primitive cube root of unity. As always for a morphism ϕ which is an inclusion, with $k = k'$, $\phi_1(P') = P' \cap F$.

11. The class of grounded fields and their morphisms clearly forms a category. The correspondences $(F,k) \to S$, $\phi \to \phi_1$ define a contravariant functor to the category of topological spaces and continuous mappings.

B. The subordinate structure sheaf of a grounded field $F|k$.

1. For each $a \in F^*$ and each $P \in S$, $\hat{a}(P)$ is defined, provided

$a \in k(P)$, as the function

$$\hat{a}(P)(\underline{O}) = \begin{cases} 1 & \text{if } a \notin \underline{O} \pmod 2 \\ 0 & \text{if } a \in \underline{O} \pmod 2 \end{cases}$$

for $\underline{O}$ any order of $k(P)$ in S . Note that

$$\text{dom } \hat{a} = \{P;\ a \in k(P)\} .$$

Thus $\hat{a}$ is a function whose values are also functions.

2. <u>dom $\hat{a}$ is an open set</u>. For if $\hat{a}$ is defined at P (dom $\hat{a}$ may be empty, as when $i = \sqrt{-1} \in F$, $\hat{1}$ has empty domain) then $\hat{a}$ is defined at $P \cap k(a)$, which is a finitely generated suborder, and at every suborder containing $P \cap k(a)$. The set of all subroders containing $P \cap k(a)$ is an open set containing P and lying in dom $\hat{a}$. Hence dom $\hat{a}$ is open.

3. Let $P_1 \in S(F|k)$. Consider the set of all nonzero $a \in k(P_1)$. The corresponding functions $\hat{a}(P_1)$, as elements of the ring of all functions from the set of all orders of $k(P_1)$ to Z_2 , generate a subring denoted by R_{P_1} ; R_{P_1} is a Boolean ring. Let $\{U_1 = \underline{O} \supset \underline{O}_1\}$ be a distinguished neighborhood of P_1 . Consider only $a \in k(\underline{O}_1)^*$, and temporarily identify $\hat{a}$ with its restriction to U_1 , which is contained in dom $\hat{a}$. The set of all $\hat{a}$ generates a ring R_{U_1} . A typical element is

$$\hat{z} = \sum_{i=1}^{n} \Pi\{\hat{a};\ a \in S_i\} ,$$

where $S_1,\ldots,S_n$ are finite subsets of $k(\underline{O}_1)^*$. The zero is $\hat{1}$, the unity is $(-\hat{1})$, and R_{U_1} is a Boolean ring; for the value of $\hat{a}$ at $P_1 \in U_1$ is a member of R_{P_1} , so the $\hat{a}$ are functions on U with values in the direct product $\Pi\{R_P;\ P \in U_1\}$, and R_{U_1} is the subring generated by the $\hat{a}$ in the ring of all functions from U_1 into $\Pi\{R_P;\ P \in U_1\}$. Since the latter is a Boolean ring so is R_{U_1}.

4. Let P be any member of $S(F|k) = S$ and let $U = \{\underline{Q};\ \underline{Q} \supset \underline{Q}_1\}$, with $\underline{Q}_1$ a finitely generated suborder of S, be a neighborhood of P; i.e., $P \supset \underline{Q}_1$. Let $\hat{y}$ and $\hat{z}$ be members of R_U, and suppose $\hat{y}(P) = \hat{z}(P)$. Then there exists a neighborhood V with $P \in V \subset U$ such that $\hat{y}$ and $\hat{z}$ are equal all over V.

Proof. By considering $\hat{y} - \hat{z}$, we see that it is enough to prove the claim in case $\hat{z}$ is zero. Thus we are given that $\hat{y}(P) = 0$. This means that $\hat{y}(P)(\underline{Q}) = 0$ for all $\underline{Q}$ (in S) which are orders of $k(P)$. Let $\hat{y}$ be expressed as above in the form

$$\hat{y} = \sum_{i=1}^{n} \Pi\{\hat{a};\ a \in S_i\},$$

where the S_i are finite subsets of $k(\underline{Q}_1)^*$. The vanishing of $y(P)(\underline{Q})$ for all orders of $k(P)$ is equivalent to the assertion that for every order $\underline{Q}$ of $k(P)$, $\underline{Q}$ contains at least one of the finite collection of finite sets of the form (union of an even number of the $-S_i$) $\cup$ (one member of each of the remaining S_i). By the Lemma below there exists a finitely generated field K between $k(\cup S_i)$ and $k(P)$ such that every order of K contains at least one of the sets. Then $P_1 = P \cap K$ is an order of K, P_1 is a finitely generated suborder and if $P_1 \subset P_2$ then every order of $k(P_2)$ intersects K in an order of K and hence contains one of the sets. Hence $\hat{y}(P_2) = 0$ for all P_2 containing P_1. Now P belongs to the open set $W = \{P_2;\ P_2 \supset P_1\}$ and so $\hat{y}$ vanishes all over the neighborhood $V = U \cap W$ of P.

Lemma. Let K be an extension of k. Let $S_1,\dots,S_n$ be finite subsets of K^*. Let K' be a subfield of $K|k$. By $C(K')$ is meant the predicate: "every order of $K'|k$ contains at least one of the S_i". If $C(K)$ is valid then there exists a finitely generated subfield $K'|k$ such that $C(K')$ is valid.

Proof. If $K|k$ has no orders then there exist positive elements p_i of k and elements t_i of K such that $-1 = \sum p_i t_i^2$. Then the field K' obtained by adjoining all S_i and all the t_i to k, satisfies $C(K')$. Let E be a subfield of $K|k$ containing all the S_i. Now $C(E)$ is equivalent to the assertion that for all x_i, and all orders P of $E|k$, if $-x_i \in P \cap S_i$ for $1 \le i \le n-1$, then S_n is contained in P. Another formulation: for all x_i, if $x_i \in S_i$ for every $i \le n-1$, then S_n is contained in the intersection of all orders of $E|k$ containing $\{-x_1,\dots,-x_{n-1}\}$. Still another: for all x_i, if $x_i \in S_i$ for all $i \le n-1$ then every element of S_n is expressible in the form

$$(1) \qquad y_j = y_j^{(x)} = a_{(e)_j}(-x_1)^{e_1^{(j)}} \cdots (-x_{n-1})^{e_{n-1}^{(j)}}, \quad (x) = (x_1,\dots,x_{n-1})$$

where $(e)_j = (e_1^{(j)},\dots,e_{n-1}^{(j)})$, each e_i is zero or 1, and $a_{(e)_j}$ has the form $\sum p_m t_m^2$, with p_m positive elements of k, t_m in E. The set of elements expressible thus is just the smallest additive and multiplicative set containing the positive elements of k, the elements $-x_1,\dots,-x_{n-1}$, and all squares of E (for a fixed sequence $x_1,\dots,x_{n-1}$), which is the intersection of all orders of $F|k$ containing the x_i. Since $C(K)$ is assumed, every $y_j \in S_n$ admits a representation (1) for each set $x_1,\dots,x_{n-1}$ $(x_i \in S_i)$. These representations are still valid in the field K' generated over k by the union of all S_i and by all the t's appearing in the expressions for the

$a_{(e)_j}$. Then K' is finitely generated over k and C(K') holds.

5. Since, as is easily verified, $U = \{\underline{O} \supset \underline{O}_1\} \supset V = \{\underline{O} \supset \underline{O}_2\}$, for finitely generated $\underline{O}_1$ and $\underline{O}_2$, if and only if $\underline{O}_1 \subset \underline{O}_2$, there is a natural restriction map from R_U to R_V, when $U \supset V$. For if $a \in k(\underline{O}_1)$ then $a \in k(\underline{O}_2)$, and the restriction of $\hat{a}|U$ to V is just $\hat{a}|V$. Thus is induced a restriction map of all of R_U to R_V . Hence, for each P the set of all R_U , with U ranging over the distinguished neighborhoods of P , together with the restriction maps, forms a direct system of rings.

6. Let L_P be the direct limit of the direct system. Let $U = \{\underline{O} \supset \underline{O}_1\}$ be a distinguished neighborhood of P , and let $\hat{z}|U$ be a member of R_U . Then P belongs to the domain of $\hat{z}|U$ so $\hat{z}$ is defined at P . The correspondence

$$\hat{z} \longmapsto \hat{z}(P) ,$$

where we have written $\hat{z}$ in place of $\hat{z}|U$, defines a ring homomorphism $R_U \to R_P$. The family of all these is a direct family. There is induced a canocical homomorphism $L_P \to R_P$, which will now be shown to be an isomorphism. An arbitrary element z of R_P is a finite sum of finite prodicts of functions of the form $\hat{a}(P)$, with a in k(P) . Adjoin all these elements a to k to get a finitely generated extension, let $\underline{O}_z$ be the intersection of P in the extension, and let U be the distinguished neighborhood $\{\underline{O} \supset \underline{O}_z\}$ of P . The $\hat{a}$ are now defined over U and the same combination that produced z from the $\hat{a}(P)$ yields an element $\hat{z}$ (supressing P in every case) of R_U . Also $\hat{z}(P) = z$ so $\hat{z}$ corresponds to z in the map from R_U to R_P . It follows that L_P maps onto R_P . Suppose x belongs to the direct limit L_P and that x maps to zero in R_P . There exists a distinguished neighborhood U of P and an element $\hat{z}$ in R_U such that the image of $\hat{z}$ in L_P by the canonical map is x . By the commutativity relations, $\hat{z}(P) = 0$. By No. 4, there is a distinguished neighborhood V of P , contained in U , such that $\hat{z}|V$ is zero.

But $\hat{z}|V$ also maps to x, so $x = 0$. <u>Hence</u> L_P <u>and</u> R_P <u>are canonically isomorphic</u>.

7. The assignment $U \to R_U$, for distinguished open sets U in $S(F|k)$, together with the restriction maps, defines a presheaf. The uniqueness axiom for sheaves is clear. If $U = \cup\{U_i; i \in I\}$, $U = \{\underline{O} \supset \underline{O}'\}$, $U_i = \{\underline{O} \supset \underline{O}_i\}$, with $\underline{O}'$ and $\underline{O}_i$ finitely generated, then, by the first remark in No. 6, U is equal to one of the U_i. Hence follows the existence axiom (considering only distinguished open sets). By extending as usual to all open sets we get a sheaf of Boolean rings, called the <u>subordinate structure sheaf of</u> $F|k$. The stalk at P is R_P. For an arbitrary open set U, the associated ring, still denoted by R_U, can be regarded as a ring of functions on U.

8. <u>Direct image</u>. Let $(F,k) \overset{\phi}{\rightarrowtail} (G,k)$ be an inclusion morphism. There is induced a continuous map $\psi\colon S(F') \to S(F)$, $(\psi = \phi_1)$ which in turn induces the direct image $\psi_*(G)$ of the subordinate structure sheaf G of (G,k); $\psi_*(G)$ is a sheaf on $S(F)$. If U is an open set in $S(F)$ then $\psi^{-1}(U)$ is just the set of all suborders of G which contain some suborder in U. For if $\underline{O}'$ belongs to $\psi^{-1}(U)$ then $\underline{O}' \cap F = \psi(\underline{O}') \in U$ and $\underline{O}'$ contains $\underline{O}' \cap F$. Conversely, if $\underline{O}'$ is a suborder of G containing a suborder $\underline{O}$ in U then $\underline{O}'' = \underline{O}' \cap F$ is a suborder of F containing $\underline{O}$. By A4, $\underline{O}''$ belongs to U, while $\psi(\underline{O}') = \underline{O}''$ so $\underline{O}'$ belongs to $\psi^{-1}(U)$. In particular, $\psi^{-1}(U)$ contains U. For a distinguished open U, say $U = \{\underline{O} \in S(F); \underline{O} \supset \underline{O}_1\}$, $\psi^{-1}(U) = \{\underline{O} \in S(G); \underline{O} \supset \underline{O}_1\}$.

9. <u>Ringed space morphisms</u>. The pair $(S(F|k), F)$ is a ringed space, where G is the subordinate structure sheaf of $F|k$. Let $\phi\colon (F,k) \to (F,k)$ be an inclusion morphism of grounded fields, let F, G, respectively, be the subordinate strucutre sheaves, and let S, T be the base spaces, $S = S(F|k)$,

$T = S(G|k)$. Let $\psi = \phi_1$ be the associated continuous mapping of T onto S . Pictorially we have

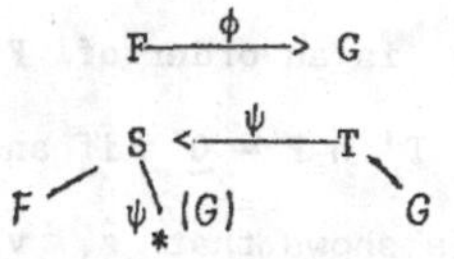

We define a morphism θ of presheaves from F to $\psi_*(G)$ as follows: let U be a distinguished open, $U = \{\underline{O} \in S;\ \underline{O} \supset \underline{O}_1\}$. An element $\hat{z}$ of $F(U)$ is a sum of products of elements $\hat{a}$, with a in $k(\underline{O}_1)$. Now $\psi^{-1}(U) = \{\underline{O} \in T;\ \underline{O} \supset \underline{O}_1\}$ by No. 9, and $\hat{z}$ can therefore be regarded as a function $\hat{z}_1$ on $\psi^{-1}(U)$ and thence as a member of $G(\psi^{-1}(U)) = \psi_*(G)(U)$. The correspondence $\hat{z} \longmapsto \hat{z}_1$ thus defines a ring homomorphism(1) from $F(U)$ to $\psi_*(G)(U)$. The commutativity requirements obviously hold. Hence θ , where $\theta_U\colon F(U) \to \psi_*(G)(U)$ is defined by $\theta_U(\hat{z}) = \hat{z}_1$, is a morphism of presheaves from F to $\psi_*(G)$, defined on the bases of distinguished neighborhoods. Induced is a morphism, still denoted by θ , of the sheaves, defined over all open sets. Thus the pair (ψ,θ) is a morphism of ringed spaces. In fact we have now got a contravariant functor from the category of grounded fields and <u>inclusion</u> morphisms to the category of ringed spaces and their morphisms.

For arbitrary grounded field morphisms the only additional difficulties are notational. For example the correspondence $\hat{z} \longmapsto \hat{z}_1$ can be defined as follows. Let U be a distinguished open in S , let $\hat{z}$ belong to $F(U)$. Then $z = \sum \Pi\, \hat{a}_{ij}$, where a_{ij} belong to $k(\underline{O}_1)$, $U = \{\underline{O} \in S;\ \underline{O} \supset \underline{O}_1\}$. Set $z_1 = \sum \Pi\, \hat{b}_{ij}$ with $b_{ij} = \phi(a_{ij})$.

10. Suppose $\hat{z} = \sum \Pi\, \hat{a}_{ij} = 0$ on the distinguished open set $U = \{\underline{O} \in S;\ \underline{O} \supset \underline{O}_1\}$, where the a_{ij} belong to $k(\underline{O}_1)$. This means that for every suborder $\underline{O} \supset \underline{O}_1$, and every order $\underline{O}'$ of $k(\underline{O})$, $\hat{z}(\underline{O})(\underline{O}') = 0$. The value of $\hat{z}(\underline{O})(\underline{O}')$

(1) It must be checked that the correspondence is a mapping. See 10 below.

is determined entirely by knowledge, for each i and j, of whether a_{ij} belongs to $\underline{O}'$ or not. Let P belong to $\psi^{-1}(U) = \{\underline{O} \in T; \underline{O} \supset \underline{O}_1\}$. Let P' be an order of $k(P)$. Then $\underline{O} = P \cap F$ is an order of $F \cap k(P)$ and it contains $\underline{O}_1$; similarly $\underline{O}' = P' \cap F$ is an order of $F \cap k(P) = k(\underline{O})$. Hence $\hat{z}(\underline{O})(\underline{O}') = 0$. But a_{ij} belongs to $P' \cap F = \underline{O}'$ if and only if a_{ij} belongs to P'. Hence $\hat{z}_1(P)(P') = 0$. This shows that $\hat{z}_1$ vanishes over $\psi^{-1}(U)$, so that $\hat{z} \longmapsto \hat{z}_1$ is a mapping.

C. Change of ground field.

1. Let $k \subset K \subset F$, where k and K are ordered with $k^+ \subset K^+$. Let $S_k = S(F|k)$, $S_K = S(F|K)$ and denote the subordinate structure sheaves by $\mathcal{F}_k$, $\mathcal{F}_K$, respectively. For any suborder P in S_k, let $J_k(P)$ be the set of all suborders containing P, and let $J_K(P)$ be the set of all suborders in S_K containing P. Observe that S_K is contained in S_k. Let

$$\psi\colon S_K \to S_k$$

be the inclusion mapping.

2. ψ is continuous. Let P_1 be any member of S_K, let $\underline{O}_1$ be a finitely generated member of S_k contained in P_1 so that $U_1 = J_k(\underline{O}_1)$ is a neighborhood of $P_1 = \psi(P_1)$ in S_k. Now $K(\underline{O}_1)$ is a finitely generated field over K and P_1 is an order of $K(P_1)$. Also P_1 contains $\underline{O}_1$ and K^+, $K(P_1)$ contains $K(\underline{O}_1)$, and hence $P_2 = P_1 \cap K(\underline{O}_1)$ is an order of $K(\underline{O}_1)$. Thus P_2 is a finitely generated suborder in S_K containing $\underline{O}_1$, and contained in P_1; $J_K(P_2)$ is a neighborhood of P_1. If P belongs to $J_K(P_2)$ then P contains $\underline{O}_1$ and hence $\psi(P) = P$ belongs to $J_k(\underline{O}_1)$. This shows that $\psi(J_K(P_2))$ is included in $U_1 = J_k(\underline{O}_1)$ and hence that ψ is continuous.

3. For a distinguished neighborhood $J_k(\underline{O}_1)$ in S_k,

$$\psi^{-1}(J_k(\underline{O}_1)) = J_K(\underline{O}_1) \subset J_k(\underline{O}_1),$$

and the set is open in S_K. The equalities and inclusion are obvious and the openness claim is just the statement that ψ is continuous.

4. There exists a canonical morphism (ψ,θ) of the ringed space $(S(F|K), \mathcal{F}_K)$ to $(S(F|k), \mathcal{F}_k)$. Let U be an open set in S_k and let $V = \psi^{-1}(U)$. In order to define a morphism from $\mathcal{F}_k$ to $\psi_*\mathcal{F}_K$ it suffices to consider the special case of a distinguished open set U, say $U = J_k(\underline{O}_1)$. Let $\hat{z}$ belong to $\mathcal{F}_k(u)$, say $\hat{z} = \sum \Pi \hat{a}_{ij}$, a_{ij} being members of $k(\underline{O}_1) \subset K(\underline{O}_1)$. Since V is contained in U, the functions $\hat{a}_{ij}$ can be taken as functions relative to S_K, defined over V, so that $\hat{z}_1 = \sum \Pi \hat{a}_{ij}$, with the changed interpretation of $\hat{a}_{ij}$, is a member[2] of $\mathcal{F}_K(V) = \psi_*\mathcal{F}_K(U)$. Explicitly, $\hat{z}_1$ is defined as follows: for every $\underline{O}$ in V (and also in S_K) and every order P of $K(\underline{O})|K$,

$$\hat{z}_1(\underline{O})(P) = \hat{z}(\underline{O})(P).$$

The correspondence $\theta_U : \hat{z} \longmapsto \hat{z}_1$ is a homomorphism then from $\mathcal{F}_k(U)$ to $\psi_{*K}(U)$. If $U_1 \subset U$ and $V_1 = \psi^{-1}(U_1)$, then the square

$$\begin{array}{ccc} \mathcal{F}_k(U) & \xrightarrow{\theta_U} & \mathcal{F}_K(V) = \psi_*\mathcal{F}_K(U) \\ \downarrow & & \downarrow \\ \mathcal{F}_k(U_1) & \xrightarrow{\theta_{U_1}} & \mathcal{F}_K(V_1) \end{array}$$

(2) Every a_{ij} belongs to $K(P_1)$, for every finitely generated P_1 in S_K belonging to V; hence $\hat{z}_1$ is defined over every distinguished open subset V' of V and represents a member of $\mathcal{F}_K(V)$, the latter being simply the inverse limit of all such $\mathcal{F}_K(V')$.

is easily seen to be commutative. The family $\theta = \{\theta_U\}$ is therefore a sheaf morphism from F_k to $\psi_* F_K$ and the pair (ψ,θ) is a morphism of ringed spaces. It is called the change of ground field.

§2. A Universal Ordered Domain

A. Extensions of ordered fields. By an order-embedding we mean a field embedding $\phi: F \to G$, where F and G are ordered fields and ϕ preserves order. If F and G are ordered fields and $F \subset G$, then G is called an order-extension of F provided the inclsuion map is an order-embedding: in other words $F^+ = G^+ \cap F$, where F^+ is the set of all non-negative elements of F. We also may say "G is ordered over F" and refer to the "order-extension $G|F$".

1. Let $\phi: F \to G$ be an order-embedding of ordered fields, let α and β be order-embeddings of F and G into their real-closures $\overline{F}$ and $\overline{G}$ respectively. Suppose that $\overline{F}/\alpha F$ and $\overline{G}/\beta G$ are algebraic. Then there exists precisely one order-embedding $\overline{\phi}: \overline{F} \to \overline{G}$ such that $\overline{\phi}\alpha = \beta\phi$.

Proof. First suppose that α and β are inclsuions. Then the algebraic closure $\overline{\phi F}$ of ϕF in $\overline{G}$ is a real-closed algebraic extension of ϕF. By the uniqueness theorem for real closures (Theorem 8, Jacobson [3]) there is exactly one order-embedding $\overline{\phi}: \overline{F} \to \overline{\phi F}$ extending ϕ. If $\mu: \overline{F} \to \overline{G}$ is any order-extension of ϕ then $\mu\overline{F}$ is a real-closed algebraic extension of ϕF also lying in $\overline{G}$. Hence $\mu\overline{F} = \overline{\phi F}$ and, by the uniqueness theorem again, $\mu = \overline{\phi}$. For the general case simply factor α and β as surjections α_1, β_1, respectively, followed by inclusions, take $\phi_1 = \beta_1\phi\alpha_1^{-1}$ extending ϕ to αF, and $\overline{\phi}_1$ as the unique extension of ϕ_1 to F, as in the first part. Then $\overline{\phi}_1\alpha = \beta\phi$. If $\gamma\alpha = \beta\phi$, and γ_1 is the restriction of γ to $\alpha(F)$, then $\gamma_1\alpha_1 = \beta_1\phi$, $\gamma_1 = \beta_1\phi\alpha_1^{-1} = \phi_1$, so $\gamma = \overline{\phi}_1$. This proves the uniqueness.

The unique $\overline{\phi}$ of the theorem will be referred to as the canonical extension of ϕ.

2. Isbell's Lemma (special case of Th. 4.1, [2]). Let j be an order-embedding of F into F' with F algebraic over $j(F)$. Let α and β be order-embeddings of F' into H. If $\alpha j = \beta j$ then $\alpha = \beta$.

Proof. The propositon is immediately deduced from the case where j is an inclusion. To prove this case let ϕ be the common restriction of α and β to F, let $\overline{F}$ be a real-closure of F, $\overline{H}$ a real-closure of H. The canonical extensions $\overline{\alpha}$ and $\overline{\beta}$ from $\overline{F}$ to $\overline{H}$, also extend ϕ, while $\overline{F}$ is a real-closure of F. Hence each of $\overline{\alpha}$ and $\overline{\beta}$ is the canonical extension of ϕ, whence $\overline{\alpha} = \overline{\beta}$, $\alpha = \beta$.

3. Let α and β be two order-embeddings of F into real-closed fields $\overline{F}$ and $\overline{G}$, respectively, and suppose that $\overline{F}/\alpha F$, $\overline{G}/\beta F$ are algebraic. Then there exists a unique order-embedding $\overline{\phi}: \overline{F} \to \overline{G}$ such that $\overline{\phi}\alpha = \beta$, namely, the canonical extension of the identity on F. Moreover, $\overline{\phi}$ is surjective.

Proof. $\overline{G}$ is an algebraic ordered extension of the real-closed field $\overline{\phi}(\overline{F})$. Hence $\overline{G} = \overline{\phi}(\overline{F})$.

4. Let α be an order-embedding of F into a real-closure $\overline{F}$ with $\overline{F}$ algebraic over αF. Let ϕ be an order-embedding of F into G with G algebraic over ϕF. The there exists a unique order-embedding $\alpha': G \to \overline{F}$ such that $\alpha'\phi = \alpha$.

Proof. Let $\overline{G}$ be a real-closure of G, let i be the inclusion $G \subset \overline{G}$,

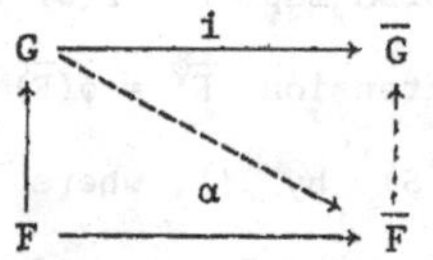

let $\overline{\phi}$ be the canonical extension of ϕ. Then $\overline{G}$ is algebraic over ϕF and also over $\overline{\phi}\alpha F$. By the last result, $\overline{\phi}$ is surjective. Let $\alpha' = \overline{\phi}^{-1} i$. Then

$\alpha'\phi = \overline{\phi}^{-1}i\phi = \overline{\phi}^{-1}\overline{\phi}\alpha = \alpha$, as required. If also $\beta\phi = \alpha$ then $\overline{\phi}\beta\phi = \overline{\phi}\alpha = i\phi$. Since G is algebraic over ϕF , ϕ can be cancelled according to Isbell's Lemma: $\overline{\phi}\beta = i$, $\beta = \overline{\phi}^{-1} i = \alpha'$. Thus the uniqueness is proved.

5. Let G be an algebraic order-extension of F and let α be an order-embedding of F into a real-closed field C . Then there exists a unique extension of α to an order-embedding of G into C .

Proof. Let $\overline{\alpha F}$ be the algebraic closure of αF in C . As just proved

$$\begin{array}{ccc} G & & \\ | & & \\ F & \xrightarrow{\alpha} & \overline{\alpha F} \subset C \end{array}$$

there is a unique extension to $G \to \overline{\alpha F}$. But any extension of α to G maps G into an algebraic extension of αF , hence into $\overline{\alpha F}$. So the uniqueness still applies.

6. Let F be an ordered field, let S be a set of independent indeterminates, let $F(S)$ be an order-extension of F , and let $\overline{F}$ be a real-closure of F . Then there exists an order of $\overline{F}(S)$ containing $F(S)^+$ = set of all nonegative elements. In other words, every order of $F(S)$ containing F^+ extends to an order of $\overline{F}(S)$.

Proof. Let $\overline{F(S)}$ be a real-closure of $F(S)$. Let $\phi\colon \overline{F} \to \overline{F(S)}$ be the canonical extension of the inclusion map $F \to F(S)$. Then S is algebraically independent over the algebraic extension $\overline{F}^{\phi} = \phi(\overline{F})$ of $F = \phi(F)$; hence $\overline{F}^{\phi}(S)$ is mapped isomorphically onto $\overline{F}(S)$ by ψ , where ψ agrees with ϕ^{-1} on $\overline{F}^{\phi}$ and fixes each element of S . We take for an order of $\overline{F}(S)$ the image P of the order in $\overline{F}^{\phi}(S)$ as ordered subfield of $\overline{F(S)}$. Since ψ reduces to the identity on $F(S)$, we see that the order of $F(S)$ is contained in P ; i.e.,

$\overline{F}(S)$ is an order-extension of $F(S)$. See the diagram below.

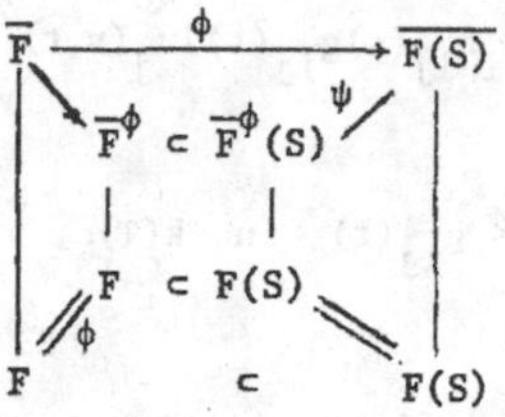

7. Artin's Theorem (Th. 13, p. 290 [1]). Let F be an ordered field, let $F(X) = F(X_1,\ldots,X_n)$ be the rational function field in n variables ordered as order-extension of F. If $f_1(X),\ldots,f_m(X)$ is a finite set of strictly positive elements of $F(X)$ then there exists a point $z = (z_1\ldots,z_n)$ in $\overline{F}^n$, where $\overline{F}$ is a real-closure of F, such that $f_i(z) > 0$, $1 \le i \le m$.

Proof. If, in Artin's proof of the case $F = \mathcal{Q}$, $\mathcal{Q}$ is replaced by the real field $\mathbb{R}$ then the proof still works and is dependent only on Sturm's theorem. Since the latter holds for all real-closed fields (loc. cit. p. 283) our claim is valid for the case $F = \overline{F}$. Let $\overline{F}(X)$ be ordered to extend the order of $F(X)$ (See §6). Thus each $f_i(X)$ is positive in $\overline{F}(X)$ so the Artin theorem applies.

8. Let F be an ordered field, let F' be an algebraic order-extension of F, let S be an independent set of indeterminates, and let $F(S)$ be an order-extension of F. Then $F'(S)$ can be ordered as an order-extension of both F' and $F(S)$.

Proof. Any real-closure $\overline{F}$ of F' is also a real-closure of F. So we take $F \subset F' \subset \overline{F}$, apply 6 to get $\overline{F}(S)$ as order-extension of $F(S)$. The order induced in $F'(S)$ makes $F'(S)$ an order-extension of both F' and $F(S)$.

9. Let k be any ordered field, let $S \cup T$ be an algebraically independent set of indeterminates and let $k(S)$ and $k(T)$ be ordered over k. Then $k(S \cup T)$ can be ordered so as to be an order-extension of both $k(S)$ and $k(T)$.

Consider the set C of all members of $k(S \cup T)$ of the form

$$\textstyle\sum_j [\sum_i f_{ij}(s) g_{ij}(t)] \pi_j(s,t)^2 ,$$

$$0 < f_{ij}(s) \text{ in } k(S) , \quad 0 < g_{ij}(t) \text{ in } k(T) , \quad \pi_j(s,t) \text{ in } k(S \cup T) .$$

This C is obviously just the smallest additively and multiplicatively closed set containing all positive members of $k(S)$ and $k(T)$, and all squares in $k(S \cup T)$. If C does not contain -1 then it is contained in an order of $k(S \cup T)$ which will satisfy the conslusion of the theorem. Suppose C does contain -1. Then there exist elements $s_1,\dots,s_m$ in S and $t_1,\dots,t_n$ in T, elements $f_{ij}(s_1,\dots,s_m)$, $g_{ij}(t_1,\dots,t_n)$, $\pi_j(s_1,\dots,s_m,t_1,\dots,t_n)$, with $0 < f_{ij}(s)$, $0 < g_{ij}(t)$, and

$$\sigma(s,t) \equiv \textstyle\sum_j (\sum_i f_{ij}(s) g_{ij}(t)) \pi_j(s,t)^2 = -1 .$$

By Artin's theorem (No. 7) there exist points $a = (a_1,\dots,a_m)$ in $\overline{k}^m$, $b = (b_1,\dots,b_n)$ in $\overline{k}^n$ such that $f_{ij}(a) > 0$, $g_{ij}(b) > 0$ for all i and j. Moreover, such a and b can be chosen so that the $\pi_j(a,b)$ are defined. But these conditions imply that $\sigma(a,b)$ is a non-negative member of $\overline{k}$, which contradicts the displayed equation.

10. Let F and G be ordered over k. There exists an ordered field H and order-embeddings of F and G into H.

Proof. Take copies of F and G sufficiently disjoint that transcendence bases S for F, T for G, can be chosen so that $S \cup T$ is algebraically independent over k.

$$
\begin{array}{ccccc}
G & \overset{h}{\subset} & G(S) & \overset{\phi}{\dashrightarrow} & \overline{F(T)} \\
| & & i\,| & & | \\
k(T) & \subset & k(S \cup T) & \subset & F(T) \\
| & & |\,j & & |\,m \\
k & \subset & k(S) & \subset & F
\end{array}
$$

First take an order of $k(S \cup T)$ so that i and j are order-inclusions (§8). Next choose orders for $G(S)$ and $F(T)$ to make h and m order-inclusions.

Let $\overline{F(T)}$ be a real-closure for the ordered field $F(T)$. Noting that $\overline{F(T)}$ is algebraic over $k(T \cup S)$, we see that the inclusion extends to an order-embedding $\phi: G(S) \to \overline{F(T)}$, by 5. Thus ϕh and m embed G and F, respectively, in $\overline{F(T)}$.

Note that the H of Proposition 10 can be chosen to be the composite of the images of F and G in it, hence the power of H can be the maximum of the powers of G and H.

11. Let $\mathcal{F}$ be a set of ordered fields over k. Then there exists an ordered field K over k in which every field in $\mathcal{F}$ can be order-embedded.

Proof. Let S be a set of power 2^n containing k, where $n = \sum\{\text{card } F;\ F \in \mathcal{F}\}$. Consider the set $\mathcal{T}$ of all triples $(\mathcal{F}_1, \Phi_1, S_1)$, where $\mathcal{F}_1$ is a nonempty subset of $\mathcal{F}$, Φ_1 consists of one mapping from each ordered field in $\mathcal{F}_1$ into S_1, and S_1 is an ordered field over k contained in S such that each $\phi \in \Phi_1$ is an order-embedding of the corresponding ordered field into S_1, while S_1 is the composite of all images by the ϕ in Φ_1. $\mathcal{T}$ is not empty. For triples in $\mathcal{T}$ we write

$$(\mathcal{F}_1, \Phi_1, S_1) \le (\mathcal{F}_2, \Phi_2, S_2)$$

to signify that $\mathcal{F}_1$ is contained in $\mathcal{F}_2$, Φ_1 is contained in Φ_2 and S_2 is an order-extension of S_1. It is easy to see that $\mathcal{T}$ is inductive. To complete the induction it suffices to observe that for any triple $(\mathcal{F}_1,\Phi_1,S_1)$, the power of S_1 is at most $n < 2^n$ = power of S; there is therefore plenty of room in S to apply (§10) to S_1 and F, where F is any member of $\mathcal{F}$ not in $\mathcal{F}_1$, to get $(\mathcal{F}_2,\Phi_2,S_2)$ strictly greater than $(\mathcal{F}_1,\Phi_1,S_1)$, where $\mathcal{F}_2 = \mathcal{F}_1 \cup \{F\}$. Thus if we choose $(\mathcal{F}',\Phi',S')$ to be any maximal element of $\mathcal{T}$ then $\mathcal{F}' = \mathcal{F}$, and every member of $\mathcal{F}$ is order-embedded in S'.

REFERENCES

1. Emil Artin, Uber die Zerlegung definiter Funkionen in Quadrate, Abh. Math. Sem. Univ. Hamburg 5(1972), 100-115.

2. John R. Isbell, Embedding two ordered rings in one ordered ring, J. Algebra 4(1966), 341-364.

3. Nathan Jacobson, Lectures in Abstract Algebra III, Van Nostrand, Princeton, 1964.

D.W. Dubois
University of New Mexico
Albuquerque, NM 87131 USA

Tomás Recio
Universidad Málaga
Málaga, SPAIN

The Extension Theorem for Nash Functions

Gustave A. Efroymson
University of New Mexico

Introduction.

The main result of this paper is Theorem 2: Let h be a Nash function $h: R^n \to R$ with non-empty zero set $h^{-1}(0)$ and let U be an open semi-algebraic set containing $h^{-1}(0)$. Let $f: U \to R$ be Nash. Then there exists $g: R^n \to R$, Nash, and $g = f$ on $h^{-1}(0)$.

This theorem would follow easily if we had a good cohomology theory with $H^1(X, O_X) = 0$, but Hubbard [H] has shown that we do not have such a theory. So it is necessary to proceed in other ways. We first prove an approximation theorem which resembles the extension theorem and is interesting in its own right.

Theorem 1. Let U and U' be open semi-algebraic sets with $\overline{U} \subset U' \subset R^n$. Suppose we are given a function $e(r) = 1/(C+r^{2m})$ where $r^2 = \sum_1^n x_i^2$, and a Nash function $f: U' \to R$. Then there exists $g: R^n \to R$ such that g is Nash and $|g-f| < e(r)$ on U.

Before proving this theorem we need some preliminary results but we can give an idea of the proof now. First cover by a union of U_i s so that on each $U_i \times R$, $p_f(z,x)$ the irreducible polynomial for f has constant Thom polynomial (i.e., the root structure of p and all its derivatives is constant). Thom's lemma [E] asserts that this can be done with a finite cover. Then we work by induction on the degree of $p_f(z,x)$ in z. Here we use, if necessary, the fact that the roots of the derivative of $p_f(z,x)$ can be approximated (by induction). We then construct a polynomial $p_{new}(z,x)$ which will have on U_i a root of g_i close to f and g_i is defined and Nash on R^n. Using a "partition of unity", we can put the g_i s together to get a Nash function g approximating f. Next we prove the extension theorem in the case where $dp_f/dz(f(x),x) \neq 0$ on U. Finally, we show that this case suffices to get the general result.

In the last part of the paper we discuss the problem of extending a Nash function off of $h^{-1}(0)$, instead of a neighborhood of $h^{-1}(0)$.

Section 1. The Approximation Theorem

We begin by recalling a result which follows from the Tarski-Seidenberg principle.

Proposition 1. Let $f\colon R^n \to R$ be a continuous semi-algebraic function which is > 0 on R^n. Then there exists a real constant C and a positive integer m such that if $e(x) = 1/(C+r^{2m})$ then $f(x) > e(x)$ for all x in R^n. Here as in the introduction $e(x) = e(r) = 1/(C+r^{2m})$, $r^2 = \sum_1^n x_i^2$.

Proof. Note that $1/f$ is also a continuous semi-algebraic function and positive. So we are reduced to proving that if f is continuous and semi-algebraic then there exists a real constant C and an integer m as above with $f(x) < C + r^{2m}$. So let $g(r) = \max(f(x)\colon |x| = r)$. Since $\{x\colon |x| = r\}$ is compact, $g(r)$ is well defined and always positive. But $g(r) = s$ if and only if for all x with $|x| = r$, $g(x) \leq s$ and there exists y with $|y| = r$ and $g(y) = s$. That means that we can apply the Tarski-Seidenberg principle to conclude that $g(r)$ is an algebraic function. But then $g(r)$ is a semi algebraic function so in particular for r large enough, g is a root of $a_d(r)z^d + \sum_{i=0}^{d-1} a_i(r)z^i$. Let $a_d(r) = \lambda \prod_{i,j} (r-\lambda_i)(r^2+b_jr+c_j)$ be the factorization of $a_d(r)$ into linear and quadratic irreducible factors. Let $M = \max\{\lambda_i\} + 1$. Then for $r > M$ we have $a_d(r) > \lambda \prod_j (c_j - b_j^2/4) > 0$. From this it follows easily that if m is the highest power of r that appears in any of the $a_i, i < d$, then for r large enough, $g(r)$ will be less than r^{2m}. And since $g(r)$ is continuous, there will exist a bound C for all smaller r. Q.E.D.

We should say now what we mean by constant Thom polynomial. We mean that $p_f(z,x)$ and all the $d^kp_f/dz^k(z,x)$ have constant multiplicity of roots as x varies on U. We would like to assume that in our cover, on each U_i, the Thom polynomial $p_f(z,x)$ is constant on the closure of U_i. Actually this does not really follow from the above. We need to note that the boundary of U_i will be of lower dimension than U_i so that we will already have an approximation for f on the boundary which will carry over into U, We will do this in detail later but first we prove the following.

Proposition 2. Let D be a connected semi-algebraic set of the form $D = \{x \in V: h_1(x),\cdots,h_s(x) \geq 0\}$ where V is $\{x \in R^n: f_1 = 0,\cdots,f_t = 0\}$. Let f be a Nash function on D with irreducible polynomial $p_f(z,x)$ such that $p_f(z,x)$ is a constant Thom polynomial for f on D. Then for any $e(r)$ as above, there exists g Nash on R^n so that $|g-f| < e(r)$ on D.

<u>Proof</u>. We work by induction on the degree in z of $p_f(z,x)$. This means that if we have a root $b(x)$ of $dp_f/dz(z,x)$, then we can approximate $b(x)$ by $b'(x)$ Nash on R^n so that $|b'(x)-b(x)| < e(r)$ on D for any $e(r)$ as above. Now let $h(x) = \sum_i(|h_i|-h_i) + \sum_i f_i^2$. Then h is semi-algebraic, $h \geq 0$ and $h = 0$ only on the closure of D. At the end of each step we will replace $|h_i|$ by $(h_i^2+e)^{1/2}$ for some suitable $e(r)$. We now consider several cases.

<u>Case 1</u>. $dp_f/dz(z,x) \neq 0$ on $R \times D$. Then dp_f/dz must have even degree and so p_f will have odd degree in z. We can take $p_f(z,x) = \sum_{i=0}^{d} a_i(x)z^i$ with $a_d(x) > 0$ on D. Now we wish to consider $p(z-a(x)) = p_1(z,x)$ so that by choosing $a(x)$ large enough, we will have, since the degree of $p = p_f$ is odd, that the new $a_0(x)$ will be < 0 everywhere on D. But if we approximate $f(x) + a(x) = f_1(x)$ closely, we will have $f(x) = f_1(x) - a(x)$ approximated closely also. So we can assume that $a_0(x) < 0$ on D. But then we can let $p_{new}(z,x) = p(z,x) + u_1hz - u_2h$ where u_1 and u_2 are chosen so that

1) $p_{new}(0,x) = p(0,x) - u_2h$ is < 0 everywhere.

2) $dp_{new}/dz = dp/dz + u_1h$ is > 0 everywhere.

Now 1) is possible since $p(0,x) < 0$ when $h = 0$ and so will be < 0 when $h < e(r)$, for some $e(r)$ found as in Proposition 1. But then choose $u_2h > p(0,x)/e$ and we will have what we want.

For 2) we use the fact that, since p is of odd degree with positive leading coefficient $a_d(x)$ on D, we can take $a_{dnew} = a_d + \lambda h$ so that a_{dnew} is > 0 everywhere.

So $p_{new}(z,x) = 0$ will have a unique root $z = f(x)$ for each x. By choosing e' small enough when we replace each $|h_i|$ by $(h_i^2+e')^{1/2}$, we can make the coefficients of p_{new} as close to those of p on D as we wish. Moreover, it is clear that then $f(x)$ is Nash. So Case 1 is finished once we show:

<u>Proposition 3</u>: Let $f(x)$ be a root of $p(z,x) = \sum a_i z^i$ on D (as above). Assume that f is continuous on D and that $dp/dz(f(x),x) > 0$ on D. Then given $e(x)$ as above, there exists $d(x)$ of the same form so that if $|a_i - b_i| < d(x)$, then $p_{new} = \sum b_i z^i$ will have a root $g(x)$ which is also Nash on D and $|f-g| < e$.

<u>Proof</u>. Let $\epsilon(x) = \min\{y: p(f(x)+y,x) > 0, p(f(x)-y,x) < 0, dp/dz(z,x) > 0$ on $[f(x) - 2y, f(x) + 2y]$ and $0 \leq y \leq 1\}$. Then clearly $\epsilon(x)$ is a semi-algebraic function of x and since $dp/dz(f(x),x) > 0$ on D, we know that $\epsilon(x) > 0$ on D. Now

let $g(x) = 1/\epsilon(x)$ for x in D, and $g(x) = 1$ off D. Then as in the proof of Proposition 1, we can find C and m so that $g(x) < C + r^{2m}$ for all x. So if we let $e_2(x) = 1/(C+r^{2m})$, we have $p(f(x) + e_2(x),x) > 0$ on D and $p(f(x) - e_2(x),x) < 0$ on D, and $dp/dz(z,x) > 0$ on $[f(x) - 2e_2(x), f(x) + 2e_2(x)]$, an interval on the z-axis. So choose d so small that $dp_{new}/dz(z,x) > 0$ on $[f(x)-e_2(x), f(x)+e_2(x)]$, $p_{new}(f(x)-e_2(x),x) < 0$, $p_{new}(f(x)+e_2(x),x) > 0$. The desired conclusion follows.

<u>Case 2</u>. $f(x)$ is the largest root of p_f on D, and there exists b, a root of $dp_f/dz(z,x)$ with $b < f$. Now find b' by our induction hypothesis so that $b < b' < f$. To do this choose e so that $f - b > e$, and then approximate $b + e/2$ closer than $e/4$. Then we have $dp_f/dz(z,x) > 0$ for $z \geq b'$ and now we let $p_{new}(z,x) = p(z,x) + u_1hz - u_2h$ where we choose u_2 so that $p_{new}(b'(x),x) = p(b'(x),x) - u_2h < 0$ and u_1 so that $dp_{new}/dz(z,x) = dp/dz)z,x) + u_1h$ is > 0 for all $z > b'$. This is done by noting that $dp/dz(z,x)$ is bounded below for $z > b'(x)$ and so we can proceed as at the end of Case 1.

<u>Case 3</u>. $f(x)$ is a root of $p_f(z,x)$ with $b_1 < f < b_2$ where b_1 and b_2 are roots of $dp_f/dz(z,x)$ on D.

Now find b'_1 and b'_2 so that $b_1 < b'_1 < f < b'_2 < b_2$ and b'_1 and b'_2 are Nash everywhere. So we can define $p_{new}(z,x) = p_f(z,x) + [h(u_2-u_1)/(b'_2-b'_1)](z-b'_1) - u_1h$ where u_1 and u_2 are chosen so that

1) $p_{new}(b'_1,x) = p_f(b'_1,x) - u_1h$ is < 0 everywhere.

2) $p_{new}(b'_2,x) = p_f(b'_2,x) + u_2h$ is > 0 everywhere.

3) $dp_{new}dz(z,x) = dp_f/dz(z,x) + h(u_2-u_1)/(b'_2 \cdot b'_1) > 0$ for $b'_1 < z < b'_2$.

This can be done as in Case 2 by noting that $dp/dz(z,x)$ is bounded from below for $b'_1 < z < b'_2$ (in terms of some $-(C+r^{2m})$). Then one checks that $p_{new}(z,x)$ will have a root $g(x)$ which is Nash everywhere and is near f on D.

Any remaining cases are similar to one of the above.

Now back to the proof of Theorem 1. We cover U' by $\bigcup_{i=1}^{m} U_i'$ where each U_i' is a connected semi-algebraic set of the form $\{x \in V_i': h_{ij} > 0 , j = 1,\cdots,s_i\}$ and $V_i' = \{x$ in $R^n: f_{ij}(x) = 0 , j = 1,\cdots,t_i\}$. And on each U_i', $p_f(z,x) = p_i(z,x)$ has constant Thom polynomial. We need only consider those U_i' with $U_i' \cap U \neq \emptyset$. So first consider the lowest dimensional of the U_i' with $U_i' \cap U \neq \emptyset$. Then the boundary of U_1' will not intersect U, so that we can replace the inequalities: $h_{ij} > 0$ by the inequalities $h_{ij} \geq 0$ to get a new set D_i on which the Thom polynomial of f is constant. So we can apply Proposition 2 to approximate f by g_i on D_i so that there $|f-g_i| < e/4m$. This inequality will hold on an open subset U_i'' of R^n. And then there exist open sets $W_i \supset V_i \supset U_i''$ on which $|f-g_i| < e/m$, $e/2m$ and $e/4m$ respectively.

Now we work by induction on the dimension of U_i' and since we have already approximated f on the U_j of lower dimension, we can remove the union of the U_j'' to get a closed set D_i which can be written as a union of sets of the correct form so Proposition 2 will apply (using the "unproved theorem" of Brumfiel, see [B-E] Proposition 5.1 for a proof). So there exists a Nash function g_i on R^n so $|g_i-f| < e/4m$ on D_i. Then, as above, this inequality will hold on an open set U_i'' of R^n, and there exist open sets $W_i \supset V_i \supset U_i''$ so that $|f_i - g_i| < e/2m$ on V_i and $< e/m$ on W_i.

We now construct for each i a global Nash function e_i so that $\sum e_i = 1$, and e_i is close to 0 outside W_i. Choose $d_j > 0$ so that $V_i = \{x: (h_{ij}+d_j) > 0,\ j = 1,\cdots,r_i\}$. Choose $e' \ll$ all d_i and $< e/m$ and so $e'|f(x)-g_i(x)| < e/m$ on U. Then take $m_i = \prod[((h_{ij}+d_{j/2})^2 + e')^{1/2} + (h_{ij} + d_{j/2})]$ and we see that $m_i > \prod d_j$'s on U_i. And $0 < m_i < \prod_j(e'/d_j)$ off W_i. So taking $e_i = m_i/\sum m_i$'s, we find that $\sum e_i = 1$ and $e_i \ll e$ outside W_i. Finally, let $g = \sum e_i g_i$. Then $|f-g| = |\sum e_i f - \sum e_i g_i| \leq \sum e_i |f-g_i|$. So if $x \in W_i$, then $|f(x) - g_i(x)| < e/m$. And if $x \notin W_i$ then $e_i |f(x) - g_i(x)| < e/m$ since $e_i < e'$ outside W_i. So $|f(x) - g(x)| \leq \sum e_i |f(x) - g_i(x)| < e$. Q.E.D.

Section 2. The Extension Theorem.

We first prove the extension theorem in the case where f is a root of some $p_f(z,x) = \sum_{i=0}^{d} a_i(x)z^i$ where the $a_i(x)$ are in $N(R^n)$, i.e., are Nash functions on R^n , and also we require both $a_d(x) \neq 0$ and $\partial p_f/\partial z(f(x),x) \neq 0$ on U . Recall that we wish to construct $g: R^n \to R$, g Nash so that $f = g$ on $S = h^{-1}(0)$.

So let $p_f = \sum a_i z^i$. As in the earlier part of the paper, we can choose $a_{dnew} = a_d + uh$ so that $a_d > 0$ on R^n . And then we can divide by the new a_d so that p_f is monic. Now apply Proposition 1 to choose $d > 0$ so that if $|a_i - b_i| < d$, then $\sum b_i z^i$ will have a unique root f' with $|f-f'| < e/2$ where $e <$ the distance from f to the nearest other root of p_f . We are only taking x in U here.

Now choose f_1 so that f_1 is Nash on R^n and $|f-f_1| < e/4$ on U . Note that $p_f(z,x) = (z-f)q(z,x)$ on U . So approximate the coefficients of $q(z,x)$ to get $q'(z,x)$ in $N(R^n)[z]$ so that $q'(f(x),x) > 0$ and $q'(f_1(x),x) > 0$. By adding some uh to the constant term of q' , we can make $q'(f_1(x),x) > 0$ everywhere.

Let $(z-f_1)q'(z,x) = \sum b_i z^i$ and we can assume that we have $|a_i - b_i| < d$ on U . There exists d' so that, if $|c_i - b_i| < d'$, $\sum c_i z^i$ has a root g with $|g-f_1| < e/4$. Choosing e' small enough, we let $e_i = a_i + (c_i - a_i)h/(h^2+e')^{1/2}$. But then $\Sigma e_i z^i$ will have a root $g(x)$ which will be Nash on R^n since, for h small, $e_i \approx a_i$ and for other x, $e_i \approx c_i$. Moreover, since $e_i = a_i$ when $h = 0$, g and f agree when $h = 0$.

The general case for the extension theorem will follow from the following.

<u>Theorem 3</u>. Let U and U' be open connected semi-algebraic sets with $U' \subset U \subset R^n$. Let f be a Nash function on U, then there exists g Nash on U' with g a root of a polynomial

$$p_g(z,x) = z^d + \sum_{i=0}^{d-1} a_i(x) z^i$$

where the $a_i(x)$ are in $N(R^n)$ and where f is in $N(R^n)[g]$ localized at U' and where $\partial p_g/\partial z(g(x),x) \neq 0$ on U'.

<u>Proof</u>: Let A be the ring $N(R^n)$ localized at U, i.e., $A = \{a(x)/b(x)$: a,b are in $N(R^n)$ and $b(x) \neq 0$ for x in $U\}$ Similarly let A' be the polynomial ring $R[x_1,\cdots,x_n]$ localized at U. Then let K' be the quotient field of $A'[f]$ and B' the integral closure of A' in K'. Let K be the quotient field of $A[f]$, and B = the integral closure of A in K. Let $p_f(z,x)$ be the irreducible polynomial for $f(x)$ over A' (i.e., multiply out denominators so that all coefficinets are in A' which can be done since A' is a U.F.D.

By multiplying $f(x)$ by a power of the leading coefficient of $p_f(z,x)$, we obtain an element $h(x)$ of B' so that for each

x in U, $p_h(z,x)$ will have d roots $h_1(x) = h(x)$, $h_2(x) \cdots h_d(x)$ in $\mathbb{C}$. We call $h_1(x),\cdots,h_d(x)$ the conjugates of $h(x)$. There is a non-singular branch of Spec B' corresponding to the factor $(z-h(x))$ of $p_h(z,x)$. Let $t_1(z,x),\cdots,t_r(z,x)$ be representatives of generators of B' over A'. So applying Zariski's Main Theorem, we find that for each $j \neq 1$, and each a in U, there exists $t_i(z,x)$ with $t_i(h_j(a),a) \neq t_i(h_1(a),a)$. Now consider the factorization of $p_h(z,x)$ in $A[z]$. (Recall A is also a U.F.D. [B-E] p. 219 Theorem 4.1.) Then, by renumbering, we can let $h_1(x),\cdots,h_c(x)$, be the conjugates of $h_1(x)$ over A. But then, for each a, there exists $\lambda_1,\cdots,\lambda_r$ so that if we define $g_a(x) = \sum_i \lambda_i t_i(h_1(x),x)$, and $g_{ja}(x) = \sum_i \lambda_i t_i(h_j(x)x)$, we find $g_{ja}(a) = \sum\lambda_i t_i(h_j(a),a) \neq \sum\lambda_i t_i(h_1(a),a) = g(a)$. So we have for each a in U, a Nash function $g_a(x)$ in B so that $p_a(z,x) = \prod_{j=i}^{c} (z-g_{ja}(x))$ is the irreducible polynomial for g_a over A and we have $dp_a/dz(g_a(a),a) \neq 0$. But then $dp_a/dz(g_a(x),x) \neq 0$ will hold over a Zariski open set in U and so by the usual dimension argument, there will exist a finite number of points $a_1,\cdots,a_s$ in U so that for all b in U, there will exist g_{a_i} with $dp_a/dz(g_{a_i}(b),b) = 0$.

Now let $g_a(x)$ denote any of the g_{a_i} as above. Then let $p_a(z,x) = (z-g_a)q_a(z,x)$ in $B[z]$. So if $g_{aj}(x)$ is as above then $q_a(g_{aj}(x),x) = 0$ if $j \neq 1$. And for all b in U, there exists q_{a_i} as above with $q_{a_i}(g_{a_i}(b),b) \neq 0$. So consider

$\sum_{i=1}^{s} q_{a_i}(g_{a_i}(x),x)^2$ which is > 0 on U and so in particular is $> e(x)$ on U' for some $e(x) = 1/(C+r^m)$ as above. Now we apply the approximation theorem. We approximate the coefficients of the $q_a(z,x)$, which are Nash functions on U, so closely on U' by elements of $N(R^n)$ that we get new $q'_a(z,x)$ such that

$$\sum_{i=1}^{s} q'_{a_i}(g_{a_i}(x),x)^2 > e(x)/2$$

and

$$\left|\sum_{i=1}^{s} q'_{a_i}(g_{a_{ij}}(x),x)^2\right| < e(x)/2 .$$

Then let

$$g(x) = \sum_{i=1}^{s} q'_{a_i}(g_{a_i}(x),x)^2$$

and we see that

$$g_j(x) = \sum_{i=1}^{s} q'_{a_i}(g_{a_ij}(x),x)^2 \neq \sum_{i=1}^{s} q'_{a_i}(g_{a_i}(x),x)^2 = g(x)$$

for all x in U' and $j \neq 1$. Since the $g_{a_i}(x)$ are Nash on U, g(x) is Nash on U. Moreover, the irreducible polynomial (over A[z]) $p_g(z,x) = \prod_{j=1}^{c}(z-g_j(x))$ will have $dp_g/dz(g(a),a) \neq 0$ for all a in U'.

All that is left is to show that f is in $A[g]$. But if $C = A[g]$ localized on U', then by the above, for each a in U', we will have $\hat{C}_{(g(a),a)} \approx N(R^n)_{(a)} \approx R[[x_1,\cdots,x_n]]$. So since $C_{(g(a),a)} = \hat{C}_{(g(a),a)} \cap K$, f will be in each $C_{(g(a),a)}$ and so in C. This completes the proof of Theorem 3.

To finish the proof of the extension theorem, consider first the case where $h^{-1}(0)$ is connected. Then we use Theorem 3 to find $g(x)$ so that $f(x) = p(g(x),x)/q(g(x),x)$ where $q(g(x),x) \neq 0$ on U', where $p(z,x)$ and $q(z,x)$ are in $N(R^n)$, and $h^{-1}(0) \subset U' \subset U$, and $g(x)$ can be extended to $G(x)$ on R^n. Then we note that since $q(G(x),x) \neq 0$ on $h^{-1}(0)$ and $h^{-1}(0)$ is connected, that $q(G(x),x)$ has constant sign on $h^{-1}(0)$ and so by the Tarski-Seidenberg argument, there exists $\lambda(x)$ polynomial on R_n so that $q(G(x),x) + \lambda(x)h(x) \neq 0$ on R^n. So $F(x) = p(G(x),x)/[q(G(x),x) + \lambda(x)h(x)]$ extends $f(x)$. For the case of non-connected $h^{-1}(0)$, just use Mostowski's separation theorem [E], page 138. The details are left as an exercise for the interested reader.

Section 3. Applications.

Let S be a non singular Nash variety in R^n, i.e., S = the zero set of Nash functions and S is non-singular. Then there exists a normal bundle B to S in R^n and one can define for any $f\colon S \to R$, an extension f_1 of f to a neighborhood U of S. To be more formal, consider for each point x in R^n, the point(s) $\pi(x)$ on S which are closest to x. Since S is non-singular, for points x sufficiently close to S, there will be a unique $\pi(x)$. Now for each real $r > 0$, there exists $e(r)$ so that if $|x| = r$ and x is a distance $< e(r)$ from S, then $\pi(x)$ is unique. So we take an $e(r)$ neighborhood of S.

Now f_1 can be defined as $f_1(x) = f(\pi(x))$ where π is the map $\pi = B \to S$, where B is the $e(r)$ neighborhood of S. Then it is clear that f_1 will be Nash since p is an algebraic function. Next, using Theorem 2, we can extend f_1 to R^n. •

Summarizing, we state the following.

Theorem 4. If S is a non-singular Nash variety in R^n and $f: S \to R$ is a Nash function on S, then there exists $g: R^n \to R$ which is Nash and $g = f$ on S.

But what if S is singular? For S = the "Whitney umbrella", $(x^2+y^2)z = x^3$, we define

$$f(x,y,z) = (z-1)^2/((z-1)^2 + x^2 + y^2),$$

and $f = 1$ on the "stem". We claim f is not extendable off S. Let S_C be the complexification of S. Then f extends to a neighborhood of S in S_C. But near $(0,0,1)$ f is not analytic on S_C since it is not even continuous there. For consider f on $\{z = 1\} \cap S_C$. There $f = 0$. Next consider f on $\{x = 0, y = 0\} \cap S_C$. There $f = 1$ at $(0,0,1)$.

The above example is non-coherent so maybe coherence is sufficient for extendability? Of course, for any non-coherent S

an example as above can be produced. The case of singular curves will be covered in a future work of the author.

Bibliography

[B-E] Bochnak, J., Efroymson, G., Real Algebraic Geometry and the Hilbert 17th Problem, Math. Ann. 251, 213-241 (1980).

[E] Efroymson, G. Substitution in Nash functions. Pacific J. Math. 63, 137-145 (1976).

[H] Hubbard, J., On the cohomology of Nash sheaves, Topology 11, 265-270 (1972).

SEPARATION DES COMPOSANTES CONNEXES REELLES

DANS LE CAS DES VARIETES PROJECTIVES

par Jean HOUDEBINE et Louis MAHÉ

Dans ce papier, on étend aux variétés projectives sur un corps réel clos, les résultats obtenus dans le cas affine [4], c'est-à-dire qu'on démontre le théorème suivant :

THEOREME : *Soit* $X \hookrightarrow \mathbb{P}^n_k$ *une variété projective sur un corps réel clos* k. *Soit* F_1 *un ouvert-fermé semi-algébrique de* X(k) *et* F_2 *son complémentaire : il existe un* X-*espace quadratique* $Q \in W(X)$ *tel que* $\hat{Q}(F_1) = 2^t$ *et* $\hat{Q}(F_2) = -2^t$ *pour un certain entier* t.

Avant tout, quelques définitions s'imposent :

1. DEFINITIONS, NOTATIONS.

1.1. *Topologie forte.*

X(k) est l'ensemble des points k-rationnels de X (on dira ici réels) et la topologie forte sur X(k) est celle qui admet pour base d'ouverts les ensembles $[U,f_1,\dots,f_m] = \{x \in U(k)/f_1(x) > 0,\dots,f_m(x) > 0\}$ où U est un ouvert affine de X et les f_i sont des fonctions régulières sur U ([6] p. 309). Notons $x_0,x_1,\dots,x_n$ un système de coordonnées de X et $s = \sum_{i=0}^{n} x_i^2$. Il est clair que l'ouvert affine $D_s = \{p \in X/s$ est inversible en $p\}$ contient tous les points réels de X et que l'immersion ouverte $D_s \hookrightarrow X$ induit un homéomorphisme des espaces $D_s(k)$ et X(k) munis de leur topologie forte respective.

1.2. *Ensembles semi-algébriques.*

Une partie S de X(k) est dite semi-algébrique si elle est combinaison booléenne (finie) d'ouverts de la forme [U,f]. La trace sur $D_s(k)$ d'une telle partie de X(k) est encore un semi-algébrique de $D_s(k)$ en ce sens (si f est régulière sur U, elle l'est aussi sur $U \cap D_s$). Réciproquement, comme D_s est lui-même un ouvert affine de X, et que l'ouvert $[U,f]$ de $D_s(k)$ a même complémentaire dans X(k) et dans $D_s(k)$, un semi-algébrique de $D_s(k)$ est donc aussi un semi-algébrique de X(k).

Reste à voir que sur l'ouvert affine $D_s(k)$, cette notion locale de semi-algébrique coïncide avec la notion usuelle qui est : combinaison booléenne finie d'ensembles de la forme $\{x \in D_s(k)/f(x) > 0\}$ pour f régulière sur D_s. Or, si U est un ouvert affine de D_s, U est de la forme $U = D_v \cap D_s$ et si g est une fonction régulière sur D_v, on a $g = \frac{u}{v^\alpha}$ avec $d^o u = d^o v^\alpha = r$.

Posons alors $f = \frac{uv^\alpha}{s^r}$, f est une fonction régulière sur D_s et $f(x) > 0 \iff g(x) > 0$ pour tout x de $D_s(k)$. (Pour ces questions, consulter [1] §6)

1.3. *Séparation sur* A_s.

Des points de vue "semi-algébrique" et "topologie forte", l'espace X(k) est donc identique à l'ouvert affine $D_s(k)$ et on en déduit donc en particulier que les composantes semi-algébriques ([4] p. 4) de X(k) sont en nombre fini et coïncident avec les composantes connexes pour la topologie forte dans le cas où $k = \mathbb{R}$.

Soit donc F_1 un ouvert-fermé semi-algébrique de X(k) et F_2 son complémentaire : on peut les considérer comme ouverts-fermés semi-algébriques de $D_s(k)$ et d'après la proposition 3.1 de [5], il existe un D_s-espace quadratique q tel que $\hat{q}(F_1) = 2^{t'}$, $\hat{q}(F_2) = -2^{t'}$.

Notons k[X] l'anneau gradué des polynômes définis sur la variété homogène associée à X, et pour f homogène dans k[X], A_f l'anneau des fonction régulières sur D_f : $A_f = (k[X]_f)^{(0)}$; q est donc un A_s-espace quadratique (libre) tel que, si $\tilde{F}_1$ et $\tilde{F}_2$ sont les constructibles de $\mathrm{Spec}_R A_s$ associés à F_1 et F_2, on ait $\hat{q}(\tilde{F}_1) = 2^t$, $\hat{q}(\tilde{F}_2) = -2^t$. q est donc représenté dans la base canonique de A_s^p par une matrice symétrique inversible dont les coefficients sont de la forme $\frac{P(\bar{x})}{s^i}$ avec $d^oP = 2i$. On peut alors trouver un entier m' tel que tous les coefficients de q et q^{-1} (on note encore q la matrice de la forme q) soient de la forme $\frac{P(\bar{x})}{s^{m'}}$. Notons alors $q = \frac{q_1'}{s^{m'}}$, $q^{-1} = \frac{q_2'}{s^{m'}}$, on a $\frac{q_1' q_2'}{s^{2m'}} = I_{A_s^p}$: on peut donc trouver $r \in \mathbb{N}$ tel que $s^{2r} q_1' q_2' = s^{2r} s^{2m'} I_{k[X]^p}$; posons alors $q_i = s^r q_i'$ et $m = r+m'$, on a $q_1 q_2 = s^{2m} I_{k[X]^p}$.

On note q_{i1} et q_{i2} les A_{x_i}-modules quadratiques définis par les matrices q_1/x_i^{2m} et q_2/x_i^{2m} dans la base de $A_{x_i}^p$ correspondant à celle de A_s^p. Si l'on développe l'expression s^{2m}, on obtient : $s^{2m} = \sum_{i=0}^{n} x_i^{4m} + \sum_{j=1}^{\ell} t_j^2$ et donc

$$q_{i1} q_{i2} = q_1 q_2 / x_i^{4m} = \frac{s^{2m}}{x_i^{4m}} I = \Big(1 + \sum_{k \neq i} \Big(\frac{x_k}{x_i}\Big)^{4m} + \sum_{j=1}^{\ell} \Big(\frac{t_j}{x_i^{2m}}\Big)^2\Big) I .$$

On convient d'appeler s_i le terme $s^{2m} - x_i^{4m}$; s_i est donc une somme de carrés dans k[X] et donc aussi $\frac{s_i}{x_i^{4m}}$ dans A_{x_i}. q_{i1} satisfait donc les conditions de la proposition 4.9 de [4] et on peut trouver Q_i, matrice symétrique inversible sur A_{x_i} telle que $Q_i \simeq 2^{t''} q_{i_1}$ sur $A_{x_i}[\Sigma^{-1}]$. Comme q_{i1} et donc Q_i, a la signature voulue sur $\tilde{F}_1 \cap D_{x_i}(k)$ et $\tilde{F}_2 \cap D_{x_i}(k)$, le seul point est de montrer qu'on peut trouver un choix cohérent des Q_i de façon à constituer un élément de W(X) (pour la définition de W(X) voir [4] Ch. I). C'est-à-dire qu'il faut trouver un cocycle (λ_{ij}) tel que $\lambda_{ij}^{*} Q_i \lambda_{ij} = Q_j$.

Avant de faire cette construction, nous allons développer

2. QUELQUES POINTS DE CALCUL MATRICIEL.

2.1. *Les matrices* $\Lambda(a_1,\dots,a_n)$ *de Karoubi.*

Etant donné une suite d'éléments (a_i) dans un anneau A (commutatif, unitaire), pour tout n il existe une matrice Λ_n d'ordre 2^{n-1} telle que $\Lambda_n\Lambda_n^* = (\sum_{i=1}^{n} a_i^2)I$ ([3] p. 384).

Ces matrices sont construites par récurrence en posant

$$\Lambda_1 = a_1 , \quad \Lambda_{n+1} = \begin{bmatrix} \Lambda_n & a_{n+1}I \\ -a_{n+1}I & \Lambda_n^* \end{bmatrix} .$$

Les matrices que nous allons utiliser ici sont un peu plus particulières :

a) On prend pour Λ_n la matrice d'ordre 2^n $\Lambda(0,a_1,\dots,a_n)$ de façon à faire jouer à chaque a_i des rôles symétriques (la diagonale a en effet un rôle spécial).

b) Les éléments $a_i \in A$ sont remplacés par des matrices scalaires d'ordre p a_iI_p, p étant un entier fixé.

Les matrices Λ_n ainsi construites sont donc d'ordre $p2^n$ et vérifient bien sûr $\Lambda_n\Lambda_n^* = (\sum_{i=1}^{n} a_i^2)I$.

2.2. *Permutations circulaires.*

Un des intérêts particuliers des matrices Λ_n ainsi construites réside dans la proposition suivante :

PROPOSITION : *Il existe une matrice* P *d'ordre* $p2^n$, *vérifiant* $P^* = P^{-1}$ *constituée de blocs identités d'ordre* p *(au signe près) et de blocs nuls de même* *ordre, et telle que*

$$P^* \Lambda_n P = \Lambda(0, a_n, a_1, \ldots, a_{n-1}) .$$

<u>PREUVE</u> : Montrons d'abord qu'il existe une matrice $S_{i,i+1}$ telle que $S^*_{i,i+1} \Lambda_n S_{i,i+1}$ soit la matrice Λ_n dans laquelle on a simplement permuté a_i et a_{i+1} : il suffit de prendre pour $S_{i,i+1}$ la matrice d'ordre $p2^n$ constituée de blocs diagonaux d'ordre $p2^{i+1}$

$$S_i = \begin{bmatrix} I & 0 & 0 & 0 \\ 0 & 0 & I & 0 \\ 0 & I & 0 & 0 \\ 0 & 0 & 0 & -I \end{bmatrix}$$ où I est l'identité d'ordre $p2^{i-1}$.

On a alors :

$$S_{i,i+1} = \begin{bmatrix} S_i & 0 & \cdots & 0 \\ 0 & S_i & \cdots & \vdots \\ \vdots & & \ddots & 0 \\ 0 & \cdots & 0 & S_i \end{bmatrix} .$$

En effet, Λ_n a la forme suivante :

$$\Lambda_n = \begin{bmatrix} V_i & X & \cdots & X \\ X & V_i^* & & \vdots \\ \vdots & \vdots & \ddots & \vdots \\ X & \cdots & & V_i^o \end{bmatrix}$$ avec $V_i^o = V_i$ ou V_i^* et X des blocs scalaires

$$\text{et } V_i = \begin{bmatrix} \Lambda_{i-1} & a_i & a_{i+1} & 0 \\ -a_i & \Lambda^*_{i-1} & 0 & a_{i+1} \\ -a_{i+1} & 0 & \Lambda^*_{i-1} & -a_i \\ 0 & -a_{i+1} & a_i & \Lambda_{i-1} \end{bmatrix}$$

où tous les blocs sont d'ordre $p2^{i-1}$.

Comme $S_i^* S_i = S_i^2 = I$, la matrice $S^*_{i,i+1} \Lambda_n S_{i,i+1}$ aura comme blocs diagonaux d'ordre $p2^{i+1}$ les $S_i^* V_i S_i$ et $S_i^* V_i^* S_i$, les autres blocs étant inchangés puisque scalaires : $S_i^*(aI)S_i = aI$.

Or le calcul montre que

$$S_i^*V_iS_i = \begin{bmatrix} \Lambda_{i-1} & a_{i+1} & a_i & 0 \\ -a_{i+1} & \Lambda_{i-1}^* & 0 & a_i \\ -a_i & 0 & \Lambda_{i-1}^* & -a_{i+1} \\ 0 & -a_i & a_{i+1} & \Lambda_{i-1} \end{bmatrix}$$

d'où le résultat.

Maintenant, toute permutation de $\{1,2,\ldots,n\}$, et en particulier le cycle $(n,n-1,n-2,\ldots,1)$, peut s'obtenir en composant des transpositions $(i,i+1)$: il existe donc une matrice P d'ordre $p2^n$, constituée de blocs identités (au signe près) d'ordre p et de blocs nuls, telle que

$$P^*\Lambda_nP = \Lambda(0,a_n,a_1,\ldots,a_{n-1})$$

et bien sûr

$$P^*\Lambda_n^*P = \Lambda^*(0,a_n,a_1,\ldots,a_{n-1})$$

(on rappelle que $\Lambda_n^* = -\Lambda_n = \Lambda(0,-a_1,\ldots,-a_n)$.

2.3. *Les matrices* D *et* T.

On note $D_n = \Lambda(0,1,0,\ldots,0)$ la matrice d'ordre $p2^n$ et $T = \begin{bmatrix} 0 & I \\ -I & 0 \end{bmatrix}$ où I est d'ordre p. D_n est donc constituée de blocs diagonaux d'ordre 2p T ou $T^* = -T$, et d'autres blocs nuls de même ordre. On aura besoin, au paragraphe 3, des calculs de $D_n^*\Lambda_nD_n$ et de $\Lambda_nD_n^* + D_n\Lambda_n^*$.

a) $D_n^*\Lambda_nD_n$.

On peut écrire Λ_n sous la forme $\Lambda_n = \Lambda(a_1T,a_2,\ldots,a_n)$. (On commence la construction par a_1T). On démontre alors par récurrence sur n que

$$D_n^*\Lambda_nD_n = \Lambda(a_1T,-a_2,\ldots,-a_n) .$$

Preuve : <u>n = 1</u> $\Lambda_1 = a_1 T$, $D_1 = T$, $D_1^* = -T$; donc $D_1^*\Lambda_1 D_1 = -a_1 T^3 = a_1 T$ (car $T^2 = -T$).

<u>n ⟶ n+1</u>

On a $\Lambda_{n+1} = \begin{pmatrix} \Lambda_n & a_{n+1} \\ -a_{n+1} & \Lambda_n^* \end{pmatrix}$, $D_{n+1} = \begin{pmatrix} D_n & 0 \\ 0 & -D_n \end{pmatrix}$

et on vérifie aisément que

$$-D_{n+1}\Lambda_{n+1}D_{n+1} = \begin{pmatrix} -D_n\Lambda_n D_n & -a_{n+1} \\ a_{n+1} & -D_n\Lambda_n^* D_n \end{pmatrix} = \Lambda(a_1 T, -a_2, \ldots, -a_{n+1}).$$

b) <u>$\Lambda_n D_n^* + D_n\Lambda_n^*$</u> .

On montre par récurrence que $\Lambda_n D_n^* + D_n\Lambda_n^* = 2a_1 I$

$n = 1$: $a_1 T(-T) + T(-a_1 T) = -2a_1 T^2 = 2a_1 I$

$n \to n+1$:

$$\begin{pmatrix} \Lambda_n & a_{n+1} \\ -a_{n+1} & \Lambda_n^* \end{pmatrix}\begin{pmatrix} D_n^* & 0 \\ 0 & D_n \end{pmatrix} + \begin{pmatrix} D_n & 0 \\ 0 & D_n^* \end{pmatrix}\begin{pmatrix} \Lambda_n^* & -a_{n+1} \\ a_{n+1} & \Lambda_n \end{pmatrix}$$

$$= \begin{pmatrix} \Lambda_n D_n^* + D_n\Lambda_n^* & \\ & \Lambda_n^* D_n + D_n^*\Lambda_n \end{pmatrix} = \begin{pmatrix} 2a_1 I & 0 \\ 0 & 2a_1 I \end{pmatrix}.$$

2.4. <u>*Matrices "microscopiques" et matrices "macroscopiques"*</u>.

Soient $p > 0$ et n deux entiers, on appellera <u>matrice micoscopique</u> une matrice de la forme

$$\overline{A} = \begin{pmatrix} A & 0 & \ldots & 0 \\ 0 & A & & \vdots \\ \vdots & & \ddots & 0 \\ 0 & \ldots & 0 & A \end{pmatrix}$$

où A est une matrice d'ordre p et $\overline{A}$ d'ordre $p2^n$ (c'est le produit des tenseurs A et I_{2^n}).

On appellera d'autre part <u>matrice macroscopique</u> une matrice de la forme

$$B = \begin{bmatrix} a_{11}I_p & \cdots\cdots & a_{12^n}I_p \\ \vdots & & \vdots \\ a_{2^n 1}I_p & \cdots\cdots & a_{2^n 2^n}I_p \end{bmatrix}$$

(C'est le produit d'un tenseur d'ordre 2^n par le tenseur d'ordre p, I_p).

On a alors le

<u>LEMME</u> : *Les matrices microscopiques et macroscopiques intercommutent.*

<u>PREUVE</u> : Récurrence sur n

$n=0$: Evident car les matrices macroscopiques sont scalaires.

$n \to n+1$: Soit B macroscopique d'ordre $p2^{n+1}$:

$$B = \begin{bmatrix} B_{11} & B_{21} \\ B_{12} & B_{22} \end{bmatrix}$$ les B_{ij} étant macroscopiques d'ordre $p2^n$

et soit $\bar{A} = \begin{bmatrix} \bar{A}' & 0 \\ 0 & \bar{A}' \end{bmatrix}$ microscopique ; on a :

$$B\bar{A} = \begin{bmatrix} B_{11}\bar{A}' & B_{21}\bar{A}' \\ B_{12}\bar{A}' & B_{22}\bar{A}' \end{bmatrix} = \begin{bmatrix} \bar{A}'B_{11} & \bar{A}'B_{21} \\ \bar{A}'B_{12} & \bar{A}'B_{22} \end{bmatrix} = \bar{A}B \,. \qquad \text{c.q.f.d.}$$

<u>Remarque</u> : On démontrerait aussi facilement que des matrices macroscopiques et microscopiques d'ordre p.n commutent.

Nous pouvons maintenant faire la

3. CONSTRUCTION DU X-ESPACE QUADRATIQUE.

Soient q_1 et q_2 les matrices du § 1, supposées d'ordre p. On notera $M = \bar{q}_1$, $N = \bar{q}_2$ les matrices microscopiques d'ordre $p2^{n+\ell}$. Changeant légèrement de notation par rapport au § 2, on notera

$$\Lambda_i = \Lambda(0, x_0^{2m}, x_{i+1}^{2m}, \ldots, x_n^{2m}, t_1, \ldots, t_\ell, x_1^{2m}, \ldots, x_{i-1}^{2m}) \quad \text{pour} \quad i \neq 0,$$

et $\Lambda_o = \Lambda(0,x_1^{2m},\ldots,x_n^{2m},t_1,\ldots,t_\ell)$. Notons $t'' = n+\ell$.

Posons $Q_i = \frac{1}{x_i^{2m}}\begin{bmatrix} M & \Lambda_i^* \\ \Lambda_i & N \end{bmatrix}$.

On a $\text{dét}\, Q_i = \text{dét}\left(\frac{MN-\Lambda_i\Lambda_i^*}{x_i^{4m}}\right) = \text{dét}\left(\frac{s^{2m}-s_i}{x_i^{4m}} I\right) = 1$ puisque $M,N,\Lambda_i,\Lambda_i^*$ inter-commutent (cf. [2] pb. 56, p. 56), les deux premières étant microscopiques, les deux dernières macroscopiques. Q_i est donc une matrice symétrique inversible à coefficients dans A_{x_i}.

Montrons que, comme A_{sx_i}-espaces quadratiques, Q_i est isométrique à $2^{t''+1}q_{1i}$: le produit $B_i^* Q_i B_i$ où $B_i = \begin{bmatrix} I - \Lambda_i^*/x_i^{2m} \\ 0 & M/x_i^{2m} \end{bmatrix}$ vaut $\frac{1}{x_i^{2m}}\begin{bmatrix} M & 0 \\ 0 & M \end{bmatrix}$ et B_i est inversible sur A_{sx_i}.

Q_i a donc la signature souhaitée sur D_{x_i} : il reste à montrer qu'on peut recoller ces espaces. Pour cela, posons pour $i \neq 0$

$$f_i = \frac{1}{x_i^{2m}\sqrt{2}}\begin{bmatrix} (\Delta_i+\Lambda_i) & N \\ M & (\Delta_i+\Lambda_i)^* \end{bmatrix} \quad \text{où } \Delta_i = x_i^{2m} D_{t''} \text{ (notation du § 2).}$$

Toujours à cause de 2.4, on a

$$\text{dét}\, f_i = \text{dét}\left[\frac{(\Delta_i+\Lambda_i)(\Delta_i+\Lambda_i)^*-MN}{2x_i^{4m}}\right] = \text{dét}\left[\left(\frac{x_o}{x_i}\right)^{2m} I\right]$$

car $\Lambda_i+\Delta_i = \Lambda(0,x_0^{2m}+x_i^{2m},x_{i+1}^{2m},\ldots,x_{i-1}^{2m})$ et donc

$$(\Lambda_i+\Delta_i)(\Lambda_i+\Delta_i)^* = \Big((x_0^{2m}+x_i^{2m})^2 + \sum_{\substack{j\neq 0\\ j\neq i}} x_j^{4m} + \sum_{j=1}^{\ell} t_j^2\Big)I = (s^{2m}+2x_0^{2m}x_i^{2m})I .$$

Calculons alors $f_i^* Q_i f_i$.

On obtient :

$$\frac{1}{2x_i^{6m}}\begin{bmatrix} \Delta_i^* M \Delta_i + x_i^{4m} M & s^{2m}\Delta_i^* + \Delta_i^* \Lambda_i^* \Delta_i^* + x_i^{4m}\Delta_i^* - s_i \Delta_i^* + \Lambda_i^* x_i^{4m} \\ s^{2m}\Delta_i + \Delta_i \Lambda_i \Delta_i + x_i^{4m}\Delta_i - s_i \Delta_i + \Lambda_i x_i^{4m} & \Delta_i N \Delta_i^* + x_i^{4m} N \end{bmatrix}$$

c'est-à-dire encore :

$$\frac{1}{2x_i^{6m}}\begin{bmatrix} 2x_i^{4m} M & 2x_i^{4m}\Delta_i^* + x_i^{4m}\Lambda_i^* - \Delta_i^*\Lambda_i^*\Delta_i \\ 2x_i^{4m}\Delta_i + x_i^{4m}\Lambda_i - \Delta_i^*\Lambda_i\Delta_i & 2x_i^{4m} N \end{bmatrix}.$$

Or, comme on l'a vu en 2.3 a), on a :

$$\Delta_i^*\Lambda_i\Delta_i = x_i^{4m}\Lambda(x_0^{2m}T, -x_{i+1}^{2m}, \dots, -x_{i-1}^{2m})$$

et

$$x_i^{4m}\Lambda_i = x_i^{4m}\Lambda(x_0^{2m}T, x_{i+1}^{2m}, \dots, x_{i-1}^{2m}).$$

D'où l'on tire que

$$2x_i^{4m}\Delta_i + x_i^{4m}\Lambda_i - \Delta_i^*\Lambda_i\Delta_i = 2x_i^{4m}\Lambda(0, x_i^{2m}, x_{i+1}^{2m}, \dots, x_{i-1}^{2m}).$$

Notant $2x_i^{4m}\Lambda_{iO}$ cette dernière matrice, on obtient que :

$$f_i^* Q_i f_i = \frac{1}{x_i^{2m}}\begin{bmatrix} M & \Lambda_{iO}^* \\ \Lambda_{iO} & N \end{bmatrix} \qquad \square$$

Si l'on note P la matrice macroscopique d'ordre $p2^{t''}$ associée au cycle $(t'', t''-1, \dots, 1)$ (2.2) et $\overline{P} = \begin{bmatrix} P & 0 \\ 0 & P \end{bmatrix}$, on a

$$\left(\frac{x_i}{x_0}\right)^m \overline{P}^{i-1*} f_i^* Q_i f_i \overline{P}^{i-1} \left(\frac{x_i}{x_0}\right)^m = \frac{1}{x_0^{2m}}\begin{bmatrix} M & \Lambda_O^* \\ \Lambda_O & N \end{bmatrix} = Q_O$$

car $P^{i-1*}{}_{iO}P^{i-1} = \Lambda_O$, et P^{i-1} étant macroscopique

on a $P^{i-1*}MP^{i-1} = MP^{i-1*}P^{i-1} = M$

$P^{i-1*}NP^{i-1} = NP^{i-1*}P^{i-1} = N.$

Sur $D_{x_i x_0}$, $f_i \overline{P}^{i-1}\left(\frac{x_i}{x_0}\right)^m$ réalise donc une isométrie entre Q_i et Q_O ($i \neq 0$). Notons λ_{iO} cette isométrie. Pour calculer son inverse, il suffit de connaître l'inverse de f_i. Or, en raison des commutations, l'inverse se calculant par blocs, on vérifie aisément que

$$f_i^{-1} = \frac{1}{x_0^{2m}\sqrt{2}} \begin{bmatrix} (D_i+\Lambda_i)^* & -N \\ -M & (D_i+\Lambda_i) \end{bmatrix} .$$

<u>*On pose alors* $\lambda_{ij} = \lambda_{i0}\lambda_{j0}^{-1}$</u> : C'est un isomorphisme a priori défini sur $D_{x_0} \cap D_{x_i} \cap D_{x_j}$; on va montrer qu'il est en fait défini sur $D_{x_i} \cap D_{x_j}$: on a

$$\lambda_{ij} = (\frac{x_i}{x_0})^m f_i \overline{P}^{-i-1} (\frac{x_0}{x_j})^m \overline{P}^{-j+1} f_j^{-1} ;$$

pour que λ_{ij} soit défini sur $D_{x_i x_j}$, il faut et il suffit que $\overline{P}^{-i-1}\lambda_{ij}\overline{P}^{j+1}$ le soit. Or

$$P^{-i-1}\lambda_{ij}\overline{P}^{j+1} = (\frac{x_i}{x_j})^m \overline{P}^{-i-1} \frac{1}{x_i^{2m}\sqrt{2}} \begin{bmatrix} (D_i+\Lambda_i) & N \\ M & (D_i+\Lambda_i) \end{bmatrix} \overline{P}^{i+1}\overline{P}^{-j-1} \frac{1}{x_0^{2m}\sqrt{2}} \begin{bmatrix} (D_j+\Lambda_j)^* & -N \\ -M & (D_j+\Lambda_j) \end{bmatrix} \overline{P}^{j+1}$$

$$= \frac{1}{2(x_i x_j)^m x_0^{2m}} \begin{bmatrix} P^{-i-1}(D_i+\Lambda_i)P^{i+1} & N \\ M & P^{-i-1}(D_i+\Lambda_i)^* P^{i+1} \end{bmatrix} \begin{bmatrix} P^{-j-1}(D_j+\Lambda_j)^* P^{j+1} & -N \\ -M & P^{-j-1}(D_j+\Lambda_j)P^{j+1} \end{bmatrix} .$$

Notons $\Lambda^i = P^{-i-1}(D_i+\Lambda_i)P^{i+1} = \Lambda(0, x_1^{2m}, x_2^{2m}, \ldots \overset{i}{(x_0^{2m} + x_i^{2m})}, \ldots, x_n^{2m}, t_1, \ldots, t\,)$, il reste alors à calculer

$$\begin{pmatrix} \Lambda^i & N \\ M & \Lambda^{i*} \end{pmatrix} \begin{pmatrix} \Lambda^{j*} & -N \\ -M & \Lambda^j \end{pmatrix} = \begin{pmatrix} \Lambda^i\Lambda^{j*} - s^{2m}I & N(\Lambda^j - \Lambda^i) \\ M(\Lambda^{j*} - \Lambda^{i*}) & \Lambda^{i*}\Lambda^j - s^{2m}I \end{pmatrix} .$$

Or, si l'on note $T_i = \Lambda(0,0,\ldots,0,\overset{i}{1},0\ldots,0)$, on a $\Lambda^i = (\Lambda_0 + x_0^{2m}T_i)$ et donc

$$\begin{aligned} \Lambda^i\Lambda^{j*} = \Lambda^{i*}\Lambda^j &= (\Lambda_0 + x_0^{2m}T_i)(\Lambda_0^* + x_0^{2m}T_j^*) \\ &= \Lambda_0\Lambda_0^* + x_0^{2m}[T_i\Lambda_0^* + \Lambda_0 T_j^* + x_0^{2m}T_iT_j^*] \\ &= s_0 I + x_0^{2m}[T_i\Lambda_0^* + \Lambda_0 T_j^* + x_0^{2m}T_iT_j^*] \end{aligned}$$

et $\Lambda^i\Lambda^{j*} - s^{2m}I = -x_o^{4m}I + x_0^{2m}[T_i\Lambda_0^* + \Lambda_0 T_j^* + x_0^{2m}T_iT_j^*]$.

D'autre part $\Lambda^j - \Lambda^i = x_0^{2m}(T_j - T_i)$; on peut donc écrire :

$$\overline{P}^{-i-1}\lambda_{ij}\overline{P}^{j+1} = \frac{1}{2(x_ix_j)^m}\begin{bmatrix} T_i\Lambda_0^*+\Lambda_0T_j^*+x_0^{2m}(T_iT_j^*-I) & N(T_j-T_i) \\ M(T_j^*-T_j^*) & T_i\Lambda_0^*+\Lambda_0T_j^*+x_0^{2m}(T_iT_j^*-I)\end{bmatrix}$$

et cette matrice est définie sur $D_{x_ix_j}$: il en est donc de même de λ_{ij}.
Il est clair que c'est un isomorphisme puisque

$$\text{dét}\lambda_{ij} = \text{dét}((\frac{x_i}{x_0})^m \times (\frac{x_0}{x_i})^{2m} I) \cdot \text{dét}((\frac{x_0}{x_j})^m (\frac{x_j}{x_0})^{2m} I)$$

$$= \text{dét}(\frac{x_j}{x_i})^m I .$$

D'autre part, la collection des (λ_{ij}) définit évidemment un cocycle puisque le calcul formel sur l'anneau des fonctions sur X nous donne

$$\lambda_{ij}\lambda_{jk} = \lambda_{io}\lambda_{jo}^{-1}\lambda_{jo}\lambda_{ko}^{-1} = \lambda_{io}\lambda_{ko}^{-1} = \lambda_{ik} .$$

Ceci termine la construction du X-espace quadratique Q, vérifiant $\hat{Q} = 2^{t''+1}\hat{q}$ et donc

$$\hat{Q}(F_1) = 2^{t'+t''+1} , \quad \hat{Q}(F_2) = -2^{t'+t''+1} . \quad \square$$

REFERENCES

[1] H. DELFS, M. KNEBUSCH : Semialgebraic topology over a real closed field II, Math. Z. 178, 175-213 (1981).

[2] P.R. HALMOS : A Hilbert space problem book, Prnceton, Van Nostrand Company (1967).

[3] M. KAROUBI : Localisation de formes quadratiques I, Ann. Sc. Ec. N. Sup. 4e série, 7, fasc. 3 (1974) 359-404.

[4] M. KNEBUSCH : Symmetric bilinear forms over algebraic varieties, in conf. on quadratic forms, Queen's papers in p. and ap. Math. n° 46 Orzech ed. (1977) 103-283.

[5] L. MAHE : Signatures et composantes connexes. Math. Annalen (à paraître).

[6] I. SHAFAREVICH : Basic Algebraic geometry, Springer Study Edition (1977).

IRMAR
Université de Rennes I
Campus de Beaulieu
35042 - RENNES-CEDEX
(FRANCE)

On Real One-Dimensional Cycles

Friedrich Ischebeck

Numerous people have proved what L. Bröcker called EPT (= Everybody Proved this Theorem). It says, that a divisor on a projective, nonsingular $\mathbb{R}$-variety, whose "real part" is $\mathbb{Z}/2$-homologous to zero, is linear equivalent to a divisor supporting no real point ([3]). Naturally one may ask, whether there are analogous results for cycles in other dimensions. As to the zero-dimensional case, it is settled in [8] and [5].

Here we consider the one-dimensional case and prove in the 2nd section: Any one-dimensional cycle on a projective, nonsingular $\mathbb{R}$-variety X, whose class in $H_1(X(\mathbb{R}),\mathbb{Z}/2)$ is zero, is rational equivalent to a cycle, whose support has only finitely many real points.

General remarks on real cycles in any dimension are made in the first chapter.

Even for zero-dimensional cycles the method used to prove our main theorem gives us a new proposition. Namely finitely many real points lying on one connected component of a nonsingular projective $\mathbb{R}$-variety can be linked by one draught (German: Zug) of a curve (chapter 3).

1. Real rational equivalence

<u>Definitions:</u> An $\mathbb{R}$-variety (i.e. an integral scheme of finite type over $\mathbb{R}$) is called real, if its function field is formally real (or equivalently, if $X(\mathbb{R})$ is Zariski-dence in X).

Analogously we define "real prime-cycles", especially "real prime-divisors".

Let $Z_m(X)$, resp. $Z_m^{\mathbb{R}}(X)$ be the abelian group, free generated by the m-dimensional prime cycles resp. by the real m-dimensional prime-cycles, and $\alpha: Z_m(X) \to Z_m^{\mathbb{R}}(X)$ the canonical epimorphism.

A real (Weil-) divisor of a normal $\mathbb{R}$-variety X is called a real principal divisor, if it is the image of a principal divisor under $\alpha: \mathrm{Div}(X) \to \mathrm{Div}^{\mathbb{R}}(X)$. ($\mathrm{Div} = Z_{n-1}$, where $n = \dim X$.)

Define subgroups $F_m(X) \subset Z_m(X)$ and $F_m^{\mathbb{R}}(X) \subset Z_m^{\mathbb{R}}(X)$ as follows:

a) F_m is generated by those $z \in Z_m(X)$ which are the direct image ([6] 1.2) of some principal divisor of a normal $\mathbb{R}$-variety V of dimension m+1 under a proper morphism $f: V \to X$. $F_m(X)$ is the group of m-cycles, rational equivalent to zero (loc.cit. 1.8.).

b) $F_m^{\mathbb{R}}$ is generated by those $z \in Z_m^{\mathbb{R}}(X)$ which are the direct image of some <u>real</u> principal divisor of a <u>real</u> normal variety V of dimension m+1 under a proper morphism $f: V \to X$.

Two real m-cycles are called real rational equivalent, if their difference is in $F_m^{\mathbb{R}}(X)$.

Call $A_m(X) := Z_m(X)/F_m(X)$ resp. $A_m^{\mathbb{R}} := Z_m^{\mathbb{R}}(X)/F_m^{\mathbb{R}}(X)$ the m-th Chow-group of X, resp. the m-th real Chow-group of X.

Let X be normal. Then the real divisor class group $C^{\mathbb{R}}(X)$ is defined to be the residue class group of $\mathrm{Div}^{\mathbb{R}}(X)$ modulo the subgroup of the real principal divisors.

There is an epimorphism $C(X) \to C^{\mathbb{R}}(X)$ induced by α.

<u>Proposition 1:</u> $2C^{\mathbb{R}}(X) = 0$.

This is Satz 1 of [8], and also an easy consequence of [9] 4.2.

Lemma: $2A_m^{\mathbb{R}}(X) = 0$ for any $\mathbb{R}$-algebraic scheme X.

Proof: Let $Z \in Z_m^{\mathbb{R}}(X)$ be a prime cycle, considered as an m-dimensional subvariety, $\nu: \tilde{Z} \to Z$ its normalization, $V = \mathbb{P}_1 \times \tilde{Z}$, $f = i \cdot \nu \cdot pr_2$, with $i: Z \to X$ the inclusion. Then f is proper and with $p \in \mathbb{P}_1(\mathbb{R})$ the divisor $h := 2(\{p\} \times \tilde{Z})$ is a real principal divisor of V, according to prop. 1. We have $f_*(h) = 2Z$.

Proposition 2: Let X be an $\mathbb{R}$-algebraic scheme of pure dimension n.

a) Generally $F_m^{\mathbb{R}} = \alpha(F_m(X)) + 2Z_m^{\mathbb{R}}(X)$.

b) If X is normal, then $F_{n-1}^{\mathbb{R}}(X) = \alpha(F_{n-1}(X))$ and $A_{n-1}^{\mathbb{R}}(X) = C^{\mathbb{R}}(X)$, i.e. $F_{n-1}^{\mathbb{R}}(X)$ consists only of real principal divisors. (According to the definition $F_{n-1}^{\mathbb{R}}(X)$ could possibly be bigger.)

c) If X is nonsingular and $m < n$, then $F_m^{\mathbb{R}}(X) = \alpha(F_m(X))$.

Proof: a) The lemma says $2Z_m^{\mathbb{R}}(X) \subset F_m^{\mathbb{R}}(X)$. Now let V be a normal $\mathbb{R}$-variety of dimension m+1, f: V → X be a proper morphism and p be a nonreal prime divisor. If f(p) is also nonreal, then $\alpha(f_*(p)) = 0$. If dim p > dim f(p), then $f_*(p) = 0$. In the remaining case, where f(p) real and dim p = dim f(p), the function field of p is of even (finite) degree over that of f(p), since an extension of odd degree over a formally real field is itself formally real. In any case we have for nonreal p, that $\alpha(f_*(p)) \in 2Z_m^{\mathbb{R}}(X)$ holds.

A nonreal V has no real prime divisor. For each field having a place into a formally real field is itself formally real. So for nonreal V and any divisor d on V one has $\alpha(f_*(d)) \in 2Z_m^{\mathbb{R}}(X) \subset F_m^{\mathbb{R}}(x)$.

Now let V be real and $h = \sum_{p \text{ real}} n_p p + \sum_{q \text{ nonreal}} n_q q$ be a principal divisor on V. Then $h' := \sum_{p \text{ real}} n_p p$ is a real principal divisor and we have $\alpha(f_*(h)) = f_*(h')+2z$ with some real cycle z. So $\alpha(f_*(h)) \in F_m^{\mathbb{R}}(X) + 2Z_m^{\mathbb{R}}(X) = F_m^{\mathbb{R}}(X)$. We have shown that $\alpha(F_m(X)) + 2Z_m^{\mathbb{R}}(X) \subset F_m^{\mathbb{R}}(X)$.

To show the converse inclusion let h' be a real principal divisor of a normal real variety V and h be a principal divisor whose image in $\mathrm{Div}^{\mathbb{R}}(X)$ is h'. Then one has (with the above notation) $f_*(h') = \alpha(f_*(h)) - 2z \in \alpha(F_m(X)) + 2Z_m^{\mathbb{R}}(X)$.

b) To prove the first equality one still has to show that $2Z_{n-1}^{\mathbb{R}}(X) \subset \alpha(F_{n-1}(X))$. This follows from prop. 1, since every real principal divisor comes from $F_{n-1}(X)$. By [6] §1.8., Cor. (2) we know, that $F_{n-1}(X)$ consists only of principal divisors. So by definition every element of $F_{n-1}^{\mathbb{R}}(X) = \alpha(F_{n-1}(X))$ is a real principal divisor.

c) We still have to show $2Z_m^{\mathbb{R}}(X) \subset \alpha(F_m(X))$.

Let Z be an m-dimensional prime cycle. Since X is nonsingular, there exists a closed subvariety Y of X of dimension $m+1$ containing X and normal in (the generic point of) Z. Let $\tilde{Y}$ be its normalization and $f: \tilde{Y} \to X$ be the induced morphism.

Now for almost all prime cycles Z' of dimension m lying on Y, especially for $Z' = Z$, the following holds: $f^{-1}(Z')$ is a prime divisor on $\tilde{Y}$ of degree 1 over Z', because Y is normal in almost all those Z', especially in Z. So there are only finitely many nonreal prime divisors of $\tilde{Y}$ whose image under f is real, say $p_1,\ldots,p_r$. Using [9] Satz 4.1 we find rational functions g_0, g_1 on $\tilde{Y}$ with the following properties: $\mathrm{ord}_{f^{-1}(Z)}(g_i) = 1$, $\mathrm{ord}_{p_j}(g_0) = 0$, $\mathrm{ord}_{p_j}(g_1) > 0$, and $\mathrm{ord}_q(g_i) \geq 0$, $\mathrm{ord}_q(g_0) \cdot \mathrm{ord}_q(g_1) = 0$ for all real prime divisors q of $\tilde{Y}$. Then we have $\mathrm{div}(g_0^2+g_1^2) = 2\cdot f^{-1}(Z) + \Sigma u_j q_j$, where the q_j are non real and distinct from the $p_1,\ldots,p_r$. Therefore $f_*(\mathrm{div}(g_0^2+g_1^2)) = 2Z$

Corollary: Let $B_m(X)$ be the subgroup of $A_m(X)$, which is generated by the classes of nonreal prime cycles. Then

$A_m^{\mathbb{R}}(X) = A_m(X)/(B_m(X) + 2A_m(X))$. If X is nonsingular and $m < n$, then $A_m^{\mathbb{R}}(X) = A_m(X)/B_m(X)$.

In other words: For nonsingular X and $m < n$ holds: An m-cycle is real rational equivalent to zero, iff it is rational equivalent to a cycle whose support contains only a lower dimensional set of real points.

Remark: The proposition 2 of 1.6 and that of 1.8 of the paper [6] hold for real cycles only modulo 2. So I don't see how to avoid the effort I had to prove prop. 2c).

2. One-dimensional cycles

For the rest of the paper let X be a projective nonsingular $\mathbb{R}$-variety of dimension n. Every real prime cycle Z of X of dimension m has a fundamental class in $H_m(Z(\mathbb{R}),\mathbb{Z}/2)$; see e.g. [1].

So one gets a canonical homomorphism $r: Z_m^{\mathbb{R}}(X) \otimes \mathbb{Z}/2 \to H_m(X(\mathbb{R}),\mathbb{Z}/2)$, (loc.cit. 5.12). (Here $X(\mathbb{R}), Z(\mathbb{R})$ carry their "strong topology" - induced by the ordering of $\mathbb{R}$. Further H_* is in loc.cit. the "homology for locally compact spaces", which is identical to the singular homology in our case, since $X(\mathbb{R})$ and $Z(\mathbb{R})$ are triangulable see [2] 5.).

Proposition 3: For real rational equivalent cycles $z_1, z_2 \in Z_m^{\mathbb{R}}$ holds $r(z_1) = r(z_2)$.

This is identical with prop. 5.13 of [1]. But perhaps I must make some remarks concerning the different vocabulary. The space V in [1] is the same as our $X_{\mathbb{C}}(\mathbb{C})$, and V^{o} is our $X(\mathbb{R})$. (Here $X_{\mathbb{C}} = X \otimes_{\mathbb{R}} \mathbb{C}$.) The really defined cycles of V are the cycles of X, whereas the cycles of V^{o} are the real cycles of X:

$$\mathfrak{Z}_{\mathbb{R}}(V;Z_2) = Z(X) \otimes \mathbb{Z}/2 ,$$
$$\mathfrak{Z}(V^{o};Z_2) = Z^{\mathbb{R}}(X) \otimes \mathbb{Z}/2 .$$

The maps α and r in [1] are the same as our α (tensored with $\mathbb{Z}/2$) and r. To prove the proposition, we know from proposition 2, that there exists a $z \in Z(X)$ being rational equivalent to 0 with $\alpha(z) = z_1 - z_2$. So prop. 5.13 of [1] gives $r(z_1-z_2) = r\alpha(z) = p(z) = 0$.

By proposition 3 we obtain natural homomorphisms:

$$\lambda_m : A_m^{\mathbb{R}}(X) \longrightarrow H_m(X(\mathbb{R}), \mathbb{Z}/2) .$$

<u>Conjecture:</u> λ_m is injective for every m.

This is true for trivial reasons for $m \geq n$. EPT [3] says, that λ_{n-1} is injective. Following [8] or [5] λ_o is bijective (even if X is only complete and not necessary projective).

The aim of this paper is to prove:

<u>Theorem:</u> λ_1 is injective.

<u>Preliminary remarks:</u> We set $X^{o} := X(\mathbb{R})$ for any $\mathbb{R}$-algebraic scheme X. To any $\mathbb{R}$-morphism $f: Y \to X$ one gets by restriction a continuous map (w.r.t. strong topology) $f^{o}: Y^{o} \to X^{o}$. "Modulo 2" we may write any real cycle on X as a sum of pairwise distinct real prime cycles. So we

identify it with a reduced closed real subscheme Y of X. Let $\tilde{Y} \to Y$ be a desingularization; then the (canonical) map $\tilde{Y}^o \to X^o$ is a so named "closed singular manifold" (analogously to the name "singular simplex" in Algebraic Topology - see [4] 2.1).

In the case $\dim Y = 1$, the class of Y in $H_1(X^o, \mathbb{Z}/2)$ is zero, iff $f^o\colon \tilde{Y}^o \to X^o$ is zero-bordant, i.e. iff there is a two-dimensional compact bounded C^∞-manifold M with $\partial M = \tilde{Y}^o$ and a continuous map $F\colon M \to X^o$ with $F|\tilde{Y}^o = f^o$, see [4] p. 13. One may as well suppose, that F is a C^∞-map according to P.E. Conner, E.E. Floyd : "Differentiable periodic maps" Thm. 9.1. So it suffices to prove the following :

Proposition 4: Let X be as above, Y be a real nonsingular projective $\mathbb{R}$-algebraic scheme of pure dimension, and $f\colon Y \to X$ be a morphism such that $Y \xrightarrow{f} f(Y)$ is birational. If $f^o\colon Y^o \to X^o$ is zero-bordant, then $f(Y)$ as a cycle on X is real rational equivalent to zero.

Proof: The zero-bordism of f^o is given by some C^∞-map $F\colon M \to X^o$, where M is a compact, bounded C^∞-manifold with $\partial M = Y^o$ and $F|Y^o = f^o$. Since X is projective, one can embed X^o into some $\mathbb{R}^p$ as an algebraic subset. (See e.g. [10]3. cor.3 of prop.2.) Further M can be C^∞-embedded into some $\mathbb{R}^{q+1}$ in such a way, that for $(x_1,\dots,x_{q+1}) \in M$ one has $x_{q+1} \geq 0$ and that Y^o is an algebraic subset of $\mathbb{R}^q \subset \mathbb{R}^{q+1}$ and $Y^o = M \cap \mathbb{R}^q$ and finally that $N := M \cup M_-$ is an unbounded C^∞-submanifold of $\mathbb{R}^{q+1}$. Here $\mathbb{R}^q$ is embedded into $\mathbb{R}^{q+1}$ by $(x_1,\dots,x_q) \longmapsto (x_1,\dots,x_q,0)$ and M_- is symmetric to M, i.e. $M_- := \{(x_1,\dots,x_{q+1}) | (x_1,\dots,x_q,-x_{q+1}) \in M\}$. Next choose two tubular neighbourhoods U', U of X^o in $\mathbb{R}^p$ with $\overline{U'} \subset U$. By Weierstraß' approximation theorem in the strong form of [10] 5. Thm. 1 one can approximate F by a polynomial map $\varphi\colon \mathbb{R}^{q+1} \to \mathbb{R}^p$ such that $\varphi(M) \subset U'$ and $\varphi|Y^o = f^o$ holds.

Now approximate N by the union V of some connected components (w.r.t. strong topology) of a nonsingular compact algebraic subset Z^o of $\mathbb{R}^{q+1}$ such that the following conditions are fulfilled:

i) V lies in a tubular neighbourhood of N and the projection gives a diffeomorphism $V \to N$;

ii) $V \cap \mathbb{R}^q = Y^o$;

iii) $\varphi(V_+) \subset U$, here $V_+ := \{(x_1,\ldots,x_{q+1}) \in V \mid x_{q+1} \geq 0\}$.

The possibility of such an approximation is shown in Tognoli's booklet [10] during the proof of 5. Thm. 2, especially in "step III" and "step IV".

Set $V_\varepsilon := \{(x_1,\ldots,x_{q+1}) \in V \mid x_{q+1} > -\varepsilon\}$. Then there exists $\varepsilon > 0$, so that even iii') $\varphi(V_\varepsilon) \subset U$ and V_ε has the same number of connected components as V, i.e. V_ε is V_+ equipped with a collar.

Now let $\pi: U \to X^o$ be the projection and $\psi := \pi \circ (\varphi|V_\varepsilon): V_\varepsilon \to X^o$. The maps π and ψ are analytic but generally not polynomial. However the graph Γ_ψ of ψ possesses an open neighbourhood S (w.r.t. the strong topology) with $\Gamma_\psi = S \cap \{(z,x) \in Z^o \times X^o \mid (\varphi(z)-x) \perp T_{x,X^o}\}$. (Here T_{x,X^o} is the tangent space of X^o in x and "$\perp$" means "orthogonal".) Now the last set $\{(z,x) \in \ldots.\}$ is defined by polynomial equations. In other words if we denote by W the closure of Γ_ψ in $Z \times_{\mathbb{R}} X$ w.r.t. the Zariski-topology, then $\Gamma_\psi = S \cap W$. Γ_ψ is a real analytic manifold, isomorphic (analytically) to V_ε. Its dimension is pure and coincides with that of W. So $\dim W = \dim \Gamma_\psi = \dim V_\varepsilon = \dim M = \dim Y+1$. Further W is regular in all points of Γ_ψ, since Γ_ψ is analytically nonsingular.

Now take a resolution of singularities $\sigma: \widetilde{W} \to W$, which does not touch its regular locus, especially leaves Γ_ψ unchanged ([7]).

We may consider Y^o as an algebraic subset of $\widetilde{W}^o$; namely we have $\varphi|Y^o = f^o$ and $\pi|f^o(Y^o) = id_{f^o(Y^o)}$ and so $Y^o \simeq \Gamma_{f^o} \subset \Gamma_\psi \simeq \sigma^{-1}(\Gamma_\psi) \subset \widetilde{W}^o$, where the isomorphisms are given by polynomial maps. The induced embedding $Y^o \to \widetilde{W}^o$ is zero-bordant by virtue of the canonical embeddings $V_+ \subset \Gamma_\psi \subset \widetilde{W}^o$. Therefore the singular manifold Y^o in $\widetilde{W}^o$ is zero-bordant, and so a fortiori its class in $H_1(\widetilde{W}^o, \mathbb{Z}/2)$ is zero. By EPT [3] the divisor Y of $\widetilde{W}$ is a real principal divisor, where $\widetilde{Y}$ is the Zariski-closure of Y^o in $\widetilde{W}$.

Set $p := pr_2 \circ \sigma : \widetilde{W} \to X$ (where $pr_2: Z \times X \to X$ is the projection). Then $p(\widetilde{Y}) = f(Y)$ and $\widetilde{Y} \xrightarrow{p} f(Y)$ is birational. Therefore $p_*(\widetilde{Y}) = f(Y)$.

3. Linking points

Proposition 5: Let $P_o, \dots, P_n$ be finitely many points, lying on one connected component (w.r.t. the strong topology) of $X(\mathbb{R})$. Then they may be linked by one draught of an $\mathbb{R}$-algebraic curve; i.e. there exist some nonsingular curve C, a morphism $f: C \to X$ and a connected component Γ (w.r.t. the strong topology) of $C(\mathbb{R})$ with $P_i \in f(\Gamma)$ for all i.

Proof: If $n = 1$, then $P_o + P_1$ is zero-bordant and the construction in the proof of prop. 4 gives the desired curve C and morphism $f: C \to X$ as $p: \widetilde{W} \to X$. For bigger n the proof is analogous: Start with a C^∞-map $c: [0,n] \to X^o$ with $c(i) = P_i$, approximate it polynomially by $\varphi: \mathbb{R} \to \mathbb{R}^p \supset X^o$, such that $\varphi(i) = P_i$ and work with some $I_\varepsilon = (-\varepsilon, n+\varepsilon) \subset \mathbb{R}$ instead of V_ε exactly as in the proof of prop. 4.

Acknowledgements: I have to thank J.-L. Colliot-Thélène (Paris) and E. Freitag (Heidelberg) for some advices.

References

1. Borel, A., Haefliger, A.: La classe d'homologie fondamentale d'un espace analytique. Bull.Soc.math.France 89 (1961)461-513

2. Borel, A., Moore, J.C.: Homology theory for locally compact spaces. Mich.math. J. 7 (1960)137-159

3. Bröcker, L.: Reelle Divisoren. Arch.d.Math. 35 (1980)140-143

4. Bröcker, T., tom Dieck, T.: "Kobordismentheorie". Lecture Notes in Math. 178, Springer-Verl. Berlin - Heidelberg - New York 1970

5. Colliot-Thêlène, J.-L., Ischebeck, F.: L'équivalence rationnelle sur les cycles de dimension zéro des variétés algébriques réelles. C.R.Acad.Sc. Paris, 292 (1981) Série I 723-725

6. Fulton, W.: Rational equivalence on singular varieties. Publ.Math. IHES 45 (1975)147-167

7. Hironaka, H.: Resolutions of singularities of an algebraic variety over a field of characteristic zero. Annals of Math. 79 (1964)109-326

8. Ischebeck, F.: Reelle Divisoren und Nullzyklen. Preprint.

9. Ischebeck, F.: Binäre Formen und Primideale. man.math. 35 (1981) 147-163

10. Tognoli, A.: "Algebraic Geometry and Nash Functions". Acad. Press, London - New York 1978

Mathematisches Institut
der Universität Münster
Einsteinstraße 62
D - 4400 Münster
BR Deutschland

Sur l'homologie des surfaces algèbriques réelles

J.J. Risler

Au cours des journées de Rennes, beaucoup de conversations ont portées sur le sujet suivant : soit T une surface algébrique réelle homéomorphe à un Tore ; l'homologie $H_1(T,\mathbb{Z}/_{2\mathbb{Z}})$ est-elle engendrée par des cycles algébriques ? (c.f. par exemple (B-K-S) où la question est posée explicitement) ainsi que (B-T)).

La note qui suit montre que la réponse est en général négative.

Je remercie A.Tognoli et J. Le Potier pour d'utiles conversations sur le sujet de cette note.

§1 - Fibrés algébriques réels

Soient $X \subset \mathbb{R}P^k$ une variété algébrique réelle lisse non vide de dimension n, $\tilde{X}$ sa complexifiée ($\tilde{X}$ est la sous-variété de CP^k définie par les mêmes équations que X).

Nous supposerons aussi que $\tilde{X}$ est lisse. Dans la suite nous poserons $\mathbb{Z}_2 = \mathbb{Z}/_{2}\mathbb{Z}$, $\mathcal{O}_X$ (resp. $\mathcal{O}_{\tilde{X}}$) désignera le faisceau des fonctions algébriques régulières sur X (resp. sur $\tilde{X}$) et $\mathcal{O}_{X,An}$ le faisceau des fonctions analytiques réelles sur X (considérée comme variété analytique).

Si Y est un sous-ensemble algébrique fermé de X de dimension n-1, Y définit un élément de $H_{n-1}(X,\mathbb{Z}_2)$ noté $[Y]$ (c.f. par exemple (B-H)) ; nous noterons $H_{n-1}^{Alg}(X,\mathbb{Z}_2)$ le sous-espace de $H_{n-1}(X,\mathbb{Z}_2)$ engendré par ces cycles algébriques, et $H^1_{Alg}(X,\mathbb{Z}_2)$ son dual pour la dualité de Poincaré ($H^1_{Alg}(X,\mathbb{Z}_2)$ est donc un sous-espace de $H^1(X,\mathbb{Z}_2)$).

Soit τ l'involution sur $\tilde{X}$ induite par la conjuguaison complexe sur CP^k, τ^* l'action qu'elle induit sur le groupe $H^1(\tilde{X},\mathcal{O}^*_{\tilde{X}})$ qui classifie les fibrés de rang 1 sur $\tilde{X}$. Si $G_{\mathbb{R}}$ désigne le groupe des classes d'isomorphisme de fibrés définis sur $\mathbb{R}$ (i.e. les fibrés dont le cocycle est défini à l'aide d'un recouvrement par des ouverts stables par τ et dont les équations correspondantes sont réelles); On a des inclusions :

$$G_{\mathbb{R}} \subset \mathrm{Ker}\ \tau^* \subset H^1(\tilde{X}, \mathcal{O}^*_{\tilde{X}})$$

<u>Remarque 1</u> : Un résultat classique (dont nous n'aurons pas besoin ici) affirme que comme $X \neq \emptyset$, on a l'égalité : $G_{\mathbb{R}} = \mathrm{Ker}\ \tau^*$.

<u>Proposition 1</u> : Il existe une application naturelle $\phi : G_{\mathbb{R}} \to H^1(X,\mathbb{Z}_2)$ dont l'image est $H^1_{Alg}(X,\mathbb{Z}_2)$.

<u>Démonstration</u> :

a) Définition de ϕ - Soit $\tilde{F}$ un fibré algébrique complexe de rang 1 sur $\tilde{X}$ défini sur $\mathbb{R}$: On lui associe canoniquement un fibré algébrique F sur X en droites réelles en définissant F sur les parties réelles des ouverts trivialisant $\tilde{F}$ avec le même cocycle, d'où un élément [F] de $H^1(X, \mathcal{O}^*_X)$.

Il faut maintenant voir que cette application se factorise par $G_{\mathbb{R}}$, i.e. que si $\tilde{F}$ est trivial (en tant que fibré complexe), F est un fibré réel trivial.

Mais si $\tilde{F}$ est trivial, il possède une section s partout non nulle, ce qui définit un isomorphisme $\psi : \Gamma^*(\tilde{F}) \to \mathcal{O}^*(\tilde{X})$ tel que $\psi(s) = 1$ ($\Gamma^*(\tilde{F})$ est l'ensemble des sections globales non nulles de $\tilde{F}$). D'autre part, X étant projective, on a un isomorphisme : $\mathcal{O}^*(\tilde{X}) \tilde{\to} \mathbb{C}^*$, d'où un isomorphisme $\psi' : \Gamma^*(\tilde{F}) \tilde{\to} \mathbb{C}^*$.

$\tilde{F}$ étant défini sur $\mathbb{R}$, il existe une involution naturelle σ sur $\Gamma^*(\tilde{F})$ définie par $\sigma(s)(x) = \bar{s}(\bar{x}))$, et il suffit pour montrer que F est trivial de trouver une section s' partout non nulle de $\tilde{F}$ qui soit "définie sur $\mathbb{R}$", i.e. telle que $\sigma(s') = s'$.

Mais l'involution σ induit (par l'isomorphisme ψ') une involution τ sur $\mathbb{C}^*$ (telle que $\tau(z) = \mu z$ avec $\mu = \psi'(s)$ et $\mu\bar{\mu}= 1$) qui possède une droite réelle de points fixes, ce qui achève de démontrer l'existence d'un morphisme : $G_{\mathbb{R}} \to H^1(X,\mathcal{O}^*_X)$.

Toute fonction algébrique régulière étant analytique, on a un morphisme canonique :

$H^1(X, \mathcal{O}^*_X) \to H^1(X, \mathcal{O}^*_{X,An})$, et un morphisme (qui est un isomorphisme) :

$H^1(X, \mathcal{O}^*_{X,An}) \to H^1(X,\mathbb{Z}_2)$

déduit de la suite exacte de l'exponentielle :

$$0 \to \mathcal{O}_{X,An} \overset{exp}{\to} \mathcal{O}^*_{X,An} \to \mathbb{Z}_2 \to 0,$$

ce qui achève de définir ϕ ($\phi([\tilde{F}]$ est donc la première classe de Stiefel-Whitney du fibré F considéré comme fibré analytique).

Il est immédiat de vérifier que ϕ est une application $\mathbb{Z}$-linéaire.

b) $\mathrm{Im}\ \phi \subset H^1_{Alg}(X,\mathbb{Z}_2)$

Si F est un fibré réel de rang 1 sur X, on lui associe de la manière habituelle un diviseur de Cartier D qui définit à son tour un "diviseur de Weil", i.e. une somme $\Sigma n_i Y_i$ où les Y_i sont des sous-variétés de X dont les complexifiées $\tilde{Y}_i$ sont irréductibles et de codimension 1 ; chaque Y_i définit un élément $[Y_i]$ de $H^{Alg}_{n-1}(X,\mathbb{Z}_2)$ (qui est nul si Y_i n'est pas de codim.1), et donc D définit un élément de $H^{Alg}_{n-1}(X,\mathbb{Z}_2)$ ne dépendant que de la classe de D, et dont le dual est l'image de F dans $H^1_{Alg}(X,\mathbb{Z}_2)$ (cf. par exemple (G-A) p.84 où l'analogue complexe est traité).

c) $\mathrm{Im}\ \phi = H^1_{Alg}(X,\mathbb{Z}_2)$

Soit $Y \subset X$ un sous-ensemble algébrique de codim 1 ; Y est défini sur chaque ouvert affine de X par une équation réelle ; on en déduit qu'il existe un sous-ensemble algébrique $\tilde{Y} \subset \tilde{X}$ (défini par les mêmes équations) de codimension 1 dans $\tilde{X}$ qui définit un fibré $\tilde{F}$ de rang 1 sur $\tilde{X}$ défini sur $\mathbb{R}$, tel que $\phi[\tilde{F}]$ soit le dual de $[Y]$.

§2 - Un exemple de surface X pour laquelle $H^{Alg}_1(X,\mathbb{Z}_2) \neq H_1(X,\mathbb{Z}_2)$.

Soit n un entier ≥0, et T_n la surface réelle compacte orientable de genre n (T_n est le "tore à n trous"). Nous allons montrer le théorème suivant :

Théorème : Pour $1 \leq n \leq 9$, il existe dans $\mathbb{R}P^3$ une surface algébrique réelle projective X de degré 4, homéomorphe à T_n, et telle que $\mathrm{rang}(H^1(\tilde{X}, \mathcal{O}^*_X)) = 1$.

De la proposition 1 ci-dessus, on déduit immédiatement le corollaire suivant :

Corollaire : Pour une telle surface X, on a $\dim H^{Alg}(X,\mathbb{Z}_2) \leq 1$ et donc :

$$H^{Alg}_1(X,\mathbb{Z}_2) \neq H_1(X,\mathbb{Z}_2).$$

Démonstration du théorème : soit n un entier tel que $1 \leq n \leq 9$.

D'après les résultats de Ut kin(U), il existe dans $\mathbb{R}P^3$ une surface algébrique réelle connexe X_1 de degré 4 telle que X_1 soit homéomorphe à T_n. Si $\tilde{X}_1$ est la complexifiée de X_1 plongée dans CP^3 par $\phi_1 : \tilde{X}_1 \to CP^3$, X_1 est une "surface K3" munie d'une involution τ antiholomorphe, et d'un élément $\beta \in H^2(\tilde{X}_1,\mathbb{Z})$, image par ϕ_1^* d'un générateur $e \in H^2(CP^3,\mathbb{Z})$, et vérifiant :

- $\tau^*(\beta) = -\beta$

- $<\beta,\beta> = 4$,

$<,>$ étant la forme bilinéaire sur $H^2(\tilde{X}_1,\mathbb{Z})$ définie par le cup produit.

La théorie des déformations de telles surfaces (cf. (Kh) lemme 1.3 et théorème 2.5 ; on pourra consulter aussi (L-P) et (R)) montre le lemme suivant :

Lemme 1 : Il existe une surface K3 $\tilde{X}'$ munie d'une involution antilinéaire τ', obtenue par "déformation équivariante" de $\tilde{X}_1$, et un élément $\beta' \in H^2(\tilde{X}',\mathbb{Z})$ tels que :

a) $<\beta',\beta'> = 4$ et $\tau'^*(\beta') = -\beta'$

b) $H^1(\tilde{X}', \mathcal{O}^*_{\tilde{X}'})$ est de rang 1, engendré par β' (ceci a un sens, car $\tilde{X}'$ étant une surface K3, $H^1(\tilde{X}', \mathcal{O}_{\tilde{X}'}) = 0$ et le morphisme : $H^1(\tilde{X}', \mathcal{O}^*_{\tilde{X}'}) \to H^2(\widetilde{X'},\mathbb{Z})$ obtenu avec la suite de l'exponentielle est injectif).

c) $X' = \mathrm{Fix}\,\tau'$ est homéomorphe à X.

Le théorème résulte maintenant immédiatement du lemme suivant ((Kh), lemme 2.2):

Lemme 2 : Pour une telle surface $\tilde{X}'$, il existe un plongement ϕ' équivariant (i.e. tel que $\overline{\phi'} = \phi' \circ \tau'$) : $\tilde{X}' \overset{\phi'}{\to} CP^3$, tel que $\phi'^*(e) = \beta'$.

Pour démontrer le théorème, il suffit en effet de remarquer que la surface $X = \phi'(X')$ répond aux conditions demandées. c.q.f.d.

Remarque 2 : Il doit être possible de montrer que l'on peut choisir une déformation telle que X soit isotope à X_1 dans $\mathbb{R}P^3$; on en déduirait (cf. (U)) que pour $1 \leq n \leq 8$, X peut être choisie homotope à un point, ou au contraire telle que le morphisme

$\phi_* : H_1(X,\mathbb{Z}_2) \to H_1(\mathbb{R}P^3,\mathbb{Z}_2)$ soit non nul.

En revanche pour une surface de degré 4 homéomorphe à T_9, le morphisme ϕ_* : $H_1(X,\mathbb{Z}_2) \to H_1(\mathbb{R}P^3,\mathbb{Z}_2)$ est nécessairement non nul (cf. (U)).

Bibliographie

(B-H) : Borel-Haefliger : la classe d'homologie fondamentale d'un espace analytique, Bull. Soc. Math. France, 89, 461-513 (1961).

(B-K-S) :Bochnak - Kucharz - Shiota : On the divisor class group... , ce volume.

(B-T) : Bennedetti-Tognoli : Remark, and counterexamples in the theory of real Algebraic vector bundles and cycles, ce volume.

(G-A) : Griffiths-Adams : Topics in Algebraic and Analytic geometry (Princeton University Press).

(Kh) : Kharlamov : The topological type of non singular surfaces in $\mathbb{R}P^3$ of degree four, Functional Analysis, vol 10 n°4 (1977) p.295-305.

(L-P) : Le Potier : Exposé au Séminaire de Géométrie Algébrique 1981-1982 de l'école Polytechnique.

(R) : Risler : Exposé au Séminaire de Géométrie Algébrique 1981-1982 de l'école Polytechnique.

(U) Utkin : Topological classification of non singular fourth-order surfaces, Dokl- Nauk. SSSR, 175, n°1, 40-43 (1967).

U.E.R. de Mathématiques

UNIVERSITE PARIS VII.

ETUDE DES COUPURES DANS LES GROUPES ET CORPS ORDONNES

par

Raymond ROLLAND

INTRODUCTION

Nous étudions les coupures, dans le cas de Groupes Abéliens Totalement Ordonnés (GATO), et dans celui des corps ordonnés, selon 3 points de vue : leurs propriétés algébriques, leurs relations entre elles et leur "forme" du point de vue de l'ordre. Pour les trois aspects, nous utilisons l'idée fondamentale qui consiste à associer à chaque coupure t, une coupure $\hat{t}$ qui estime la "largeur" de t (cf. I 1).

Le point de vue algébrique nous conduit à étudier le rang des GATO et des corps ordonnés. Nous donnons une caractérisation des GATO de rang donné (II 2) et des corps ordonnés de rang donné et de corps résiduel donné (III 2). Dans les deux cas, la partie difficile est le plongement du GATO ou du corps dans un objet maximum correspondant. C'est en fait le théorème de plongement de Hahn déjà étudié sous des formes diverses par de nombreux auteurs. Nous en donnons ici une nouvelle démonstration en termes de coupures.

Le second point de vue consiste à chercher dans quelles coupures t' viennent s'ajouter les nouveaux éléments lorsqu'on construit une extension d'un GATO divisible ou d'un corps réel clos, en plaçant un élément X dans une coupure t. Cela définit en fait une relation d'équivalence notée $t \sim t'$ dans le cas des GATO (cf. I 3) et $t \equiv t'$ dans le cas des corps réel clos (cf. I 6). Le résultat le plus important est le théorème de III 1 : Dans un corps réel clos K si $t = \hat{t}$ et $t' \equiv t$, alors t' est de la forme $a + c\,t^q$ avec $a, c \in K$ et $q \in \mathbb{Q}$. Les relations d'équivalences permettent aussi d'introduire naturellement une classification des coupures qui est constamment utilisée.

Le troisième aspect nous conduit à étudier le caractère initial et le caractère final des coupures (cf. II 3 et III 3). En fait on peut voir que le rang détermine essentiellement les types de coupures qui peuvent exister dans le GATO ou le corps ordonné, ce qui est exploité dans les applications (IV) : en IV 1 nous trouvons une caractérisation des GATO η_α, en IV 2, nous étudions l'existence de corps ordonnés comprenant des coupures de types donnés puis l'existence de fonctions continues non bornées sur un intervalle fermé borné d'un corps ordonné.

On pourra noter le parallélisme remarquable entre l'étude des GATO et celle des corps ordonnés. D'autre part on exploite les résultats obtenus sur les GATO dans le cas d'un corps ordonné en remarquant qu'un corps ordonné est un GATO pour l'addition, et que ses éléments strictement positifs forment un GATO pour la multiplication, mais aussi qu'on peut lui associer d'une manière naturelle un groupe de valuation (cf. I 7).

Pour plus de détails, le lecteur pourra se reporter à ma thèse de 3ème cycle [9].

PRELIMINAIRES

DEFINITION. E *étant un ensemble totalement ordonné, une coupure* t *de* E *est la donnée de sous-ensembles* S_t, S'_t *de* E *tels que* $S_t < S'_t$ *et* $S_t \cup S'_t = E$.
On note T(E) *l'ensemble des coupures de* E.

Exemples. Les coupures $+\infty$, $-\infty$, a_+, a_- ($a \in E$) sont définies par :

$S_{+\infty} = E$, $S_{-\infty} = \emptyset$, $S_{a+} = \{x \in E / x \leq a\}$, $S_{a_-} = \{x \in E / x < a\}$.

On note $T_0(E) = \{a_+, a_- / a \in E\}$.

Si E' est une extension de E, on notera

$y \in t$ si $S_t < y < S'_t$

$y < t$ si $\exists x \in S_t$, $y \leq x$

$y > t$ si $\exists x \in S'_t$, $y \geq x$

Remarquons que T(E) est également, d'une manière naturelle, un ensemble totalement ordonné.

I - NOTIONS FONDAMENTALES ET PROPRIETES IMMEDIATES

1) $\hat{t}$

DEFINITION. *Soit G un GATO (Groupe Abélien Totalement Ordonné), à* $t \in T(G)$ *on associe* $\hat{t}$ *définie par :*

$$S_{\hat{t}} = \{y \in G/(\forall x \in G)\ (x < t \Rightarrow x + y < t)\}$$

$\hat{t}$ *mesure la "largeur" de* t *relativement à l'addition.*

Exemples. On a $\hat{a}_+ = \hat{a}_- = 0_+$ en $\widehat{+\infty} = \widehat{-\infty} = +\infty$ $(a \in G)$. On voit aisément que $\hat{t} > 0$ et $2\hat{t} = \hat{t}$ (c'est-à-dire $x < \hat{t}$ si et seulement si $2x < \hat{t}$).

On note $\hat{T}(G) = \{\hat{t}/t \in T(G)\}$. Il est facile de voir que $\hat{\hat{t}} = \hat{t}$ et d'en déduire que :

$$\hat{T}(G) = \{t \in T(G)/t = \hat{t}\} = \{t \in T(G)/\ t > 0 \text{ et } 2t = t\}.$$

DEFINITION. *Un sous groupe* H *de* G *est convexe si* $x \in H$ *et* $|y| < |x| \Rightarrow y \in H$. *(où* $|x| = \sup(x, -x)$*) si* $t \in \hat{T}(G)$, $G_t = \{x \in G/\ |x| < t\}$ *est un sous-groupe convexe de* G, *ce qui définit un isomorphisme entre* $\hat{T}(G)$ *(ordonné naturellement) et l'ensemble des sous-groupes convexes de* G *ordonné par l'inclusion. Dans cet isomorphisme, les* $t \in \hat{T}(G)$ *qui ont un prédécesseur dans* $\hat{T}(G)$ *correspondent aux sous groupes principaux (sous groupes convexes engendrés par un élément) de* G.

2) Rang d'un GATO G

DEFINITION. *On note* $\Theta(G)$ *l'ensemble des éléments de* $\hat{T}(G)$ *qui ont un prédecesseur dans* $\hat{T}(G)$.

On appelle rang de G *le type d'ordre de* $\Theta(G)$.

Un GATO G est archimédien si et seulement si $\hat{T}(G) = \{0_+, +\infty\}$ c'est-à-dire si et seulement si G est de rang 1.

PROPOSITION. *Si* $a \in G$ *et* $a > 0$, *il existe* Θ_a *et* Θ'_a *appartenant à* $\hat{T}(G)$ *tels que* $\Theta'_a < a < \Theta_a$ *et* Θ_a *est le successeur de* Θ'_a *dans* $\hat{T}(G)$ *(donc* $\Theta_a \in \Theta(G)$ *)*

Démonstration. Θ_a est définie par : $x < \Theta_a$ si $(\exists n \in \mathbb{N})$ $(x \leq n\,a)$

Θ'_a est définie par :

$0 \leq x < \Theta'_a$ si $(\forall n \in \mathbb{N})$ $(n\,x < a)$.

3) La relation $t \sim t'$.

Dans ce paragraphe, Γ désigne un QEVTO ($\mathbb{Q}$-Espace Vectoriel Totalement Ordonné).

DEFINITION. *Si* $t \in T(\Gamma)$, *on note* $\Gamma_{x,t} = \{q\,x + \gamma /\ q \in \mathbb{Q},\ \gamma \in \Gamma\}$ *ordonné par :* $q\,x + \gamma > 0$ *si* $q > 0$ *et* $-\frac{\gamma}{q} < t$ *ou si* $q < 0$ *et* $-\frac{\gamma}{q} > t$, *l'ordre de* $\Gamma_{x,t}$ *prolongeant celui de* Γ

Ainsi ordonné, $\Gamma_{x,t}$ est le QEVTO extension de Γ engendré par x, avec $x \in t$.

PROPOSITION. *Si* $y \in \Gamma_{x,t}$ *et si* $t' \in T(\Gamma)$ *est telle que* $y \in t'$, *alors* $\Gamma_{y,t'}$ *est isomorphe à* $\Gamma_{x,t}$ *par un* Γ*-isomorphisme (c.a.d. un isomorphisme de QEVTO laissant invariants les éléments de* Γ*).*

Démonstration. Le sous QEVTO de $\Gamma_{x,t}$ engendré par $\Gamma \cup \{y\}$ est isomorphisme à $\Gamma_{y,t'}$ et est égal à $\Gamma_{x,t}$.

COROLLAIRE. *La relation* $t' \cap \Gamma_{x,t} \neq \emptyset$ *est une relation d'équivalence entre* t *et* t' *notée* $t \sim t'$.

Exemples. $T_0(\Gamma)$ est la classe d'équivalence de 0_+, $\{+\infty, -\infty\}$ est celle de $+\infty$.

Dans les deux cas, si $x \in t$, $q\,x$ appartient à t ou à $-t$ suivant le signe de q (si $q \neq 0$), il en est ainsi chaque fois que $t \in \hat{T}(\Gamma)$. D'où la proposition suivante :

PROPOSITION. *Si* $t \in \hat{T}(\Gamma)$, *la classe d'équivalence de* t *est*

$$\overline{\{t\}} = \{\gamma \pm t\ /\ \gamma \in \Gamma\}$$

Si $T \subseteq T(\Gamma)$, on notera $\overline{T} = \bigcup_{t \in T} \overline{\{t\}}$, ainsi $\overline{\hat{T}(\Gamma)} = \{\gamma \pm t\ /\ \gamma \in \Gamma \text{ et } t \in \hat{T}(\Gamma)\}$.

4) <u>$\varphi(t)$ et $\Psi(t)$</u>

DEFINITION. *E étant un ensemble totalement ordonné et t une coupure de E, on note $\varphi(t)$ (ou $\varphi_E(t)$ s'il y a risque de confusion) le caractère final de S_t. On note de même $\Psi(t)$ le caractère initial de S'_t.*

Rappelons que $\varphi(t)$ et $\Psi(t)$ sont des cardinaux réguliers.

Exemples. Si G est un GATO et $a \in G$, $\varphi(a_+) = \Psi(a_-) = 1$, $\Psi(a_+) = \varphi(a_-) = \Psi(0_+)$, $\varphi(+\infty) = \Psi(-\infty)$.

5) <u>Rang d'un corps ordonné K.</u>

DEFINITIONS. *On note ^+K le QEVTO correspondant à K pour l'addition et $K^{\bullet}$ le GATO formé des éléments strictement positifs de K pour la multiplication.*

Si $t \in \hat{T}(K)$ et si $t^{\bullet}$ est la coupure correspondante dans $T(K^{\bullet})$, on note $\tilde{t} = \hat{t}^{\bullet}$.

Ainsi $\tilde{t}$ mesure la "largeur" de t relativement à la multiplication et $\tilde{t}$ peut encore être définie par :

$0 \leq y < \tilde{t}$ si $(\forall x \in K)$ $(x < t \Rightarrow xy < t)$.

On a donc $\tilde{t} > 1$ et $\tilde{t}^2 = \tilde{t}$ (c.a.d. pour $x \geq 0$, $x < \tilde{t}$ si et seulement si $x^2 < \tilde{t}$).

On note $\mathcal{C}(K) = \{t \in \hat{T}(K) \ / \ t = \hat{t}\} = \{t \in \hat{T}(K) \ / \ t > 1$ et $\tilde{t} \leq t\}$.

Exemples. $+\infty \in \mathcal{C}(K)$ et c'est évidemment son plus grand élément, mais $\mathcal{C}(K)$ a aussi un plus petit élément τ défini par :

$$x < \tau \text{ si } (\exists n \in \mathbb{N}) \ (x \leq n).$$

DEFINITION. *On note J(K) l'ensemble des éléments de $\mathcal{C}(K)$ qui ont un prédécesseur dans $\mathcal{C}(K)$.*

On appelle rang de K le type d'ordre de J(K).

PROPOSITION. *L'application qui à* t *associe* ${}^{+}K_t = \{x \in K \ / \ |x| < t\}$ *est un isomorphisme de* $\mathcal{C}(K)$ *sur l'ensemble des anneaux de valuations réels (convexes) de* K *ordonnés par l'inclusion. Dans cet isomorphisme, les éléments de* J(K) *correspondent aux anneaux de valuations principaux (anneaux de valuations réels engendrés par un élément) de* K.

La démonstration est immédiate à l'aide du lemme suivant :

LEMME. *Si* $a > \tau$, $\exists t_a, t'_a \in \mathcal{C}(K)$, $t'_a < a < t_a$ *et* t_a *est le successeur de* t'_a *dans* $\mathcal{C}(K)$ *(donc* $t_a \in J(K)$ *).*

t_a est définie par : $x < t_a$ si $(\exists n \in \mathbb{N})$ $(x \leq a^n)$ et t'_a par : $0 < x < t'_a$ si $(\forall n \in \mathbb{N})$ $(0 < x^n < a)$.

Un corps K est archimédien si et seulement si $\mathcal{C}(K) = \{+\infty\}$ c'est-à-dire si et seulement si le rang de K est nul.

6) La relation $t \equiv t'$

Dans ce paragraphe, K désigne un corps réel clos, les polynômes irréductibles de K[x] sont donc de degré 1 ou 2. Il y a donc un seul ordre possible sur K[x] avec $x \in t$ $(x - a > 0 \Leftrightarrow a < t$ pour $a \in K$ et $x^2 + c\,x + d > 0$ s'il n'a pas de racine dans K). Donc il y a un seul ordre possible sur K(x) avec $x \in t$, d'où une clôture réelle notée K_x^t de cet ordre, qui est unique à K(x)-isomorphisme près.

PROPOSITION. *Si* $y \in K_x^t$ *et* $t' \in T(K)$ *est tel que* $y \in t'$, *alors* $K_y^{t'}$ *est isomorphe à* K_x^t *par un* K-*isomorphisme.*

Démonstration. Evidemment, $K_y^{t'}$ est isomorphe un sous corps réel clos de K_x^t engendré par $K \cup \{y\}$ (par un K-isomorphisme). y est algébrique sur K[x] donc x est algébrique sur K[y] et le sous corps réel clos engendré par $K \cup \{y\}$ est K_x^t.

COROLLAIRE. *La relation* $t' \cap K_x^t \neq \emptyset$ *est une relation d'équivalence entre* t *et* t' *notée* $t \equiv t'$.

Exemple : Pour cette relation, la classe de 0_+ est

$$\overline{\{0_+\}} = T_0(K) \cup \{\pm \infty\}$$

Si $T \subseteq T(K)$ on note encore $\overline{T} = \bigcup_{t \in T} \overline{\{t\}}$, on prendra garde aux risques de confusion, les notations étant identiques à celles de la relation $\sim$ sur ${}^+K$.

7) Valuation naturelle d'un corps ordonné K.

On note $A_\tau = \{x \in K \ / \ |x| < \tau\}$ l'anneau de valuation correspondant à τ, son idéal est $M_\tau = \{x \in K \ / \ |x| < \frac{1}{\tau}\}$. $U_\tau = \{x \in K \ / \ \frac{1}{\tau} < |x| < \tau\}$ est un sous groupe multiplicatif de $K^* = K - \{0\}$.

Le groupe $\Gamma_\tau = K^*/U_\tau$, noté additivement, est le groupe de valuation de K la valuation étant la surjection canonique $v_\tau : K^* \to \Gamma_\tau$ prolongée par $v_\tau(o) = -\infty$. v_τ satisfait les propriétés :

(i) $v_\tau(a) = -\infty$ si et seulement si $a = 0$

(ii) $|a| \leq |b| \Rightarrow v_\tau(a) \leq v_\tau(b)$

(iii) $v_\tau(ab) = v_\tau(a) + v_\tau(b)$

(iv) $v_\tau(a + b) \leq \sup (v_\tau(a), v_\tau(b))$ (et on a l'égalité si $v_\tau(a) \neq v_\tau(b)$).

Remarque. Il ne faut pas confondre le rang de K et le rang de ${}^+K$, en fait le rang de K serait plutôt le rang du rang de ${}^+K$:

PROPOSITION. *$\Theta({}^+K)$ est isomorphe à Γ_τ.*

A $t \in \Theta({}^+K)$ on fait correspondre $v_\tau(x) \in \Gamma_\tau$ avec $\theta_{|x|} = t$.

DEFINITION. *On appelle corps des restes ou corps résiduel de K le corps $K_\tau = A_\tau/M_\tau$.*

Il est clair que K_τ est un corps archimédien, il est donc isomorphe à un sous corps de $\mathbb{R}$.

PROPOSITION. *Si K est réel clos, K_τ est réel clos ; de plus il est isomorphe à un sous corps de A_τ.*

II - ETUDE DES GATO

1) Classification de T(G)

Si G est un QEVTO nous pouvons classer les éléments de T(G) à l'aide de $\hat{t}$ et de $\sim$ de la façon suivante :

$T_0(G) = \{t \in T(G) \,/\, \hat{t} = 0_+ \text{ et } t \sim 0_+\} = \{a_+, a_- \,/\, a \in G\}$

$T_1(G) = \{t \in T(G) \,/\, \hat{t} = 0_+ \text{ et non } t \sim 0_+\} = \{t \in T(G) \,/\, \hat{t} = 0_+ \text{ et } t \notin T_0(G)\}$

$T'_0(G) = \{t \in T(G) \,/\, \hat{t} \neq 0_+ \text{ et } t \sim \hat{t}\} = \{a \underline{+} t \,/\, t \in \hat{T}(G) \text{ et } t \neq 0_+, a \in G\}$

$T'_1(G) = \{t \in T(G) \,/\, \hat{t} \neq 0_+ \text{ et non } t \sim \hat{t}\} = \{t \in T(G) \,/\, \hat{t} \neq 0_+ \text{ et } t \notin T'_0(G)\}$

Les expressions finales permettent de définir ces classes dans le cas des GATO car elles n'utilisent pas la relation $\sim$.

Remarque.

1) Si $\hat{t} = +\infty$, $t = \pm\infty$.

2) $t \in T_1(G)$ est équivalent à $t \notin T_0(G)$ et :

$\forall \varepsilon \in G, \varepsilon > 0 \ \exists x \in G, x < t < x + \varepsilon \ (\hat{t} = 0_+)$.

G étant un GATO, muni de la topologie l'ordre, on peut voir que le complété de G (par les ultra filtres de Cauchy sur G) s'obtient en ajoutant un élément exactement dans chaque $t \in T_1(G)$.

Exemples

1) $T(\mathbb{Q}) = T_0(\mathbb{R}) \cup T(\mathbb{Q})$

2) Si $G = \mathbb{Q}[x]$ avec $x \in 0_+$, les éléments de $T_1(G)$ correspondent aux séries formelles qui ne sont pas des polynômes.

- Soit $t_0 \in T(G)$ définie par $P < t_0$ si $\forall n \in \mathbb{N}, P < \frac{1}{n}$.

On a $t_0 \in \hat{T}(G)$ et donc $1 + t_0 \in T'_0(G)$.

- Si $r \in \mathbb{R} - \mathbb{Q}$, soit $t \in T(G)$ définie par $P < t$ si $\exists q \in \mathbb{Q}, q < r$ et $P < q$ alors $\hat{t} = t_0$ et $t \in T'_1(G)$.

2) GATO de rang donné

THEOREME. J *étant le type d'ordre d'un ensemble totalement ordonné, il existe des*

GATO Γ *et* Γ' *de rang* J *tels que :*

(i) Γ' *est un QEVTO extension de* Γ

(ii) *Un GATO* G *est de rang* J *si et seulement si il existe des plongements* $f : \Gamma \to G$ *et* $f' : G \to \Gamma'$ *tels que* $f' \circ f$ *est l'injection canonique* i *de* Γ *dans* Γ'

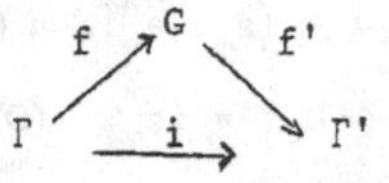

<u>Démonstration</u>. On prend pour Γ' le produit de Hahn ([3])$\Gamma' = \underset{j \in J^*}{\longmapsto\!\dashv} R_j$ où $R_j = \mathbb{R}$ et J^* est l'ordre opposé à celui de J. $\Gamma' = \{\gamma : J^* \to \mathbb{R}$ / γ a un support bien ordonné$\}$. On prend pour Γ la somme directe $\Gamma = \coprod_{j \in J^*} Z_j$ avec $Z_j = \mathbb{Z}$. On note γ_j l'élément de Γ' défini par $\gamma_j(j') = \delta_{j,j'}$, on considère que $\Gamma \subseteq \Gamma'$, alors $\gamma_j \in \Gamma$ et Γ est le sous groupe de Γ' engendré par $\{\gamma_j \ / \ j \in J\}$.

Notons $\theta_j = \theta_{\gamma_j}$, l'application qui a $j \in J$ associe θ_j est un isomorphisme de J sur $\Theta(\Gamma)$ et donc Γ est de rang J.

On peut voir aisément que si G est un GATO tel qu'il existe des plongements f et f' satisfaisant les conditions du théorème, alors G est de rang J. En particulier, Γ' est de rang J.

Si G est un GATO de rang J, soit $\Theta(G) = \{t_j \ / \ j \in J\}$ et soit t'_j le prédecesseur de t_j dans $\hat{T}(G)$ et soit $a_j \in G$ tel que $t'_j < a_j < t_j$. Il est facile de construire un plongement $f : \Gamma \to G$ tel que $f(\gamma_j) = a_j$.

Si G est un QEVTO f se prolonge en un plongement $\bar{f}$ de la clôture divisible $\bar{\Gamma}$ de Γ dans G. Il reste à définir f' pour cela on utilise le lemme suivant :

<u>LEMME</u>. *Si* G *est un QEVTO et* $\Gamma \subseteq G \subseteq \Gamma'$, *et si* $t \in T(G)$, *alors* $t \cap \Gamma' \neq \emptyset$ *si et seulement si* $t \notin \overline{\hat{T}(G)}$.

<u>Démonstration du lemmme</u>.

- On montre aisément que si $t \cap \Gamma' \neq \emptyset$, $t \notin \overline{\hat{T}(G)}$.

- Si $t \in \overline{\hat{T}(G)}$ et si $\hat{t}$ a un sucesseur t' dans $\hat{T}(G')$, soit γ_j tel que $\hat{t} < \gamma_j < t'$ et soit $a \in G$ tel que $a<t<a + \gamma_j$. Soit $r = \sup\{q \in \mathbb{Q} \ / \ a + q\,\gamma_j < t\}$ on peut alors montrer que $a \pm r\gamma_j \in t \cap \Gamma'$.

- Si $t \in \overline{\hat{T}(G)}$ et $\hat{t}$ n'a pas de sucesseur dans $\hat{T}(G)$, soit $(j_\xi)_{\xi < \lambda}$ strictement décroissante dans J telle que $(\gamma_{j_\xi})_{\xi < \lambda}$ soit coinitiale à $S'_{\hat{t}}$.

On peut définir une suite $(C_\xi)_{\xi < \lambda}$ dans G de sorte que $\forall_\xi < \lambda$,

$C_\xi < C_\xi + \gamma_{j_\xi} < t < C_\xi + 2\gamma_{j_\xi}$.

Soit $A_\xi = \{j \in \text{supp}(C_\xi) \ / \ j > j_\xi \text{ dans } J\}$, si $\xi < \xi'$ A_ξ est une section commençante de $A_{\xi'}$, et si $j \in A_\xi$, $C_\xi(j) = C_{\xi'}(j)$. On définit γ par $\text{supp}(\gamma) = \bigcup_{\xi<\lambda} A_\xi$ et si $j \in A_\xi$, $\gamma(j) = \gamma_\xi(j)$, on peut alors montrer que $\gamma \in t \cap \Gamma'$.

Pour montrer l'existence de f' on suppose d'abord que G est un QEVTO et on prend $h : H \to \Gamma'$ maximal avec H sous-QEVTO de G et $h \circ f = i$, injection canonique de Γ dans Γ'. Si $H \neq G$ soit $x \in G - H$ et $t \in T(H)$ tel que $x \in t$, soit $H = h(H)$ et t' l'image de t. On montre aisément que $t \notin \overline{\hat{T}(H)}$ donc que $t' \notin \overline{\hat{T}(H')}$ et alors $t' \cap \Gamma' \neq \emptyset$ d'où un prolongement de h.

Si G est un GATO de rang J, $\overline{G}$ est un QEVTO de rang J et il suffit de plonger $\overline{G}$ dans Γ' comme ci-dessus.

Remarques

1) Le plongement f' ainsi défini n'est pas unique en général.

2) Un plongement $h ; G \to G'$ est une Θ-extension si $h(\Theta(G)) = \Theta(G')$, on peut montrer que Γ' est maximum, parmi les GATO de rang J, vis à vis des Θ-extensions, pour cela on dira que Γ' est le GATO fortement complet de rang J. On peut montrer que $\overline{\hat{T}(\Gamma')} = T(\Gamma')$ et que Γ' est le seul GATO de rang J qui satisfait cela.

3) Calcul de $\varphi(t)$ et $\Psi(t)$

Le théorème suivant permet de se ramener à $\varphi(\hat{t})$ et $\Psi(\hat{t})$.

THEOREME. *Si G est un GATO et* $t \in T(G)$,

- *Si* $t = a + \hat{t}$, $\varphi(t) = \varphi(\hat{t})$ *et* $\Psi(t) = \Psi(\hat{t})$
- *Si* $t = a - \hat{t}$, $\varphi(t) = \Psi(\hat{t})$ *et* $\Psi(t) = \varphi(\hat{t})$
- *Si* t *n'est pas de la forme* $a \pm \hat{t}$ *avec* $a \in G$, *alors* $\varphi(t) = \Psi(t) = \Psi(\hat{t})$.

Démonstration. Les deux premières assertions sont triviales, pour montrer la dernière, on prend $(a_\xi)_{\xi < \Psi(\hat{t})}$ coinitiale à $S'_{\hat{t}}$. Pour chaque $\xi < \Psi(\hat{t})$, soit $x_\xi \in G$ tel que $x_\xi < t < a_\xi < x_\xi$. On montre alors que $\{x_\xi \; / \; \xi < \Psi(\hat{t})\}$ est cofinal à S_t et que $\{x_\xi + a_\xi \; / \; \xi < \Psi(\hat{t})\}$ est coinitial à S'_t ce qui prouve que $\varphi(t) \leq \Psi(\hat{t})$ et que $\Psi(t) \leq \Psi(\hat{t})$.

Si $(x_\xi)_{\xi < \varphi(t)}$ est cofinale à S_t, on peut construire par récurrence transfinie $(x'_\xi)_{\xi < \varphi(t)}$ de sorte que :

$$\forall \xi < \varphi(t), \; x_\xi < x'_\xi < x'_\xi + \hat{t} < x'_{\xi+1} < t$$

(en effet, $t \notin T_0(G)$ donc S_t n'a pas de borne supérieure et, d'autre part, si $x < t$, $x + \hat{t} < t$).

On peut alors montrer que $\{x'_{\xi+1} - x_\xi / \xi < \varphi(t)\}$ est coinitial à S'_t ce qui prouve que $\Psi(\hat{t}) \leq \varphi(t)$. On peut montrer de façon analogue que $\Psi(\hat{t}) \leq \psi(t)$.

PROPOSITION. *On suppose que* G *est un GATO et que* $t \in \hat{T}(G)$ *(d'où* t_+, $t_- \in T_0(\hat{T}(G))$ *)*

1) Si $t \neq 0_+$, $\varphi_G(t) = \sup\,(\omega, \varphi_{\hat{T}(G)}(t_-))$

2) a) Si $\Psi_{\hat{T}(G)}(t_+) > 1$, $\Psi_G(t) = \Psi_{\hat{T}(G)}(t_+)$

b) Si t *a un successeur* t^+ *dans* $\hat{T}(G)$, G_{t^+} / G_t *est isomorphe à un sous-groupe additif de* $\mathbb{R}$. *S'il est discret* $\Psi_G(t) = \varphi_G(t)$, *sinon* $\Psi_G(t) = \omega$.

La démonstration ne présente pas de grosses difficultés.

III - CORPS ORDONNES

1) Classification de T(K)

THEOREME. *Si* K *est un corps réel clos et si* $t \in \hat{T}(K)$,

$$\overline{\{t\}} = \{a + c\, t^q \; / \; a \in K, \; c \in K^*, \; q \in \mathbb{Q}^*\}$$

Soit $t \in \hat{T}(K)$, on pose ${}^+K_{\tilde{t}} = \{y \in K_x^t \; / \; |y| < \tilde{t}\}$; $U_{\tilde{t}} = \{y \in K_x^t \; / \; \frac{1}{\tilde{t}} < |y| < \tilde{t}\}$ en est un sous groupe multiplicatif convexe, d'où un GATO $(K_x^t)^* / U_{\tilde{t}}$ que l'on note additivement. Soit μ la surjection canonique de $(K_x^t)^*$ sur ce GATO, avec $\mu(0)=-\infty$.

La démonstration du théorème utilise une série de lemmes :

LEMME 1. 1- *Si* $y \in K_x^t$ *et* $\mu(y) = \mu(x)$ *alors* $|y| \in t$

2- $\mu(x) \notin \mu(K)$

La démonstration de ce lemme n'est pas difficile.

LEMME 2. *Si* $y \in K_x^t$ *et* $\mu(y) \notin \mu(K)$, *alors* $\exists\, c \in K^*$ *et* $q \in \mathbb{Q}^*$ *tels que* $|y| \in c\, t^q$ (*où* $c\, t^q$ *est défini de façon naturelle*).

Démonstration. Si $\mu(y) \notin \mu(K)$, y satisfait une égalité de la forme $u_o + u_1 y + \ldots + u_n y^n = 0$ avec $u_i = \sum_j a_{i,j} y^j$ ($a_{i,j} \in K$, la somme est finie). L'ensemble des couples (i,j) tels que $\mu(a_{i,j}\, x^j y^i)$ est maximum a au moins deux éléments (i,j) et (i',j'). De $\mu(y) \notin \mu(K)$ on déduit que $i \neq i'$ et de $\mu(x) \notin \mu(K)$ on déduit que $j \neq j'$. On obtient alors une égalité du type $\mu(y) = \mu(c) + q\, \mu(x)$ avec $c \in K^*$ et $q \in \mathbb{Q}^*$ d'où le résultat (d'après le lemme 1).

LEMME 3. *Si* $y \in K_x^t - K$, $P \in K[x]$ *et* $\mu(P(y)) \notin \mu(K)$ *alors* $\exists a \in K$, $\exists c \in K^*$, $\exists q \in \mathbb{Q}^*$ *tels que* $y \in a + c\, t^q$ *avec* $a=0$ *ou* $\mu(y) = \mu(a) > \mu(y-a)$.

Démonstration. On étend μ à la clôture algébrique $K_x^t(i)$ de K_x^t en posant $\mu(z) = \mu(|z|)$. On suppose que $\mu(y) \in \mu(K)$ et on décompose P en facteurs du premier degré sur K(i). Comme $\mu(P(y)) \notin \mu(K)$, il existe une racine z de P telle que $\mu(y-z) \notin \mu(K)$ et on prend pour a la partie réelle de z. Le théorème est alors une conséquence immédiate du lemme suivant :

LEMME 4. *Si* $y \in K_x^t - K$, *il existe* $a \in K$, $c \in K^*$, $q \in \mathbb{Q}^*$ *tels que* $y \in a + c\, t^q$ *avec* $a=0$ *ou* $\mu(y) = \mu(a) > \mu(y-a)$.

Démonstration. Comme y est algébrique sur K[x], x est algébrique sur K[y] d'où une relation :

$$P_o(y) + x\, P_1(y) + \ldots + x^m\, P_m(y) = 0$$

avec $P_i(y) \in K[y]$. Si $(\exists i)$ $(\mu(P_i(y)) \notin \mu(K))$ alors le résultat découle immédiatement du lemme 3, sinon il suffit de prendre i et j distincts tels que $\mu(P_i(y)) = \mu(P_j(y))$ soit maximum. On a alors $\mu(x) = \dfrac{\mu(P_j(y)) - \mu(P_i(y))}{i-j} \in \mu(K)$ ce qui est absurde.

COROLLAIRE. $\overline{\hat{T}(K)} = \{a \pm t \ / \ a \in K, \ t \in \hat{T}(K)\}$

En effet si $t \in \hat{T}(K)$, $c \in K^*$ et $q \in \mathbb{Q}^*$, $\pm c\, t^q \in \hat{T}(K)$. Nous venons en fait de montrer que $\overline{\hat{T}(K)} = \overline{\hat{T}(^+K)}$ ce qui permet de reprendre pour les corps ordonnés la classification des GATO : $T_1(K) = T_1(^+K)$, $T'_0(K) = T'_0(^+K)$, $T'_1(K) = T'_1(^+K)$. Dans le cas où K est réel clos, on a encore :

$$T_0(K) = \{t \in T(K) \ / \ \hat{t} = 0_+ \text{ et } t \equiv 0_+\}$$

$$T_1(K) = \{t \in T(K) \ / \ \hat{t} = 0_+ \text{ et non } (t \equiv 0_+)\}$$

$$T'_0(K) = \{t \in T(K) \ / \ \hat{t} \neq 0_+ \text{ et } t \equiv \hat{t}\}$$

$$T'_1(K) = \{t \in T(K) \ / \ \hat{t} \neq 0_+ \text{ et non } (t \equiv \hat{t})\}$$

Comme $T_1(K) = T_1(^+K)$, comme pour les GATO le complété d'un corps K s'obtient en ajoutant un élément exactement dans chaque $t \in T_1(K)$.

2) Corps de rang donné et de corps résiduel donné

THEOREME. *Soit* J *le type d'ordre d'un ensemble totalement ordonné et* k *un sous-corps réel clos de* $\mathbb{R}$. *Il existe des corps ordonnés* L *et* L' *de rang* J *et de corps résiduel* k *tels que :*

(i) L' *est réel clos et est une extension de* L.

(ii) *Un corps ordonné* K *est de rang* J *et de corps résiduel* k *si et seulement si il existe des plongements* $g : L \hookrightarrow K$ *et* $g' : K \hookrightarrow L'$ *tels que* $g' \circ g = i$, *injection canonique de* L *dans* L':

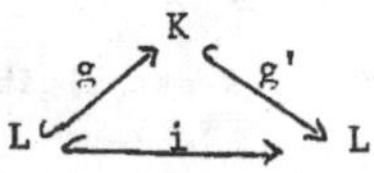

<u>Démonstration</u>. Γ' étant le GATO de rang J défini en II 2, on pose $L' = \coprod_{\gamma \in \Gamma'^*} R_\gamma$ avec $R_\gamma = k$, c'est-à-dire $L' = \{u : \Gamma'^* \to k /$ supp (u) est bien ordonné$\}$. L' est muni de la multiplication habituelle : si u, v $\in$ L' et $\gamma \in \Gamma'$, $uv(\gamma) = \sum_{\gamma_1+\gamma_2=\gamma} u(\gamma_1)v(\gamma_2)$. L' est un corps (cf. [5]) noté également $k((\Gamma'))$ et L' est réel clos (cf [6]).

Si $\gamma \in \Gamma'$ soit $x_\gamma \in L'$ défini par $x_\gamma(\gamma') = \delta_{\gamma,\gamma'}$, et soit L le sous corps de L' engendré par $k \cup \{x_{\gamma_j} / j \in J\}$. Γ' étant de rang J, L' est de rang J, le groupe de valuation de L est $\Gamma = \coprod_{j \in J^*} Z_j$ $(Z_j = \mathbb{Z})$ qui est également de rang J, donc L est de rang J.

On montre aisément que s'il existe des plongements g et g' satisfaisant aux conditions de l'énoncé, K est de rang J et de corps résiduel k.

Si K est un corps ordonné de rang J et de corps résiduel k, soit G son groupe de valuation et $v_\tau : K \to G \cup \{-\infty\}$. Soit $f : \Gamma \hookrightarrow G$ l'application définie en II 2 (G est de rang J), avec $a_j = f(\gamma_j)$ et soit $c_j \in K$, $c_j > 0$ tel que $v_\tau(c_j) = a_j$. Il existe un plongement $g : L \hookrightarrow K$ tel que $g(x_{\gamma_j}) = c_j$.

<u>LEMME</u>. *Si* K *est un corps réel clos et* $L \subseteq K \subseteq L'$, *et* $t \in T(K)$, *alors* $t \cap L' \neq \emptyset$ *si et seulement si* $t \notin \overline{\mathcal{C}(K)}$ *et* $(K_x^t)_\tau$ *est isomorphe à* k.

<u>Démonstration</u>. On voit aisément que si $t \in \overline{\mathcal{C}(K)}$, $t \cap L' = \emptyset$ et que si $t \cap L' \neq \emptyset$, $(K_x^t)_\tau$ est isomorphe à k.

Si $t \notin \overline{\mathcal{C}(K)}$ et $t \in T'_0(K)$, alors $t \equiv \hat{t}$ donc $\hat{t} \notin \overline{\mathcal{C}(K)}$. On en déduit non $(v_\tau(\hat{t}) \sim v_\tau(\hat{\tilde{t}}))$ or $v_\tau(\tilde{\hat{t}}) = \widehat{v_\tau(\hat{t})}$ donc $v_\tau(\hat{t}) \notin \overline{\hat{T}}(G)$. D'après le lemme de II 2 on a donc $v_\tau(\hat{t}) \cap \Gamma' \neq \emptyset$ et si $\gamma \in v_\tau(\hat{t}) \cap \Gamma'$, $x_\gamma \in \hat{t} \cap L'$. Comme $t \equiv \hat{t}$, on a $t \cap L' \neq \emptyset$.

Si $t \in T_1(K) \cup T'_1(K)$, on voit d'après la démonstration du lemme de II 2 que si $\hat{t}$ a un successeur t contient un élément de la forme $y = a + r x_\gamma$ avec $a \in K$, $\gamma \in G$ et $r \in \mathbb{R}$. On montre aisément que si $(K_x^t)_\tau = k$, $r \in k$ et donc $y \in K$ ce qui est absurde. Ainsi si $(K_x^t)_\tau = k$, $\hat{t}$ n'a pas de successeur, la démonstration du lemme de II 2 permet alors de trouver un élément dans $t \cap L'$.

L'existence de g' se montre alors par un argument en tous points analogue à celui utilisé en II 2. A nouveau, g' n'est pas unique en général.

Remarques. 1) L'hypothèse k réel clos est indispensable : voici un contre exemple dans le cas où $k = \mathbb{Q}$.

Soit $\mathcal{P}$ l'ensemble des nombres premiers, et pour $p \in \mathcal{P}$ soit $x_p = (p\,x)^{1/p} \in \overline{\mathbb{Q}}((\mathbb{Q}))$. Soit K le sous corps de $\overline{\mathbb{Q}}((\mathbb{Q}))$ engendré par $\{x_p / p \in \mathcal{P}\}$, son groupe de valuation est un sous groupe de $\mathbb{Q}$, donc K est de rang 1, et on peut montrer que son corps résiduel est isomorphe à $\mathbb{Q}$. $\mathbb{Q}[x]$ est bien contenu dans K, mais on ne peut pas plonger K dans $\mathbb{Q}((\mathbb{R}))$.

2) Un plongement $h : K \hookrightarrow K'$ est une J-extension si $h(J(K)) = J(K')$, on peut montrer que $\mathbb{R}((\Gamma'))$ est maximum parmi les corps ordonnés de rang J, vis à vis des J-extensions. On dira que $\mathbb{R}((\Gamma'))$ est le corps fortement complet de rang J. On peut montrer qu'il est le seul corps K de rang J a vérifier $T(K) = \overline{\mathcal{C}(K)}$.

3) Calcul de $\varphi(t)$ et $\Psi(t)$

Si K est un corps ordonné, ^{+}K est un QEVTO et le calcul de $\varphi(t)$ et $\Psi(t)$ se ramène à celui de $\varphi(\hat{t})$ et $\Psi(\hat{t})$ (cf. théorème de II 3). De même $K^{\bullet}$ est un GATO et on a les propositions suivantes :

PROPOSITION. *Si* $t \in \hat{T}(K)$, $a \in K$ *et* $a > 0$

1) $\varphi(a\tilde{t}) = \varphi(\tilde{t})$ *et* $\Psi(a\tilde{t}) = \Psi(\tilde{t})$

2) $\varphi(a/\tilde{t}) = \Psi(\tilde{t})$ *et* $\Psi(a/\tilde{t}) = \varphi(\tilde{t})$

3) *Si* t *n'est pas de la forme* $a\tilde{t}$ *ou* $a/\tilde{t}$ *pour* $a \in K$, *alors* $\varphi(t) = \Psi(t) = \Psi(\tilde{t})$.

PROPOSITION. *Si* $t \in \mathcal{C}(K)$,

1) $\varphi_K(t) = \sup(\omega, \varphi_{\mathcal{C}(K)}(t_-))$

2) a) *Si* $\Psi_{\mathcal{C}(K)}(t_+) > 1$, $\Psi_{\tilde{K}}(t) = \Psi_{\mathcal{C}(K)}(t_+)$

b) *Si* $\Psi_{\mathcal{C}(K)}(t_+) = 1$, *soit* t^+ *le successeur de* t *dans* $\mathcal{C}(K)$ *et soit* $K^{\bullet}_t = \{x \in K \,/\, \frac{1}{t} < x < t\}$.

Si $K^{\cdot}_{t^+}/K^{\cdot}_t$ *est discret, on a* $\Psi_K(t) = \varphi_K(t)$*, sinon* $\Psi_K(t) = \omega$.

4) Résultats complémentaires

THEOREME. (J. HOUDEBINE - R.ROLLAND).

Si K *est réel clos et* t, t' $\in$ T(K) *et* $t \equiv t'$, *alors* $\hat{t}' \equiv \hat{t}$.

Démonstration. Si $x \in t \cap K^t_u$ et $y \in t' \cap K^t_u$, on peut définir $f_{x,y}$ bijection de $t \cap K^t_u$ sur $t' \cap K^t_u$ de sorte que si y est la $k^{\text{ème}}$ racine du polynôme $Q(x,y) \in K[x,y]$, alors $f_{x,y}(z)$ est la $k^{\text{ème}}$ racine de $Q(z,y)$. On montre alors que $f_{x,y}$ est strictement monotone, et que son sens de variation ne dépend pas du choix de x ou de y (Il y a donc deux "orientations" possibles pour les éléments t' de $\overline{\{t\}}$ suivant le sens de variation des $f_{x,y}$ avec $x \in t$ et $y \in t'$).

Soit $K' = K^t_u$, $\hat{t} \cap K' = \hat{t}' \cap K' = \emptyset$, on montre que dans $K'^{\hat{t}}_z$, $|f_{x,y}(x+z)-y| \in \hat{t}'$, on en déduit que $\hat{t}' = c\, \hat{t}^q$ avec $c \in K'^*$ et $q \in \mathbb{Q}^*$ (d'après le théorème de III 1). Or il existe $b \in K$ tel que $v_\tau(b) = v_\tau(c)$ et alors $\hat{t}' = b\, \hat{t}^q$ donc $\hat{t}' \equiv \hat{t}$.

COROLLAIRE. *Avec les mêmes hypothèses,*

1) $\tilde{\hat{t}} = \tilde{\hat{t}}'$

2) $(\varphi(t'), \Psi(t'))$ *est égal à* $(\varphi(t), \Psi(t))$ *ou a* $(\Psi(t), \varphi(t))$.

IV - APPLICATIONS

1) GATO η_α ($\alpha > 0$)

Un ensemble totalement ordonné E est η_α si pour toutes parties X,Y de E telles que X < Y et card X + card Y < ω_α, $\exists x \in E$, X < x < Y. Cette définition est équivalente à :

$$\forall t \in T(E),\ \varphi(t) \geq \omega_\alpha \quad \text{ou} \quad \Psi(t) \geq \omega_\alpha .$$

PROPOSITION. *Un GATO* G *de rang* J *est* η_α *si et seulement si :*

(i) J *est un ensemble* η_α *et*

(ii) $\forall t \in T'_1(G)$, $\Psi(\hat{t}) \geq \omega_\alpha$

Cette proposition découle immédiatement des résultats de II 3 on peut s'en servir pour retrouver les exemples de Alling de groupes et de corps η_α ([1]). Par exemple, il est immédiat que le GATO fortement complet Γ' de rang J est un GATO η_α lorsque J est un ensemble η_α car $T'_1(\Gamma') = \emptyset$.

2) Applications de III 3.

PROPOSITION. *Si* I *est un ensemble totalement ordonné et si* $\forall i \in I$, λ_i *et* μ_i *sont des cardinaux réguliers infinis, il existe un corps réel clos fortement complet* K *et pour tout* $i \in I$ *il existe* $t_i \in \mathcal{C}(K)$ *tels que* $\varphi(t_i) = \lambda_i$ *et* $\Psi(t_i) = \mu_i$ *et* $\{t_i \ / \ i \in I\}$ *est isomorphe à* I.

Démonstration. On pose $E = \lambda_i + \mu_i^*$ où μ_i^* est l'ordre opposé à μ_i, et $J = \coprod_{i \in I} E_i$; on vérifie alors la proposition pour $K = \mathbb{R}((\Gamma'))$ corps fortement complet de rang J.

PROPOSITION. *Si* K *est un corps ordonné tel qu'il existe* $t \in T(K) - T_0(K)$ *tel que* $\Psi(t) = \Psi(0_+)$ *et* $t \neq -\infty$, *alors* $\forall a, b \in K$, $a < b$, *il existe une fonction continue sur* $[a,b]$, *à valeurs dans* K *et non majorée sur* $[a,b]$.

Démonstration. Clairement, on peut supposer $a < t < b$, comme $\Psi(t) = \varphi(+\infty)$ et comme S'_t n'a pas de borne inférieure dans K, il est facile de construire une fonction continue sur $[a,b]$, non majorée "au voisinage de t".

PROPOSITION. *La condition de la proposition précédente est réalisée dans les 3 cas suivants :*

1) $\Psi(0_+) = \omega$ *et* K *n'est pas isomorphe à* $\mathbb{R}$

2) $T_1(K) \neq \emptyset$

3) $\exists\, t \in \mathcal{C}(K) - \{+\infty\}$, $\varphi_{\mathcal{C}(K)}(t_-) \geq \Psi(0_+)$ *ou* $\Psi_{\mathcal{C}(K)}(t_+) \geq \Psi(0_+)$.

Démonstration.

1) est évident si K est archimédien, s'il ne l'est pas on prend $t = \frac{1}{\tau}$.

2) On utilise le théorème de II 3.

3) Si $\Psi(0_+) > \omega$, on trouve aisément $t' \in \mathcal{C}(K) - \{+\infty\}$ tel que $\varphi_{\mathcal{C}(K)}(t'_-) = \Psi_K(0_+)$ ou $\Psi_{\mathcal{C}(K)}(t'_+) = \Psi_K(0_+)$.

Dans le premier cas on a $\Psi_K(-t') = \Psi_K(0_+)$, et dans le second $\Psi_K(t') = \Psi_K(0_+)$.

Ainsi il reste à examiner le cas où K satisfait la condition :

(A) $\lambda = \Psi(0_+) > \omega$ et $T_1(K) = \emptyset$ et $\forall t \in \mathcal{C}(K) - \{+\infty\}$, $\varphi_{\mathcal{C}(K)}(t_-) < \lambda$ et $\Psi_{\mathcal{C}(K)}(t_+) < \lambda$.

Soit (B) la condition :

(B) Il existe une application f continue et non bornée sur l'intervalle [0,1] de K, à valeurs dans K.

Si E est totalement ordonné, nous noterons (C_λ) la condition :

(C_λ) card (E) = λ et $\forall t \in T(E)$, $\varphi_E(t) < \lambda$ et $\Psi_E(t) < \lambda$.

<u>PROPOSITION</u>. *Si* K *est un corps satisfaisant* (A), K *satisfait* (B) *si et seulement si il existe un sous ensemble* E *de* K *qui satisfait* (C_λ).

<u>Démonstration.</u> Si K satisfait (B), on peut trouver un sous ensemble E de K de cardinal λ tel que f est non bornée sur tout sous ensemble de E de cardinal λ. On montre alors que E satisfait (C_λ).

Si E est une partie de K qui satisfait (C_λ) avec $E = \{x_\xi \ / \ \xi < \lambda\}$ et si $\{b_\xi \ / \xi < \lambda\}$ est cofinal à K on construit f continue sur [0,1], telle que $f(x_\xi) = b_\xi$ (c' est facile si on remarque qu'on peut prendre E densément ordonné).

Rappelons q'un cardinal λ est faiblement compact si et seulement si pour tout ensemble E totalement ordonné de cardinal λ, il existe une partie bien ordonnée ou anti- bien ordonnée de E de cardinal λ(cf[2]). Ceci est équivalent, dans le cas où λ est régulier, à dire que (C_λ) n'est satisfaite par aucun ensemble E.

<u>COROLLAIRE</u>. *Si* λ *est faiblement compact,* (A) *équivaut à non* (B).

On suppose maintenant que λ n'est pas faiblement compact.

Nous allons donner un exemple de corps satisfaisant (A) et (B) : c'est L'_λ le corps fortement complet de rang isomorphe à λ, où λ est un cardinal régulier et $\lambda > \omega$. En utilisant III 3) on voit que L'_λ satisfait (A). λ n'est pas faiblement compact, donc il existe un ensemble totalement ordonné E qui satisfait (C_λ). On plonge alors E dans L'_λ (on utilise un argument de [4]).

Voici maintenant un exemple de corps qui satisfait (A) et non (B) : c'est $L_\lambda = \mathbb{Q}(X_{\gamma_\xi})_{\xi < \lambda}$ sous corps ordonné de L'_λ. Si $x \in L'_\lambda$ il peut s'écrire $x = \sum_{\eta < \ell(x)} a^x_\eta X_{\delta^x_\eta}$ où $\ell(x)$ est un ordinal ($\ell(x) < \lambda$), $a^x_\eta \in \mathbb{R}^*$ et $\delta^x_\eta \in \Gamma'_\lambda$ (groupe de valuation de L'_λ). On note $y \twoheadrightarrow x$ si $\ell(y) \leq \ell(x)$ et $\forall \eta < \ell(y)$, $a^x_\eta = a^y_\eta$ et $\delta^x_\eta = \delta^y_\eta$.

On montre d'abord que si λ est régulier et $\lambda > \omega$, $\forall x \in L_\lambda$, $\ell(x) < \omega \times \omega$ et que si $y \twoheadrightarrow x$, alors $y \in L_\lambda$.

On suppose que L_λ satisfait (B), alors il existe un sous ensemble E de $[0,1] \cap L_\lambda$ tel que $0 \notin E$ et f est non bornée sur tout sous ensemble de E de cardinal λ On montre alors par récurrence transfinie que

$(\forall \xi < \omega \times \omega)$ $(\exists\, \mu_\xi < \lambda)$ $(\forall x \in E)$ $(\delta^x_\xi > -\gamma_{\mu_\xi})$

On prend $\mu = \sup_{\xi < \omega\times\omega} \mu_\xi$ alors $\mu < \lambda$ et $\mathrm{card}(L_\mu) < \lambda$ donc $\exists x \in E$, $\ell(x) \geq \omega \times \omega$ ce qui est absurde.

BIBLIOGRAPHIE

[1] N.L. ALLING : A characterization of abelian η_α - groups in terms of their natural valuation. Proc. Nat. Acad. Sciences U.S.A. 47 (1961) 711-713.

[2] M.A. DICKMANN : Large infinitary languages. Studies in logic, vol. 83 (1975).

[3] H. HAHN : Uber dic nicht archemedischen grössen systeme. Sitz der Acad. der wiss. (Vienna) 116 (1907) 601-653.

[4] I. FLEISCHER : Embedding linearly ordered sets in real lexicographic products. Fund. Math. 49 (1961) 147-150.

[5] L. FUCHS : Teilweise geordnete algebraische structuren-Göttingen (1966).

[6] S. MAC LANE : The universality of power series fields. Bull. Amer. Math. Soc. 45 (1939) 888-890.

[7] P. RIBENBOIM : Sur les groupes totalement ordonnés et l'arithmétique des anneaux de valuations. Summa Brasiliensis mathematicae, Vol. 4, fasc. 1 (1958).

[8] P. RIBENBOIM : Théorie des valuations. Montréal (1964).

[9] R. ROLLAND : Thèse de 3ème cycle. Rennes (1981).

IRMAR
Université de Rennes I
Campus de Beaulieu
35042 - RENNES-CEDEX
(FRANCE)

FAISCEAU STRUCTURAL SUR LE SPECTRE RÉEL

ET FONCTIONS DE NASH

par

Marie-Françoise ROY

(Université de Paris-Nord et Université de Niamey)

Notre but est de décrire et d'étudier à la main le faisceau structural sur le spectre réel. Ce faisceau structural avait été introduit par des méthodes de théorie des topos dans [5] et [3]. Cette nouvelle description qui suit de près la description d'Artin et Mazur des fonctions de Nash [1], nous permet de montrer un nouveau résultat obtenu avec Michel Coste et fondamental pour notre théorie, l'idempotence du spectre réel. Dans le cas de l'anneau $\mathbb{R}[X_1,\dots,X_n]$, une partie de ce résultat se trouve être équivalent au lemme de substitution de Bochnak et Efroymson, qui est un outil technique essentiel pour l'étude des fonctions de Nash ([2] ou [6]).

§ 1. Description d'Artin-Mazur des fonctions de Nash.

Soit U un ouvert de $\mathbb{R}^n$. Les fonctions de Nash sur U, notées $\mathcal{N}(U)$ sont les fonctions anlytiques-algébriques, c'est-à-dire analytiques réelles qui vérifient localement une équation algébrique à coefficients polynomiaux.

Dans le cas où U est un ouvert semi-algébrique, ces fonctions sont les fonctions semi-algébriques (i.e. de graphe semi-algébrique) et analytiques.

L'anneau des germes de fonctions de Nash en un point $\vec{a}$ de $\mathbb{R}^n$ est la hensilisation de l'anneau $\mathbb{R}[X_1,\ldots,X_n]_{\vec{a}}$ des germes de fonctions régulières au point $\vec{a}$: c'est le plus petit anneau contenant les germes de fonctions régulières en $\vec{a}$ et vérifiant le théorème des fonctions implicites ([8], [7] théorème 4.4.1 ou [7']).

Le lien entre morphismes étales et fonctions de Nash est précisé dans l'article d'Artin-Mazur [1] :

Considérons, pour chaque ouvert semi-algébrique U, le système filtrant I_U dont les éléments sont les couples (V,s) avec V variété affine réelle étale sur $\mathbb{A}^n$ (l'espace affine à n dimension), et s section continue de l'homéomorphisme local de $V(\mathbb{R})$ sur $\mathbb{R}^n$ au-dessus de U

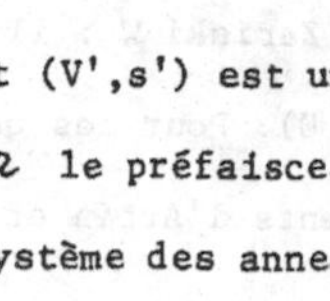

;

un morphisme entre (V,s) et (V',s') est un morphisme de variété $f : V \longrightarrow V'$ tel que $f(\mathbb{R})\circ s = s'$. Notons $\mathcal{Q}$ le préfaisceau obtenu en prenant pour chaque U la limite inductive pour ce système des anneaux de coordonnées $\mathbb{R}[V]$.

Artin et Mazur montrent :

THEOREME : *$\mathcal{A}$ est un faisceau qui coïncide avec le faisceau $\mathcal{N}$ des fonctions de Nash.*

Il n'est peut-être pas inutile de rappeler leur argumentation. Tout d'abord il est clair que $\mathcal{A}(U)$ s'envoie dans $\mathcal{N}(U)$: au représentant $(V,s,P \in \mathbb{R}[V])$ d'un élément de $\mathcal{A}(U)$ on associe $P \circ s$.

La fibre de $\mathcal{A}(U)$ au point $\vec{a}$, $\varinjlim_{U \ni \vec{a}} \ \varinjlim_{\substack{V \text{ étale sur } \mathbb{A}^n \\ \text{au-dessus de } U}} \mathbb{R}[V]$ est la limite inductive des $\mathbb{R}[X_1,\dots,X_n]_{\vec{a}}$ - algèbres locales étales équirésiduelles, c'est une autre description du hensélisé de $\mathbb{R}[X_1,\dots,X_n]_{\vec{a}}$ [9].

Il suffit donc pour conclure de montrer que $\mathcal{A}$ est un faisceau.

LEMME (Représentation canonique) : *Si l'élément* f *de* $\mathcal{A}(U)$ *est représenté par le triplet* (V,s,P), *il existe un représentant canonique* (V_o,s_o,P_o) *de* f *qui ne dépend que de* $P \circ s$, *et qui se comporte bien par restriction : si on considère un ouvert* $U' \subset U$, *il suffit pour avoir le représentant canonique de* $f_{\restriction U'}$ *de restreindre* s_o *à* U'.

Preuve : On a le diagramme commutatif suivant :

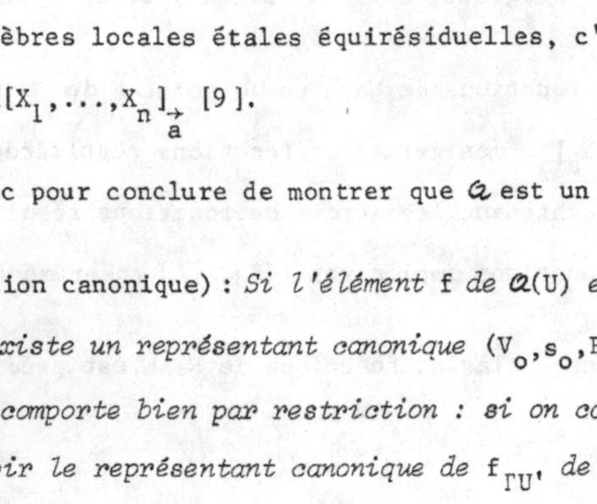

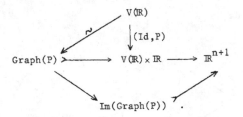

Soit W la clôture de Zariski de Im(Graph(P)), W est irréductible et de dimension n. D'autre part, Graph($P \circ s$) a pour clôture de Zariski W : il est inclus dans W, et sa dimension réelle est n (il est isomorphe à U). Pour ces questions de dimension, voir 4 prop. 8.14, ou utiliser les arguments d'Artin et Mazur.

On considère maintenant la normalisation W' de W. Puisque V est étale sur $\mathbb{A}^n$ et donc normale, on obtient un morphisme de V dans W'. Il se trouve que W' est étale sur $\mathbb{A}^n$ en tout point de l'image de V : il suffit de montrer que si y est l'image de x, les complétés $\widehat{\mathbb{R}[V]}_x$ et $\widehat{\mathbb{R}[W']}_y$ sont isomorphes. Mais puisque V est étale sur $\mathbb{A}^n$, $\widehat{\mathbb{R}[V]}_x$ et $\widehat{\mathbb{R}[X_1,\ldots,X_n]}_{\vec{a}}$ sont isomorphes. Considérons le triangle

$$\begin{array}{ccc} \widehat{\mathbb{R}[V]}_x & \longrightarrow & \widehat{\mathbb{R}[W']}_y \\ & & \downarrow i \\ & & \widehat{\mathbb{R}[X_1,\ldots,X_n]}_{\vec{a}} \end{array}$$

i est nécessairement épi. Il est aussi mono puisque $\widehat{\mathbb{R}[W']}_y$ est intègre (Main theorem de Zariski [11]), et que le noyau q de i, qui est un idéal premier, est par un argument de dimension évident égal à 0. ∎

Avec le lemme, il est clair qu'une famille f_i de $\mathcal{Q}(U_i)$ définie sur un recouvrement ouvert U_i et compatible se recolle, puisque tous les f_i ont le même représentant canonique.

Remarque : Dans [7], H. Delfs a généralisé la définition du préfaisceau $\mathcal{Q}$ au cas de l'anneau de coordonnées d'une variété V sur un corps réel clos et montré que si V est lisse, $\mathcal{Q}$ est encore un faisceau.

§ 2. Généralisation au spectre réel d'un anneau.

Contrairement à la caractérisation des fonctions de Nash comme analytiques-algébriques, la définition d'Artin-Mazur est susceptible de généralisation. Un ingrédient essentiel de leur définition est le fait, immédiat pour $\mathbb{R}^n$, qui, si V est une variété étale sur A^n, $V(\mathbb{R})$ est un espace étalé sur $\mathbb{R}^n$.

Nous allons montrer le résultat suivant.

THEOREME 2.1 : *Si* f : A ⟶ B *est un morphisme étale,* $\mathrm{Spec}_R f : \mathrm{Spec}_R B \longrightarrow \mathrm{Spec}_R A$ *est un homéomorphisme local.*

2.2. Rappels.

1. [4] $\mathrm{Spec}_R A$, le spectre réel d'un anneau a pour éléments les $\alpha = (p, \leq)$ où p désigne un idéal premier de A et $\leq$ un ordre total sur $k(p)$, le corps résiduel de p.

On note $k(\alpha)$ la clôture réelle de $k(p)$ pour l'ordre $\leq$ et π_α le morphisme de A dans $k(\alpha)$. Un <u>ouvert élémentaire</u> de $\mathrm{Spec}_R A$ est de la forme $D_{a_1,\ldots,a_n} = \{\alpha | \pi_\alpha(a_1)>0$ et ... et $\pi_\alpha(a_n) > 0\}$. Cette construction est fonctorielle : si on a $f : A \longrightarrow B$ et β dans $\mathrm{Spec}_R B$, $\mathrm{Spec}_R f(\beta)$ est le couple formé du noyau p' de $A \xrightarrow{f} B \xrightarrow{\pi_\beta} k(\beta)$ et de l'ordre induit par $k(\beta)$ sur $k(p')$. $\mathrm{Spec}_R f$ est une application continue. Un morphisme f de A dans un corps réel clos k définit un point de $\mathrm{Spec}_R A$: p est le noyau de f et $\leq$ l'ordre induit par k sur $k(p)$.

Les ouverts de $\mathrm{Spec}_R A$ sont <u>stables par générisation</u> : si $\alpha' \subset \alpha$ (i.e. $\{a | \pi_\alpha(a) \leq 0\} \subset \{a \mid \pi_{\alpha'}(a) \leq 0\}$) et si U est ouvert dans $\mathrm{Spec}_R A$, $\alpha \in U \Longrightarrow \alpha' \in U$. Un <u>constructible</u> de $\mathrm{Spec}_R A$ est une combinaison booléenne d'ouverts élémentaires.

Si $A = \mathbb{R}[V]$ est l'anneau de coordonnées d'une variété algébrique rélle affine, un point de $V(\mathbb{R})$ s'identifie à un point de $\mathrm{Spec}_R \mathbb{R}[V]$ (on prend le couple formé de l'idéal premier associé au point et du seul ordre possible sur $\mathbb{R}$). Soit $i : V(\mathbb{R}) \hookrightarrow \mathrm{Spec}_R \mathbb{R}[V]$ cette inclusion. La topologie induite par $\mathrm{Spec}_R \mathbb{R}[V]$ sur $V(\mathbb{R})$ est la topologie euclidienne, et i définit une bijection entre les constructibles (resp. les ouverts quasi-compacts) de $\mathrm{Spec}_R \mathbb{R}[V]$ et les semi-algébriques (resp. les ouverts semi-algébriques) de $V(\mathbb{R})$. On note $\tilde{X}$ le constructible associé au semi-algébrique X dans cette bijection.

2. Si $f : A \longrightarrow B$ est un morphisme étale, il existe un recouvrement de B pour la topologie de Zariski formé d'anneaux de type (*) : $B' = A[h^{-1}][X]/p\ (Q^{-1})$ avec P et Q polynômes à coefficients dans A, P de coefficient dominant une puissance de h, et P' divisant Q - [9]. On pourra toujours dans la suite se ramener à des A-algèbres étales du type (*).

3. Notons que dans [3] le résultat 2.1 apparaissait comme un corollaire de notre résultat principal. Nous le démontrons ici directement, avec des idées et des techniques de même nature.

PROPOSITION 2.3 : *Soit $\phi(x_1,\ldots,x_n)$ une formule du langage des corps ordonnés.*

Considérons les conditions

1) $\{\alpha | k(\alpha) \models \phi(\pi_\alpha(a_1,\ldots,\pi_\alpha(a_n))\}$ *est un ouvert de* $\mathrm{Spec}_R A$ *où* $(a_1,\ldots,a_n)$ *sont dés éléments de* A.

2) $\{(x_1,\ldots,x_n) \in \mathbb{R}^n | \phi(x_1,\ldots,x_n)\}$ *est un ouvert de* $\mathbb{R}^n$.

3) $\phi(x_1,\ldots,x_n)$ *est équivalente dans la théorie des corps réels clos à une combinaison positive d'inégalités polynomiales strictes portant sur* $(x_1,\ldots,x_n)$.

2) *et* 3) *sont des conditions équivalentes et impliquent la condition* 1).

Preuve : [4], proposition 4.4.

PROPOSITION 2.4 [5] : *Si* $f : A \longrightarrow B$ *est un morphisme étale,* $\mathrm{Spec}_R f : \mathrm{Spec}_R B \longrightarrow \mathrm{Spec}_R A$ *est une application ouverte.*

Preuve : Montrons d'abord que l'image de $\mathrm{Spec}_R B$ est ouverte dans $\mathrm{Spec}_R A$. D'après 2.2 2), on peut supposer B du type (*), $B = A[h^{-1}][X]/P[Q^{-1}]$. P est défini par une puissance de h et un n-uple, et Q par un m+1 uple, $\{(h,P,Q) \in \mathbb{R}^{n+m+2} | h \neq 0$ et $\exists x\ (P(x) = 0$ et $Q(x) \neq 0$ et $P'(x) \neq 0)\}$ est un ouvert de $\mathbb{R}^{n+m+2}$ (voir note ci-dessous). Donc $U = \{\alpha | k(\alpha) \models \pi_\alpha(h) \neq 0$ et $\exists\, x\ \pi_\alpha(P)(x) = 0$ et $\pi_\alpha(P')(x) \neq 0$ et $\pi_\alpha(Q)(x) \neq 0\}$ est un ouvert de $\mathrm{Spec}_R A$ (2.3). Il est clair que $U = \mathrm{Im}\ \mathrm{Spec}_R B$.

Tout ouvert élémentaire $D_{b_1,\ldots,b_n}$ de $\mathrm{Spec}_R B$ est l'image de $\mathrm{Spec}_R C$ dans $\mathrm{Spec}_R B$ avec

$$C = B[X_1,\ldots,X_n]/(X_1^2-b_1,\ldots,X_n^2-b_n)\ [2X_1^{-1},\ldots,2X_n^{-1}]$$

extension étale de B. L'image de $D_{b_1,\ldots,b_n}$ dans $\mathrm{Spec}_R A$, qui coïncide avec l'image de $\mathrm{Spec}_R C$ dans $\mathrm{Spec}_R A$ est donc ouverte. ∎

Note : On recontrera dans cet article d'autres affirmations de ce style (notamment dans la preuve de la proposition 2.5). Avec un peu d'habitude, on est amené à les considérer comme évidentes. Pour bien montrer le caractère extrêmement élémentaire des mathématiques nécessaires à leur explicitation complète, je donne dans ce premier cas une démonstration détaillée.

Soit (h_o,P_o,Q_o) un point de $X = \{(h,P,Q) | h \neq 0$ et $\exists\, x(P(x) = 0$ et $P'(x) \neq 0$ et $Q(x) \neq 0\}$.

On va montrer que X contient un ouvert V contenant (h_o,P_o,Q_o).

Considérons dans $\mathbb{R}^{1+n+m+2}$ les ouverts

$$U_1 = \{(x,h,P,Q) \mid h \neq 0 \quad \text{et} \quad Q(x) \neq 0 \quad \text{et} \quad P'(x) \neq 0\}$$

$$U_2 = \{(x,h,P,Q) \mid P(x) > 0\}$$

$$U_3 = \{(x,h,P,Q) \mid P(x) < 0\}.$$

Soit x_o un réel tel que $P_o(x_o) = 0$, $P'_o(x_o) \neq 0$, $Q_o(x_o) \neq 0$.

On a $(x_o,h_o,P_o,Q_o) \in U_1$, c'est-à-dire que U_1 contienne une boule ouverte B de centre (x_o,h_o,P_o,Q_o).

On a aussi $(x_o,h_o,P_o,Q_o) \in \overline{U}_2$

$(x_o,h_o,P_o,Q_o) \in \overline{U}_3$.

Donc $U = B \cap U_2$, $U' = B \cap U_3$ sont non vides.

Considérons $V = \pi(U) \cap \pi(U')$ où π désigne la projection de $\mathbb{R}^{1+n+m+2}$ dans $\mathbb{R}^{n+m+2}$.

V est ouvert puisque π est une application ouverte.

Soit (h_1,P_1,Q_1) un point de V, il existe des réels x_1 et x'_1 tels que

$$(x_1,h_1,P_1,Q_1) \in U \quad \text{i.e.} \quad P_1(x_1) > 0 \quad \text{et} \quad (x_1,h_1,P_1,Q_1) \in B$$

$$(x'_1,h_1,P_1,Q_1) \in U' \quad \text{i.e.} \quad P_1(x'_1) < 0 \quad \text{et} \quad (x_1,h_1,P_1,Q_1) \in B.$$

D'après le théorème des valeurs intermédiaires, P_1 s'annule en $y_1 \in [x_1,x'_1]$ et puisque (x_1,h_1,P_1,Q_1) et (x'_1,h_1,P_1,Q_1) sont contenus dans B qui est convexe, on a aussi $(y_1,h_1,P_1,Q_1) \in B$, donc $P'_1(y_1) \neq 0$ et $Q_1(y_1) \neq 0$.

On a donc bien $(h_1,P_1,Q_1) \in X$. ∎

PROPOSITION 2.5 : *Soient* k *un corps réel clos,* $P = X^n + b_1X^{n-1}+\ldots+b_n$ *un polynôme de degré* n, *unitaire à coefficients dans* k *et* a *une racine simple de* P *dans* k ; *il existe un entier* p, *des éléments* $a_o = a, a_1,\ldots,a_p$ *de* k *et une formule* $\theta_a(x_o,\ldots,x_p,y_1,\ldots,y_n)$ *du langage des corps ordonnés qui décrit* a *au sens suivant :*

a) $\theta_a(a_o,\ldots,a_p,b_1,\ldots,b_n)$ *est valide dans* k.

b) $[\forall y_1\ldots\forall y_n \forall x_o \forall x'_o\ldots\forall x_p \forall x'_p \;\; \theta_a(x_o,x_p,y_1,\ldots,y_n)$ *et*

$\theta_a(x'_o,\dots,x'_p,y_1,\dots,y_n) \longrightarrow x_o = x'_o$ *et ... et* $x_p = x'_p]$ *est vraie dans la théorie des corps réels clos.*

c) *De plus si* $Q = z_o X^m + z_1 X^{m-1} + \dots + z_m$, *les formules* $[\exists x_o \dots \exists x_p$ $\theta_a(x_o,\dots,x_p,y_1,\dots,y_n)$ et $Q(x_o) \neq 0]$ (*resp.* $[\exists x_o \dots \exists x_p\ \theta_a(x_o,\dots,x_p,y_1,\dots,y_n)$ et $Q(x_o) > 0]$) *et* $[P(x) = 0 \longrightarrow \exists x_1 \dots \exists x_p\ \theta_a(x,x_1,\dots,x_p,y_1,\dots,y_n)]$ *sont des combinaisons positives d'inégalités polynomiales strictes portant sur les variables* $(y_1,\dots,y_n,z_o,\dots,z_n)$ *et* $(x,y_1,\dots,y_n)$.

Preuve : Les éléments $a_1,\dots,a_p$ de k sont construits comme suit : a_1 et a_2 sont des racines de P' telles que a soit la seule racine de P sur $[a_1,a_2]$ (il peut arriver que P' n'ait pas de racine à gauche (resp. à droite) de a, dans ce cas $a_1 = -\infty$ ($a_2 = +\infty$) par convention). a_1, s'il existe, est racine simple de $P^{(i_1)}$ (resp. a_2 de $P^{(i_2)}$). On s'arrête au bout d'un nombre fini d'étapes à x_p, racine simple de $P^{(i_p)}$ et seule racine de $P^{(i_p)}$ dans k. Soit alors $\theta_a(x_o,\dots,x_p,y_1,\dots,y_n)$ la formule du langage des corps ordonnés qui décrit cette situation : "x_o est racine simple de $P = X^n + y_1 X^{n-1} + \dots + y_n$ et seule racine de P sur $[x_1,x_2]$ et x_1 est racine simple de $P^{(i_1)}$ et ... et x_p est racine simple de $P^{(i_p)}$ et seule racine de $P^{(i_p)}$".

a) Est clair par définition de θ_a.

b) Provient du fait que $x_p = x'_p$, puisque x_p est la seule racine de $P^{(i_p)}$, et qu'on peut montrer de proche en proche $x_i = x'_i$ jusqu'à $x_o = x'_o$.

c) Dans $\mathbb{R}^{n+m+1}$

$$\{(y_1,\dots,y_n,z_o,\dots,z_m) \mid \exists x_o \dots \exists x_p\ \ \theta_a(x_o,\dots,x_p,y_1,\dots,y_n) \text{ et } z_o x_o^m + z_1 x_o^{m-1} + \dots + z_m \neq 0\}$$

$$(\text{resp. } \{(y_1,\dots,y_n,z_o,\dots,z_m) \mid \exists x_o \dots \exists x_p\ \ \theta_a(x_o,\dots,x_p,y_1,\dots,y_n) \text{ et } z_o x_o^m + z_1 x_o^{m-1} + \dots + z_m > 0\})$$

est un ouvert : l'existence d'un p uple $(x_o,\dots,x_p)$ dans la situation décrite par θ_a avec $Q(x_o) \neq 0$ (resp. $Q(x_o) > 0$) si $Q = z_o X^m + z_1 X^{m-1} + \dots + z_m$ n'est pas affectée par de petits changements dans les coefficients $(y_1,\dots,y_n,z_o,\dots,z_m)$ de P et Q : être une racine simple d'un polynôme, être la seule racine d'un polynôme sur un intervalle fermé et ne pas annuler un polynôme (resp. rendre un polynôme strictement positif) sont des situations stables. On conclut par 2.3.

De même $\{(x,y_1,\dots,y_n) \mid P(x) = 0 \longrightarrow \exists x_1 \dots \exists x_p \ \theta_a(x,x_1,\dots,x_p,y_1,\dots,y_n)\}$ est un ouvert de $\mathbb{R}^{n+1}$: si x reste racine de P, on vient de voir que de petits changements dans les coefficients de P n'affectent pas la situation décrite par θ_a, et $\{(x,y_1,\dots,y_n) \mid P(x) \neq 0\}$ est un ouvert de $\mathbb{R}^{n+1}$. On conclut encore par 2.3. ∎

DEFINITION 2.6 : $a_o = a,a_1,\dots,a_p$ *forment l'arbre de* a *et la formule* θ_a *est la description de l'arbre de* a.

Preuve du théorème 2.1 : Soit f : A ⟶ B un morphisme étale. On peut se ramener à $B = A[h^{-1}][X]/_P\,[Q^{-1}]$ comme dans 2.2 2).

Soit β_o dans $\mathrm{Spec}_R B$ au-dessus de α_o (i.e. $\mathrm{Spec}_R f(\beta_o) = \alpha_o$), on veut montrer qu'il existe un ouvert U de $\mathrm{Spec}_R A$ et une section continue s de $\mathrm{Spec}_R f$ au-dessus de U avec $s(\alpha_o) = \beta_o$, et s(U) ouvert de $\mathrm{Spec}_R B$.

Puisque f est étale et que β_o est au-dessus de α_o, on a $k(\beta_o) \simeq k(\alpha_o)$ ($k(\beta_o)$ est réel clos et est une extension algébrique du corps réel clos $k(\alpha_o)$), et l'image de X dans $k(\alpha_o)$, a_o, est une racine simple de $\pi_{\alpha_o}(P)$. On lui associe θ_{a_o} comme en 2.5. $U = \{\alpha \mid k(\alpha) \models \pi_\alpha(h) \neq 0$ et $\exists x_o \dots \exists x_p \ \theta_{a_o}(x_o,\dots,x_p,\pi_\alpha(P))$ et $\pi_\alpha(Q)(x_o) \neq 0\}$ ($\pi_\alpha(P)$ et $\pi_\alpha(Q)$ désignent les n et m+1-uples des coefficients de $\pi_\alpha(P)$ et $\pi_\alpha(Q)$) est un ouvert de $\mathrm{Spec}_R A$ d'après 2.3 et 2.5 c). On va relever $\mathrm{Spec}_R f$ au-dessus de cet ouvert U.

A α de U, on associe le point s(α) de $\mathrm{Spec}_R B$ qui correspond au morphisme de B dans k(α) obtenu en envoyant X sur l'unique a de k(α) vérifiant $\exists x_1 \dots \exists x_p \ \theta_{a_o}(a,x_1,\dots,x_p,\pi_\alpha(P))$ (2.5 b)). Il est clair que

$s(U) = \{\beta \mid k(\beta) \models \pi_\beta(h)$ inversible et $\exists x_1 \dots \exists x_p \ \theta_a(\pi_\beta(X),x_1,\dots,x_p,\pi_\beta(P))$ et $\pi_\beta(Q(X)) \neq 0\}$, s(U) est un ouvert de $\mathrm{Spec}_R B$ d'après la dernière partie de 2.5 c) (puisque $\pi_\beta(X)$ est racine de $\pi_\beta(P)$) et 2.3.

Reste à montrer que s est un homéomorphisme de U sur s(U). $\mathrm{Spec}_R f \circ s$ est l'inclusion de U dans $\mathrm{Spec}_R A$, donc s est injective. s^{-1} est continue puisque $\mathrm{Spec}_R f$ l'est (2.2 1)), et s est continue puisqu'un ouvert de s(U) est ouvert dans $\mathrm{Spec}_R B$, et que $\mathrm{Spec}_R f$ est ouverte (2.4). ∎

On peut alors proposer la définition suivante du faisceau structural sur le spectre réel.

DEFINITION 2.7 : *Soient* U *un ouvert quasi compact de* $\mathrm{Spec}_R A$ *et* I_U *le système filtrant suivant : ses objets sont les couples* (B,s) *où* B *est une* A*-algèbre étale, et* s *une section continue de l'homéomorphisme local de* $\mathrm{Spec}_R B$ *dans* $\mathrm{Spec}_R A$ *au-dessus de* U*, un morphisme entre* (B,s) *et* (B',s') *est un morphisme* f : B ⟶ B' *de* A*-algèbre avec* $\mathrm{Spec}_R f \circ s' = s$.

On définit $\mathcal{Q}(U) = \varinjlim_{(B,s)\in I_U} B$.

Cette définition est en tout point analogue à celle du § 1. Toutefois $\mathcal{Q}$ n'a plus en général de raison d'être un faisceau, car on ne voit aucun candidat raisonnable à être un représentant canonique comme dans le lemme.

Le faisceau structural sur le spectre réel, $\mathcal{N}_{\mathrm{Spec}_R A}$ est le faisceau associé au préfaisceau $\mathcal{Q}$.

PROPOSITION 2.8 : U ⟶ $\mathcal{Q}$(U) *est un préfaisceau séparé.*

Preuve : Soit a un élément de $\mathcal{Q}$(U) représenté par (b,s) avec b ∈ B, A-algèbre étale, et s section continue de $\mathrm{Spec}_R B \longrightarrow \mathrm{Spec}_R A$ au-dessus de U. Soient U_i un recouvrement ouvert de U et B ⟶ B_i des A algèbres étales telles que $a_{\wedge u_i}$ soit représenté par $(s_{\wedge u_i}, 0)$. On va montrer que a est nul.

L'union des images des Spec B_i dans Spec B (spectre de Zariski) est un ouvert $D_{f_1} \cup \ldots \cup D_{f_p}$ de Spec B_i, soit $B' = B[(f_1^2+\ldots+f_p^2)^{-1}]$. On peut trouver $s' : U \longrightarrow \mathrm{Spec}_R B'$ et b' ∈ B' avec (s',b') représentant de a : si $s(\alpha)=(\rho,\underline{<})$, ρ est dans UIm(Spec B_i) et est un idéal réel, il ne contient pas $f_1^2+\ldots+f_p^2$ et définit un idéal ρ' de B' : on pose alors $s'(\alpha)=(\rho',\underline{<})$ et on prend pour b' l'image de b dans B'.

La famille $(B' \longrightarrow B'_i = B' \otimes_B B_i)_{i\in I}$ est couvrante pour la topologie étale, donc $B' \hookrightarrow \prod_{i \in I} B'_i$ [9], on a donc b' = 0 d'où a = 0. ∎

PROPOSITION 2.9 : *Toute section de $\mathcal{N}_{\mathrm{Spec}_R A}$ au-dessus de U est une union finie de sections, dites élémentaires, représentées par des (s_i,b_i) avec $b_i \in B_i = A[h_i^{-1}][X]/_{P_i}[Q_i^{-1}]$ comme dans 2.2 2), et s_i section de $\mathrm{Spec}_R B_i \longrightarrow \mathrm{Spec}_R A$ au-dessus d'un ouvert U_i vérifiant la propriété suivante : il existe α_o de U_i tel que pour tout α de U_i les $\pi_{s_i(\alpha)}(X)$, racines simples de $\pi_\alpha(P_i)$ dans $k(\alpha)$, vérifient la formule $\exists x_1 \ldots \exists x_m\ \Theta_{\pi_{s_i(\alpha_o)}}(X)$.*

Preuve : D'après 2.8, toute section de $\mathcal{N}_{\mathrm{Spec}_R A}$ au-dessus de U est union finie (par compacité de U[4]) de sections de $\mathcal{Q}(U_i)$ (2.2 2) et la preuve du théorème 2.1 montre qu'on peut se ramener à des (s_i,b_i) comme indiqué. ∎

PROPOSITION 2.10 : *Si $A = \mathbb{R}[X_1,\ldots,X_n]$, $\mathcal{N}_{\mathrm{Spec}_R A}$ est l'image directe du faisceau de fonctions de Nash sur $\mathbb{R}^n$ par l'inclusion* $i : \mathbb{R}^n \hookrightarrow \mathrm{Spec}_R \mathbb{R}[X_1,\ldots,X_n]$ (2.2.1).

Preuve : Soient U un ouvert semi algébrique de $\mathbb{R}^n$ et $\tilde{U}$ l'ouvert quasi compact de $\mathrm{Spec}_R \mathbb{R}[X_1,\ldots,X_n]$ correspondant (2.2 1). $\mathcal{Q}(\tilde{U})$ coïncide avec $\mathcal{Q}(U)$ définie au § 1. ∎

La fibre de $\mathcal{N}_{\mathrm{Spec}_R \mathbb{R}[X_1,\ldots,X_n]}$ en un point $\vec{a}$ de $\mathbb{R}^n$ est l'anneau des germes de fonctions de Nash en $\vec{a}$. En un point α de $\mathrm{Spec}_R \mathbb{R}[X_1,\ldots,X_n]$ qui n'appartient pas à $\mathbb{R}^n$, c'est la limite inductive pour $\tilde{U}$ contenant α des anneaux de fonctions de Nash sur U. On peut encore parler de "germes de fonctions de Nash en α". Par exemple dans $\mathrm{Spec}_R \mathbb{R}[X]$, considérons le point 0_+ obtenu en mettant X juste à droite de 0 (on décide que X est infiniment petit positif), la fibre de $\mathcal{N}_{\mathrm{Spec}_R \mathbb{R}[X]}$ en 0_+ est $\varinjlim_{\varepsilon>0} \mathcal{N}(]0,\varepsilon[)$.

Remarque : L'application $\sim$ permet de montrer de même que notre faisceau coïncide avec celui construit par H. Delfs dans [7] pour le cas K[V] avec K réel clos.

Si $V(\mathbb{R})$ est une variété algébrique réelle affine dans $\mathbb{R}^n$, l'anneau de fonctions de Nash sur un ouvert U de $V(\mathbb{R})$ peut être défini ainsi : si U' est un ouvert de $\mathbb{R}^n$ contenant U, on identifie dans l'anneau des fonctions de Nash sur U' (dans $\mathbb{R}^n$) les fonctions qui coïncident sur U, et on prend la limite inductive de ces anneaux pour le système inductif des ouverts de $\mathbb{R}^n$ contenant U.

On peut s'interroger sur les rapports entre ces fonctions de Nash sur V(ℝ) et le faisceau de $\mathcal{N}_{\mathrm{Spec}_R \mathbb{R}[V]}$.

PROPOSITION 2.11 : *Si* V(ℝ) *est de dimension* d, *soit* x *un point régulier de* V(ℝ) ; *les germes de fonctions de Nash sur* V(ℝ) *en* x *coïncident avec la fibre en* $\vec{x}$ *de* $\mathcal{N}_{\mathrm{Spec}_R \mathbb{R}[V]}$.

Preuve : Pour un ouvert U suffisamment petit, l'application de V(ℝ) dans son espace tangent en $\vec{x}$ est étale. Les hensélisés de $\mathbb{R}[V]_x$ et de $\mathbb{R}[X_1 \ldots X_d]_{\vec{0}}$ sont isomorphes, ainsi que les germes de fonctions de Nash sur V(ℝ) en $\vec{x}$ et les fonctions de Nash sur $\mathbb{R}^d$ en $\vec{0}$, et le résultat est vrai pour $\mathbb{R}^d$ en $\vec{0}$. ∎

Par contre, dans le cas d'une cubique V à point réel isolé x, la fibre de $\mathcal{N}_{\mathrm{Spec}_R \mathbb{R}[V]}$ en x ne coïncide pas avec les germes de fonctions de Nash sur V en x : x étant à lui seul un ouvert, les germes de fonctions de Nash en x sont ℝ, et la fibre de $\mathcal{N}_{\mathrm{Spec}_R \mathbb{R}[V]}$ en x n'est pas un corps, puisque qu'elle a un idéal premier minimal correspondant aux branches complexes conjuguées de V passant par x [9].

On peut raisonnablement conjecturer que les points de V(ℝ), pour lesquels la fibre de $\mathcal{N}_{\mathrm{Spec}_R \mathbb{R}[V]}$ en x (qui est le hensélisé des germes de fonctions régulières en x) coïncide avec les germes de fonctions de Nash en x sur V(ℝ) sont les points quasi-réguliers : ceux tels que les idéaux premiers minimaux du hensélisé de $\mathbb{R}[V]_x$ sont tous réels (pour une définition équivalente [10], page 51).

§ 3. Localisations strictes réelles.

La notion de localisation stricte réelle va nous permettre d'étudier les fibres du faisceau $\mathcal{N}_{\mathrm{Spec}_R A}$.

DEFINITION 3.1 : *Une localisation stricte réelle de* A *est une* A*-algèbre locale,* *ind-étale (i.e. limite inductive de morphismes étales), hensélienne de corps résiduel réel clos.*

Exemple 3.2 : Les germes de fonctions de Nash en un point $\vec{a}$ de $\mathbb{R}^n$ forment une localisation stricte réelle de $\mathbb{R}[X_1,\ldots,X_n]$: le hensélisé de $\mathbb{R}[X_1,\ldots,X_n]_{\vec{a}}$ est une $\mathbb{R}[X_1,\ldots,X_n]$ algèbre locale, ind étale, de corps résiduel $\mathbb{R}$.

PROPOSITION 3.3 : *Les localisations strictes réelles de* A *sont en bijection avec les points du spectre réel de* A.

Preuve : A une localisation stricte réelle B de A, on associe le point du spectre réel de A, défini par le morphisme de A dans le corps résiduel K de B.

Inversement, au point $\alpha = (\rho,\leq)$ de $\mathrm{Spec}_R A$, on associe la limite inductive A_α du système suivant : les objets sont les A-algèbres (f,B) qui factorisent π_α, un morphisme de (f,B) dans (f',B') étant un morphisme g : B $\longrightarrow$ B' de A-algèbres ; A_α est une A-algèbre locale, ind étale hensélienne, de corps résiduel $k(\alpha)$, et on note a_α l'image de a dans A_α. Le point de $\mathrm{Spec}_R A$ défini par A_α est évidemment α.

Inversement, montrons que A_α est isomorphe à B. K est isomorphe à $k(\alpha)$: K est un corps réel clos et une extension algébrique de $k(p)$, il induit sur $k(p)$ l'ordre $\leq$, c'est la clôture réelle de $k(p)$ pour l'ordre $\leq$. B s'envoie dans A_α puisqu'il est limite inductive de A-algèbres étales qui factorisent π_α, et ce morphisme est bijectif puisque B, étant hensélien, n'admet pas d'extension locale étale équirésiduelle. ∎

PROPOSITION 3.4 : *La fibre de* $\mathcal{N}_{\mathrm{Spec}_R A}$ *en* α *est la localisation stricte réelle* A_α.

Preuve : La fibre de $\mathcal{N}_{\mathrm{Spec}_R A}$ en α est la limite inductive des A-algèbres étales B qui ont un point de leur spectre réel au-dessus de α, c'est-à-dire des A-algèbres étales qui factorisent π_α. ∎

Remarques 3.5 :

1. Le résultat 3.4 nous autorise à dire que $\mathcal{N}_{\mathrm{Spec}_R A}$ est un faisceau qui recolle continûment les localisations strictes réelles de A.

2. Dans le cas particulier de $\mathbb{R}[X_1,\ldots,X_n]$, le résultat 3.4 a pour corollaire un "théorème des fonctions implicites avec paramètre" pour les fonctions de Nash sur $\mathbb{R}^n$.

Si $\alpha = (\rho,\leq)$ est un point de $\mathrm{Spec}_R\,\mathbb{R}[X_1,\ldots,X_n]$, ρ définit une sous variété affine de $\mathbb{R}^n$, V. π_α factorise par $\mathbb{R}[V]$, et on note encore α le point correspondant de $\mathrm{Spec}_R\,\mathbb{R}[V]$.

Théorème des fonctions implicites à paramètre : (le paramètre désigne un élément de $V(\mathbb{R})$).

Soit $P(\vec{x},y)$ un polynôme à n+1 variables tel que

$\exists U'$ ouvert semi algébrique de $V(\mathbb{R})$ composé de points réguliers avec $\alpha \in \tilde{U}'$

$\exists f$ fonction de Nash sur U' telle que $\forall x \in U'$, $P(x,f(x)) = 0$

$$\frac{\partial P}{\partial y}(x,f(x)) \neq 0$$

alors $\exists U''$ ouvert semi algébrique de $V(\mathbb{R})$ avec $\alpha \in \tilde{U}''$, contenu dans U',

$\exists U$ ouvert semi algébrique de $\mathbb{R}^n$ $U \supset U''$,

$\exists \varphi$ fonction de Nash sur U telle que

$$\forall \vec{x} \in U \quad P(\vec{x},\varphi(\vec{x})) = 0$$

$$\forall \vec{x} \in U'' \quad \varphi(\vec{x}) = f(\vec{x}) .$$

Preuve : L'anneau des germes de fonctions de Nash en α dans $\mathbb{R}^n$ est hensélien, et a pour corps résiduel l'anneau des germes de fonctions de Nash en α sur $V(\mathbb{R})$. ∎

Par exemple, dans $\mathbb{R}[X,Y]$, soit 0_+ le couple formé de l'idéal X et de l'ordre 0_+ sur $\mathbb{R}(X)$. Si une fonction de Nash sur $]0,\varepsilon[$ vérifie une équation polynomiale $P(x,0,f(x)) = 0$ telle que $\frac{\partial P}{\partial y}(x,0,f(x)) \neq 0$, on peut la prolonger en une fonction de Nash sur un ouvert U de $\mathbb{R}^2$ contenant $]0,\varepsilon'[$ avec $0 < \varepsilon' \leq \varepsilon$ telle que $\forall x \in U \quad P(x,y,\varphi(x,y)) = 0$.

PROPOSITION 3.6 : *La duale de la catégorie des localisations strictes réelles (les morphismes sont les morphismes de A-algèbres), coïncide avec la catégorie des points de* $\mathrm{Spec}_R A$ *ordonnés par inclusion.*

Preuve :

a) Si on a un morphisme f de A-algèbres entre A_α et $A_{\alpha'}$, tout élément a de A tel que $\pi_\alpha(a) > 0$ vérifie $\pi_{\alpha'}(a) > 0$: en effet $\pi_\alpha(a)$ a une racine carrée inversible dans $k(\alpha)$, donc a_α a une racine carrée inversible dans A_α qui est hensélien, et cette situation est préservée par f, c'est-à-dire que $\pi_{\alpha'}(a)$ a une racine carrée inversible dans $k(\alpha')$, donc $\pi_{\alpha'}(a) > 0$.

b) Deux morphismes de A-algèbres de A_α dans $A_{\alpha'}$ sont égaux. Nous reprenons ici la preuve de [3]. Soient donc f et f' deux morphismes de A-algèbres de A_α dans $A_{\alpha'}$. Puisque tout élément de A_α provient d'une A-algèbre étale du type (*) (2.2 2)), il suffit de montrer que f et f' coïncident pour tout élément a de A_α, qui est racine simple d'un polynôme unitaire à coefficients dans $A[h^{-1}]$, P.

Appelons $\bar{a}$ l'image de X dans $k(\alpha)$ et considérons la formule $\theta_{\bar{a}}(x_o,\dots,x_p,\bar{b}_1,\dots,\bar{b}_n)$ qui décrit $\bar{a}$ (proposition 2.5). Les éléments $\bar{a},a_1',\dots,a_p'$ de $k(\alpha)$ qui vérifient $\theta_{\bar{a}}(\bar{a},a_1',\dots,a_p',\bar{b}_1,\dots,\bar{b}_n)$ sont des racines simples de $P,P^{(i_1)},\dots,P^{(i_p)}$, et se relèvent en des éléments $a,a_1,\dots,a_p$ de A_α. Les inégalités strictes vérifiées par a_i' et a_j' dans $k(\alpha)$ sont encore vérifiées par a_i et a_j dans A_α (on dit que $a > b$ dans A_α si a-b a une racine carrée inversible), ainsi que par $f(a_i)$ et $f(a_j)$ (resp. $f'(a_i)$ et $f'(a_j)$) dans $A_{\alpha'}$. De même, à cause de 2.5 c) et puisque $\alpha' \subset \alpha$, $k(\alpha') \models \exists x_o \dots \exists x_p \; \theta_{\bar{a}}(x_o,\dots,x_p,\overline{f(b_1)},\dots,\overline{f(b_n)})$. On a $\overline{f(b_i)} = \overline{f'(b_i)}$ puisque $b_i \in A[h^{-1}]$. On montre alors, en commençant par p, que $\overline{f(a_i)} = \overline{f'(a_i)}$ pour tout i : $\overline{f(a_p)}$ et $\overline{f'(a_p)}$ coïncident puisqu'elles sont racines de $f(P^{(i_p)})$, et que $f(P^{(i_p)})$ a une seule racine dans $k(\alpha')$, et on peut remonter jusqu'à $\overline{f(a)} = \overline{f'(a)}$. Puisque $A_{\alpha'}$ est hensélien, on en déduit $f(a) = f'(a)$.

c) Reste à montrer qu'on a bien un morphisme de A_α dans $A_{\alpha'}$ si $\alpha' \subset \alpha$. Il suffit pour cela d'envoyer le système inductif qui définit A_α dans le système inductif qui définit $A_{\alpha'}$: si $A \xrightarrow{f} B$ est une A-algèbre étale qui factorise $\pi_{\alpha'}$, d'après le théorème 2.1 il existe des points β' et β de $\mathrm{Spec}_R B$ avec $\beta' \subset \beta$, β au-dessus de α, β' au-dessus de α' : le morphisme de B dans $k(\alpha)$ définit un point β au-dessus de α, et $\mathrm{Spec}_R f$ restreint à un ouvert V convenable contenant β est un homéomorphisme ; enfin V et $\mathrm{Spec}_R f(V)$ sont stables par générisation (2.2 1)).

Si on a des morphismes étales avec

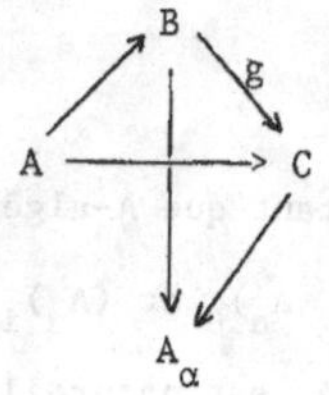

commutatif, on peut compléter le diagramme par des flèches

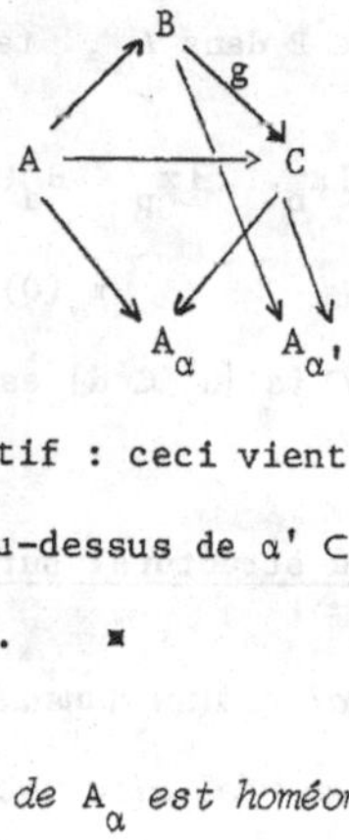

rendant le diagramme commutatif : ceci vient du fait que, d'après le théorème 2.1, il existe un unique $\beta' \subset \beta$ au-dessus de $\alpha' \subset \alpha$ et un unique $\gamma' \subset \gamma$ au-dessus de $\alpha' \subset \alpha$, donc $\mathrm{Spec}_R g(\gamma') = \beta'$. ∎

PROPOSITION 3.7 : *Le spectre de* A_α *est homéomorphe à* $\{\alpha' | \alpha' \subset \alpha\}$ *muni de la topologie induite par* $\mathrm{Spec}_R A$.

<u>Preuve</u> : Une localisation stricte réelle de A_α, $(A_\alpha)_\beta$ définit une localisation stricte réelle $A_{p(\beta)}$ de A, avec $p\beta \subset \alpha$ puisque le composé de deux morphismes ind-étales est ind-étale. Réciroquement, à tout $\alpha' \subset \alpha$ correspond un morphisme de A_α dans $A_{\alpha'}$, qui est une localisation stricte réelle $(A_\alpha)_{i(\alpha')}$ de A : puisque $A \longrightarrow A_\alpha$ ainsi que $A \longrightarrow A_{\alpha'}$ sont ind-étales, $A_\alpha \longrightarrow A_{\alpha'}$ l'est aussi.

Il est clair que $p(i(\alpha')) = \alpha'$.

Inversement on a le diagramme commutatif suivant

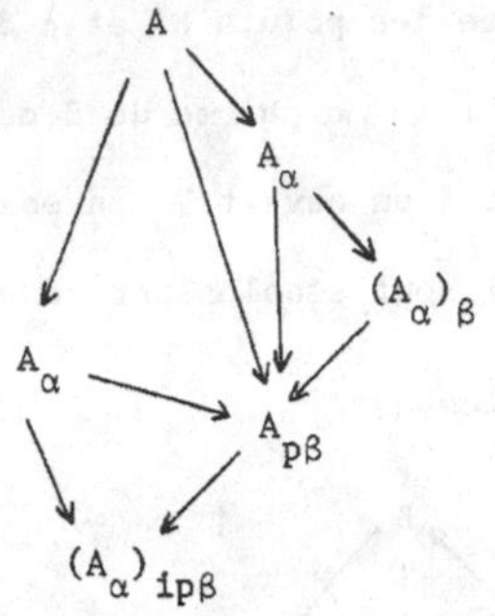

$A_{p\beta}$, $(A_\alpha)_\beta$ et $(A_\alpha)_{ip\beta}$ sont isomorphes en tant que A-algèbres, les deux flèches de A_α dans $A_{p\beta}$ sont égales d'après 3.6, donc $(A_\alpha)_\beta$ et $(A_\alpha)_{ip\beta}$ sont isomorphes en tant que A_α-algèbres. La trace de D_a sur $\mathrm{Spec}_R A_\alpha$ est naturellement un ouvert de $\mathrm{Spec}_R A_\alpha$.

Soit b un élément de A_α. Il existe une A-algèbre étale du type (*) (2.2 2)) $B = A[h^{-1}][X]/_p\ [Q^{-1}]$, et une racine a de P dans A_α, telle que $b = \dfrac{R(a)}{Q^n(a)}$.

Soit $U = \{\alpha \mid k(\alpha) \models \pi_\alpha(h) \neq 0$ et $\exists x_o \ldots \exists x_p \quad \theta_a(x_o,\ldots,x_p,\pi_\alpha(P))$ et

$\pi_\alpha(Q)^n(x_o)\pi_\alpha(P)(x_o) > 0\}$

U est un ouvert de $\mathrm{Spec}_R A$ (2.5 c)) et $U \cap \{\alpha' \mid \alpha' \subset \alpha\}$ est égal à D_b. ∎

§ 4. La propriété universelle du faisceau structural sur le spectre réel.

DEFINITION 4.1 : *Un anneau réel clos local est un anneau local hensélien de corps résiduel réel clos. Un élément d'un anneau réel clos local est strictement positif si son image dans le corps résiduel est strictement positive.*

Etre un élément strictement positif dans un anneau réel clos local s'exprime par une formule du langage des anneaux : $\exists x\ (\exists y\ x.y = f)$ et $a = x^2$ signifie que a est le carré d'un élément inversible, c'est-à-dire que a est strictement positif.

Un morphisme entre deux anneaux réels clos locaux préserve les éléments strictement positifs.

Exemple 4.2 : Une localisation strictre réelle de A est un anneau réel clos local.

PROPOSITION 4.3 : *Pour tout morphisme de* A *dans un anneau réel clos local* B*, il existe une unique localisation stricte réelle* A_α *et un unique morphisme local* $g : A_\alpha \longrightarrow B$ *tel que* $g = f\circ(\ )_\alpha$.

<u>Preuve</u> : Soit $\alpha = (\rho,\leq)$ le point de $\mathrm{Spec}_R A$ défini par le morphisme $\overline{f} : A \xrightarrow{f} B \longrightarrow K_B$ où K_B où K_B désigne le corps résiduel de B.

$\overline{f}$ factorise par $k(\rho)$, et $k(\alpha)$, clôture réelle de $k(\rho)$ pour l'ordre induit par K_B est inclus dans K_B

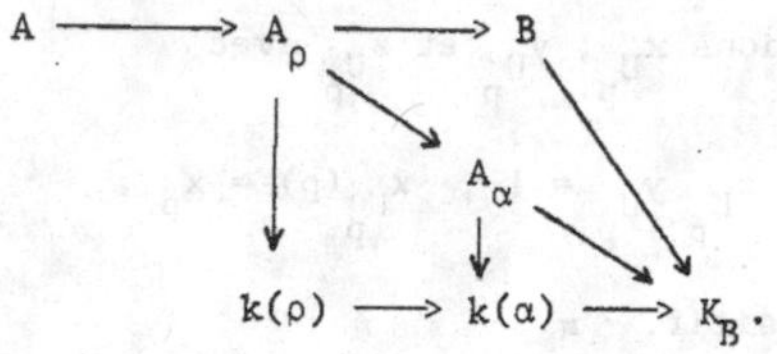

Le morphisme de A_α dans K_B se relève en un morphisme local g de A_α dans B puisque A_α est Ind-étale, et B est hensélien.

Inversement, soit (α',g') tel que $A \xrightarrow{(\ )_{\alpha'}} A_{\alpha'} \xrightarrow{g'} B$, $g\circ(\ )_{\alpha'} = f$, on a $k(\alpha') \subset K_B$ donc $\alpha' = \alpha$, $g' = g$. ∗

DEFINITION 4.4 : *Un espace annelé en anneaux réels clos locaux* (X,B) *est un faisceau d'anneaux dont les fibres sont réelles closes locales. On note* B_p *la fibre en* p *de* B, B(U) *les sections de* B *au-dessus de l'ouvert* U *et* K_p *le corps résiduel en* p.
On dit qu'un élément de B(U) *est strictement positif s'il est le carré d'un élément inversible.*

<u>Exemple 4.5</u> : $\mathcal{N}_{\mathrm{Spec}_R A}$ est un espace annelé en anneaux réels clos locaux : on a vu que la fibre A_α en α est un anneau réel clos local.

PROPOSITION 4.6 : *Soient* P *un polynôme à coefficient dans* B(U), p *un point de* U *et* x_p *une racine simple de* P_p *dans* B_p *; il existe un ouvert* U_p *contenant* p, $U_p \subset U$ *et une section* x_{U_p} *de* B *au-dessus de* U_p *avec* $x_{U_p}(p) = x_p$ *et* x_{U_p} *racine simple de* P_{U_p} *dans* $B(U_p)$.

Preuve : On a dans B_p un élément y_p avec $P_p'(x_p) \times y_p = 1$, on peut donc trouver un ouvert U_p contenant p, et des sections x_{U_p} et y_{U_p} avec $x_{U_p}(p) = x_p$, $y_{U_p}(p) = y_p$, $P_{U_p}(x_{U_p}) = 0$ et $P'_{U_p}(x_{U_p}) \times y_{U_p} = 1$. ∎

PROPOSITION 4.7 : *Si* x_p *est un élément strictement positif de* B_p, *il existe un ouvert* U_p, $p \in U_p$ *et une section* x_{U_p} *avec* $x_{U_p}(p) = x_p$ *et* x_{U_p} *strictement positif.*

Preuve : On a dans B_p un élément y_p et un élément z_p avec $x_p = y_p^2$ et $z_p y_p = 1$, d'où un ouvert U_p, des sections x_{U_p}, y_{U_p} et z_{U_p} avec

$$x_{U_p} = y_{U_p}^2 \ , \quad z_{U_p} y_{U_p} = 1 \ , \quad x_{U_p}(p) = x_p \ ,$$

donc x_{U_p} est strictement positif. ∎

THEOREME 4.8 (Propriété universelle de $\mathcal{N}_{\mathrm{Spec}_R A}$) : *Soient* X *un espace topologique,* B *un espace annelé en anneaux réels clos locaux et* $f : A \longrightarrow \Gamma(B)$ *un morphisme d'anneaux* ($\Gamma(B)$ *est l'anneau des sections globales de* B); *il existe un unique couple* (φ, g) *avec* φ *application continue de* X *dans* $\mathrm{Spec}_R A$ *et* g *morphisme local (i.e. morphisme d'espace annelés avec* g_p *local pour chaque* p *de* X) *de* $\varphi^*(\mathcal{N}_{\mathrm{Spec}_R A})$ *dans* B *tel que* $\Gamma(g) \circ \eta_A = f$, *où* η_A *désigne le morphisme canonique* $A \longrightarrow \Gamma(\mathcal{N}_{\mathrm{Spec}_R A})$.

Preuve : D'après 4.3, $\varphi(p)$ est nécessairement le point de $\mathrm{Spec}_R A$ défini par $\overline{f}_p : A \xrightarrow{f} B \longrightarrow B_p \longrightarrow K_p$ et g_p le morphisme de factorisation de $\overline{f}_p$ par $A_{\varphi(p)}$.

φ est bien une application continue :

$$\varphi^{-1}(D_a) = \{p \mid \varphi(p) \in D_a\} = \{p \mid K_p \models \overline{f_p(a)} > 0\}$$

est un ouvert d'après la proposition 4.7. Les g_p se recollent en un morphisme d'espace annelé : on va définir, par chaque U quasi-compact de $\mathrm{Spec}_R A$, un morphisme $\widehat{g}_U$ de $\mathcal{N}_{\mathrm{Spec}_R A}(U)$ dans $B(\varphi^{-1}(U))$, ce qui, après vérification de la naturalité, définira par adjonction $g : \varphi^* \mathcal{N}_{\mathrm{Spec}_R A} A \longrightarrow B$.

Il suffit de définir $\hat{g}$ pour des ouverts U munis de sections (s,b) élémentaires

(2.2) définies sur U. On a $b = \frac{R[X]}{Q^n[X]} \in A[h^{-1}][X]_{/p}[Q^{-1}]$ et on note a l'image de X dans $\mathcal{N}_{\mathrm{Spec}_R A}(U)$. On note x_α le morphisme canonique de $\mathcal{N}_{\mathrm{Spec}_R A}(U)$ dans A_α.

Soient $\alpha_o \in U$ comme dans 2.9 et $\Theta_{\overline{r_{\alpha_o}(a)}}$ la formule qui décrit la racine simple $\overline{r_{\alpha_o}(a)}$ de $\pi_{\alpha_o}(p)$. Pour chaque $p \in \varphi^{-1}(U)$, $r_{\varphi(p)(a)}$ vérifie $\exists x_1 \ldots \exists x_m\ \Theta_{\overline{r_{\alpha_o}(a)}}$, et on peut trouver des éléments $\overline{a}_o = r_{\varphi(p)}(a)$, $\overline{a}_1,\ldots,\overline{a}_m$ vérifiant $\Theta_{\overline{r_{\alpha_o}(a)}}(\overline{a}_o,\overline{a}_1,\ldots,\overline{a}_m,\overline{r_{\varphi(p)}(a)})$, ces éléments se relèvent en des a_j de $A_{\varphi(p)}$ dont les images c_j dans B_p par le morphisme g_p sont des racines simples de polynômes $f_p(P^{(i_j)})$ et vérifient les mêmes inégalités strictes que les $\overline{a}_j$.

On a d'après 4.6 et 4.7 un ouvert U_p et des éléments c_{jp} de $B(U_p)$, c_{jp} racine simple de $f_{U_p}(P^{(i_j)})$, les c_{jp} vérifiant les mêmes inégalités strictes que les $\overline{a}_j$. Les c_{jp} sont des familles de sections compatibles ; si $p \in U_{p_1} \cap U_{p_2}$, les $c_{jp_1}(p)$ (resp. $c_{jp_2}(p)$) sont des racines simples de $f_p(P^{(i_j)})$, et vérifient les mêmes inégalités strictes que les $\overline{a}_j$. Puisque

$$K_p \models \exists x_o \ldots \exists x_m\ \ \theta_{\overline{r_{\alpha_o}(a)}}(x_o,\ldots,x_m,f_p(p)) ,$$

on montre de proche en proche en commençant par m que $\overline{c_{jp_1}}(p) = \overline{c_{jp_2}}(p)$, et on en déduit $c_{op_1}(p) = c_{op_2}(p)$.

On note c la section de $B(\varphi^{-1}(U))$ obtenue en recollant les c_{op} et on définit $\widehat{g}_U(s,b) = \frac{R(c)}{Q^n(c)}$.

Il ne reste plus qu'à vérifier la naturalité de $\widehat{g}$, qui est immédiate.

Remarque 4.9 :

1. La propriété universelle de $\mathcal{N}_{\mathrm{Spec}_R A}$ est très proche de celle du faisceau structural du spectre de Zariski : on remplace anneau local par anneau réel clos local dans l'énoncé.

2. Le théorème 4.8 est une version continue de 4.3.

Rappel 4.10 : Le topos étale réel.

La topologie étale réelle sur $\mathrm{Spec}_R A$ est la topologie de Grothendieck engendrée par les recouvrements $(B \longrightarrow B_i)_{i \in I}$ où B et B_i sont des A-algèbres étales et $(\mathrm{Spec}_R B_i)_{i \in I}$ une famille surjective sur $\mathrm{Spec}_R B$.

Le topos étale réel de A, formé des faisceaux pour la topologie étale réelle, est un topos cohérent (les familles couvrantes sont engendrées par les familles couvrantes finies) : ceci résulte de la compacité du spectre réel [4]. Ses points sont les localisations strictes réelles de l'anneau A [5]. Par un théorème général de théorie des topos ([5] Annexe 1), "un topos cohérent est un topos de faisceaux sur un espace topologique, si et seulement si la catégorie de ses points est un ensemble ordonné", et la proposition 3.6 on démontre

THEOREME 4.11 : *Le topos étale réel est le topos de faisceaux sur* $\mathrm{Spec}_R A$ [3].

En effet, le topos étale réel de A est muni d'un faisceau d'anneau qui vérifie la même propriété universelle que $\mathcal{N}_{\mathrm{Spec}_R A}$ ([5] et 4.8), ces deux faisceaux d'anneaux sont donc isomorphes. En particulier, si on considère une variété algébrique réelle, la cohomologie semi-algébrique (on s'intéresse à la topologie de Grothendieck dont les recouvrement sont formés de familles couvrantes finies d'ouverts semi-algébriques) qui coïncide avec la cohomologie du spectre réel (2.2 1)) est égale à la cohomologie étale réelle. H. Delfs a montré ce résultat indépendemment dans sa thèse [7].

§ 5. L'idempotence du spectre réel.

Contrairement à la situation classique (spectre de Zariski ou spectre étale), les sections globales du faisceau structural sur le spectre réel ne redonnent pas l'anneau de départ : les fonctions de Nash globales sur $\mathbb{R}^n$ par exemple, sont plus nombreuses que les polynômes. Tout n'est cependant pas perdu, puisque quand on itère la construction, on retouve le même espace topologique muni du même faisceau. Plus précisément.

THEOREME 5.1 : *Soit* U *un ouvert quasi compact de* $\mathrm{Spec}_R A$; U *muni de la restriction du faisceau structural est isomorphe à* $\mathrm{Spec}_R N$ *muni de son faisceau structural* (N *désigne l'anneau* $\mathcal{N}_{\mathrm{Spec}_R A}(U)$).

Notations 5.2 :

1. On note r_α le morphisme de $N = \mathcal{N}_{\mathrm{Spec}_R A}(U)$ dans la fibre en α de $\mathcal{N}_{\mathrm{Spec}_R A}$, A_α.

2. On note $i : U \longrightarrow \mathrm{Spec}_R N$ l'application, qui à α de U, associe le point de $\mathrm{Spec}_R N$ défini par $N \xrightarrow{r_\alpha} A_\alpha \longrightarrow k(\alpha)$.

On note $p = \mathrm{Spec}_R N \longrightarrow \mathrm{Spec}_R A$ l'application continue $\mathrm{Spec}_R \eta$ où η désigne le morphisme canonique de A dans $N = \mathcal{N}_{\mathrm{Spec}_R A}(U)$.

Il est immédiat que $p \circ i = \mathrm{Id}_U$.

LEMME 5.3 : i *est continue.*

Preuve : Soit $f \in N$. On a un recouvrement U_i de U avec f représentée sur U_i par un couple (s_i, b_i) comme dans 2.9 avec $b_i = \dfrac{R_i[X]}{Q_i^n[X]} \in A[h_i^{-1}][X]/_{P_i}\,[Q_i^{-1}]$.

Soient $\alpha \in U_i$ et a_i l'image de X dans $k(\alpha)$.

$$U_i \cap j^{-1}(D_f) = \{\alpha \in U_i \mid \overline{r_\alpha(f)} > 0\} = \{\alpha \in U_i \mid k(\alpha) \models \pi_\alpha(h_i) \text{ inversible et } \exists x_o \ldots \exists x_p\ \theta_{a_i}(x_o, \ldots, x_p, \pi_\alpha(P_i)) \text{ et } Q_i^n(x_o) R_i(x_o) > 0\}$$

C'est un ouvert quasi compact de U d'après 2.5 c : on note $\{f \mid f \leq 0\}$ la partie de $\mathrm{Spec}_R A$ définie par $\{\alpha \mid \overline{r_\alpha(f)} \leq 0\}$; c'est une partie constructible de $\mathrm{Spec}_R A$ d'après 2.2.

Avant de prouver 5.1, nous avons besoin de quelques propriétés de N.

PROPOSITION 5.4 : *Si* f *est un élément totalement positif de* N, *i.e.* $\forall \alpha \in U$ $\overline{r_\alpha(f)} > 0$, f *admet une racine carrée totalement positive dans* N.

<u>Preuve</u> : $\overline{r_\alpha(f)}$ a dans $k(\alpha)$ une racine carrée strictement positive qui se relève en un élément g_α de A_α avec g_α strictement positif et $g_\alpha^2 = r_\alpha(f)$. Les propositions 4.6, 4.7 et l'unicité de g_α permettent de conclure. ∎

PROPOSITION 5.5 : *Si* f *est un élément totalement positif de* N, f *a dans* N *un inverse totalement positif.*

<u>Preuve</u> : L'inverse de $\overline{r_\alpha(f)}$ dans $k(\alpha)$ se relève dans A_α en un élément g_α tel que $r_\alpha(f) \times g_\alpha = 1$, ce qui définit d'après 4.7 et l'unicité de l'inverse un élément g de N inverse de f, et totalement positif. ∎

PROPOSITION 5.6 : *Soit* $\beta \in \mathrm{Spec}_R N$, $\bigcap_{f \in \beta} \{f \leq 0\}$ *est une partie non vide de* $\mathrm{Spec}_R A$.

<u>Preuve</u> : $\{f \leq 0\}$ est une partie constructible, donc par compacité [4], il suffit de montrer $\bigcap_{i=1}^{n} \{f_i \leq 0\} \neq \emptyset$ pour toute liste finie $f_1,\ldots,f_n$ d'éléments de β.

D'après [2] ou [6], III si $D = \{(x_1,\ldots,x_n) \in \mathbb{R}^n \mid x_1 > 0$ ou ou $x_n > 0\}$ il existe une fonction g strictement positive sur D nulle sur $\mathbb{R}^n - D$, construite à partir de polynômes à coefficients entiers par un nombre fini des opérations suivantes : prendre l'inverse d'une somme de carrés partout non nulle sur D, prendre la racine carrée positive d'une fonction strictement positive sur D.

Si $\bigcap_{i=1}^{n} \{f_i \leq 0\} = \emptyset$, on peut définir d'après 5.4 et 5.5, l'élément $g(f_1,\ldots,f_n)$ totalement positif de N.

Puisque $\pi_\beta(f_1) \leq 0,\ldots,\pi_\beta(f_n) \leq 0$ on a $g(\pi_\beta(f_1),\ldots,\pi_\beta(f_n)) = 0$, or $g(\pi_\beta(f_1),\ldots,\pi_\beta(f_n)) = \pi_\beta(g(f_1,\ldots,f_n))$ a une racine carrée inversible dans $k(\beta)$, contradiction. ∎

PROPOSITION 5.7 : p *est une bijection de* $\mathrm{Spec}_R N$ *dans* U, *d'inverse* i.

<u>Preuve</u> :

a) $p\beta \in U$.

On prend $\alpha \in \bigcap_{f \in \beta} \{f \leq 0\}$; on a $f \in \beta \Longrightarrow r_\alpha(f) \leq 0 \Longrightarrow f \in i(\alpha)$ donc $\beta \subset i\alpha$, donc $p\beta \subset \alpha$ et $p\beta \in U$ puisque $\alpha \in U$ et que U est ouvert, donc stable par générisation (2.2 1)).

b) $ip\beta = \beta$.

On a le diagramme commutatif

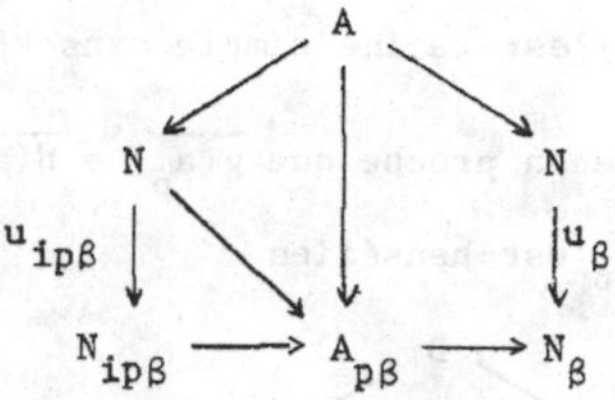

on va montrer

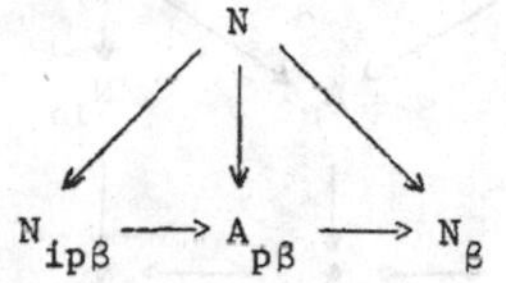

commutatif, ce qui montrera $ip\beta = \beta$, les morphismes $N_{ip\beta} \longrightarrow A_{p\beta}$ et $A_{p\beta} \longrightarrow N_\beta$ étant locaux. On note $v : N_{ip\beta} \longrightarrow A_{p\beta} \longrightarrow N_\beta$.

Soit $\alpha \in \bigcap_{f \in \beta} \{f \leq 0\}$, on a

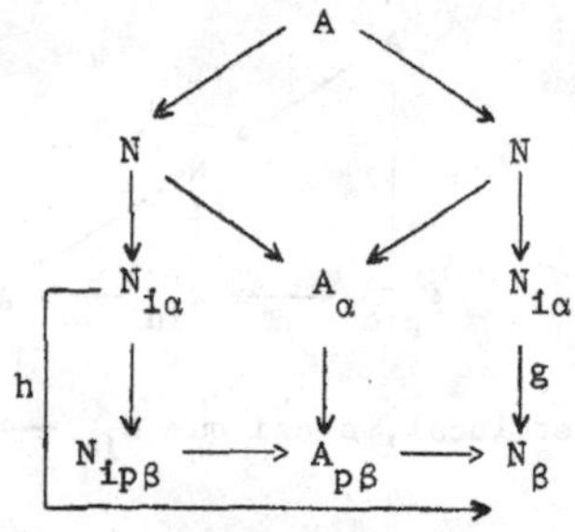

commutatif puisque $\beta \subset i\alpha$ et $ip\beta \subset i\alpha$.

Soit $f \in N$. f est une union finie de sections représentées par des (s_i, b_i) comme dans 2.9. Supposons $\alpha \in U_1$, domaine de définition de s_1 , où $b_1 = \dfrac{R_1[X]}{Q_1^n[X]} \in B_1 = A[h_1^{-1}][X]/_{p_1}[Q_1^{-1}]$, et notons $\bar{a}$ l'image de $X \in B_1$ dans $k(i\alpha)$;

$\bar{a}$ est une racine simple de (l'image de) P_1, et $k(i\alpha) \models \exists x_o \ldots \exists x_p \; \theta_{\bar{a}}(x_o, x_1, \ldots, x_p, P_1)$.

D'après 2.5 c), et puisque $\beta \subset i\alpha$, on a $k(\beta) \models \exists x_o \ldots \exists x_p \; \theta_{\bar{a}}(x_o, x_1, \ldots, x_p, P_1)$.

L'arbre de $\bar{a}$ $(\bar{a}, \bar{a}_1, \ldots, \bar{a}_p)$ de $k(i\alpha)$ se relève en des éléments de $N_{i\alpha}(a, a_1, \ldots, a_p)$; les $\bar{a}_j$ sont des racines simples de polynômes $P_1^{(i_j)}$. Les inégalités strictes vérifiées par $\bar{a}_i$ et $\bar{a}_j$ sont encore vérifiées par $\overline{g(a_i)}$ et $\overline{g(a_j)}$ (resp. $\overline{h(a_i)}$ et $\overline{h(a_j)}$) dans $k(\beta)$ et $g(a_i)$ (resp. $h(a_j)$) est racine simple dans $k(\beta)$ de (l'image de) $P_1^{(i_j)}$. On peut montrer de proche en proche que $\overline{g(a_p)} = \overline{h(a_p)}$, puis $\overline{g(a)} = \overline{h(a)}$. On en déduit $g(a) = h(a)$ puisque N_β est hensélien.

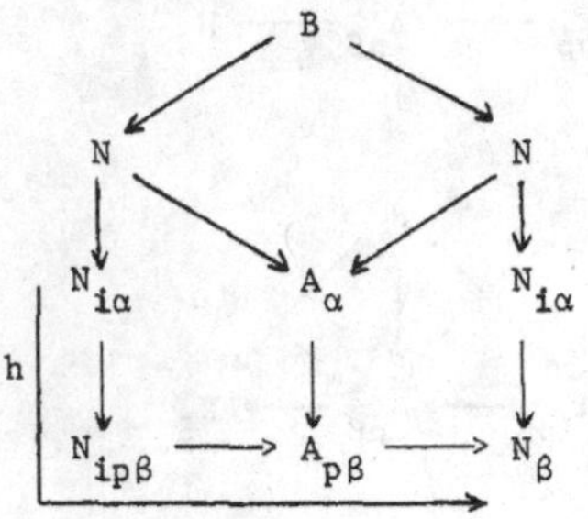

est commutatif et $v(u_{ip\beta}|f)) = u_\beta(f)$.

<u>Preuve du théorème 5.1</u> : Après 5.6 et 5.3, il reste seulement à montrer que A_α et $N_{i\alpha}$ sont isomorphes.

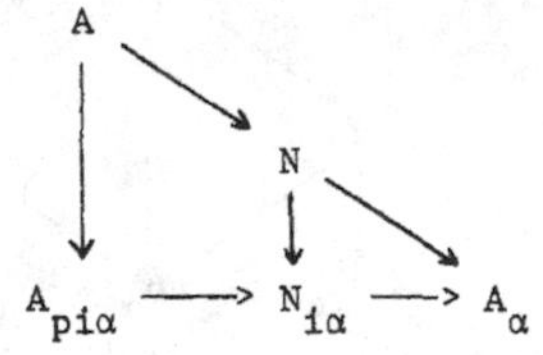

est commutatif $A_{pi\alpha} \longrightarrow N_{i\alpha}$ est local, ainsi que $N_{i\alpha} \longrightarrow A_\alpha$ et $A_{pi\alpha} \simeq A_\alpha$, donc $N_{i\alpha}$ et A_α sont isomorphes.

PROPOSITION 5.8 : *Dans le cas où* $A = \mathbb{R}[X_1, \ldots, X_n]$ *la proposition* 5.7 *est équivalente au lemme de substitution de Bochnak et Efroymson* [2] *ou* [6].

<u>Rappels 5.9</u> : Si K et L sont deux corps réels clos avec $K \subset L$, X un semi algébrique

de K^n (resp. g une fonction semi-algébrique de K^n dans K), on notera X_L (resp. g_L) l'ensemble semi-algébrique (resp. la fonction semi-algébrique) défini(e) dans L par la même formule que dans K. La correction de ces définitions -par exemple le fait que g_L est encore une fonction- sont des conséquences immédiates du principe de Tarski-Seidenberg.

<u>5.10 Le lemme de substitution</u> : *Soient* V *un ouvert semi-algébrique de* $\mathbb{R}^n$, $A = \mathbb{R}[X_1,\ldots,X_n]$, N *l'anneau des fonctions de Nash sur* U *et* φ *un morphisme de* N *dans un corps réel clos* L

1. $(\varphi(X_1),\ldots,\varphi(X_n)) \in U_L$,
2. $\varphi(g) = g_L(\varphi(X_1),\ldots,\varphi(X_n))$.

<u>Preuve de la proposition 5.8</u> : Montrons 5.7 à l'aide du lemme de substitution

a) $p(\mathrm{Spec}_R N) = \tilde{U}$.

Soient $\beta \in \mathrm{Spec}_R N$, $(\pi_\beta(X_1),\ldots,\pi_\beta(X_n)) \in U_{k(p\beta)}$ d'après 5.10 1), donc $p\beta \in \tilde{U} = \{\alpha | (\pi_\alpha(X_1),\ldots,\pi_\alpha(X_n)) \in U_{k(\alpha)}\}$.

b) p est injective.

Soient β et β' avec $p(\beta) = p(\beta') = \alpha$. Puisque N est dans ce cas une extension ind-étale de $\mathbb{R}[X_1,\ldots,X_n]$ on a $k(\beta) \simeq k(\beta') \simeq k(\alpha)$. π_β et $\pi_{\beta'}$ coïncident sur $\mathbb{R}[X_1,\ldots,X_n]$ et sont donc égaux d'après 5.10 2), or $p(ip\beta) = p\beta$, donc $ip\beta = \beta$.

Réciproquement, $\varphi : N \longrightarrow L$ définit un point β de $\mathrm{Spec}_R N$, d'où d'après 5.7 un point $\alpha \in \tilde{U}$ tel que φ factorise par $k(\alpha)$, $(\pi_\alpha(X_1),\ldots,\pi_\alpha(X_n)) \in U_{k(\alpha)}$ et donc $(\varphi(X_1),\ldots,\varphi(X_n)) \in U_L$.

φ et φ' telle que $\varphi'(g) = g_L(\varphi(X_1),\ldots,\varphi(X_n))$ définissent d'après 5.7 a) le même point β de $\mathrm{Spec}_R B$ puisqu'elles coïncident sur $\mathbb{R}[X_1,\ldots,X_n]$ et $\varphi = \varphi' = \ell \circ \pi_\beta$ où ℓ désigne l'inclusion de $k(\beta)$ dans L. ∎

<u>Remarque 5.11</u> : Les ingrédients de la preuve de 5.7, sont sous une forme plus abstraite, essentiellement les mêmes que ceux des preuves directes du lemme de substitution ([2] ou [6]).

REFERENCES

[1] ARTIN et MAZUR : *On periodic points*. Annals of mathematics n° 81, 1965.

[2] BOCHNAK et EFROYMSON : *Real algebraic geometry and the* 17th *Hilbert problem* Math. Annalen n° 251 ou
BOCHNAK et EFROYMSON : *Introduction to Nash Functions*, ce volume.

[3] COSTE M. et COSTE-ROY M.-F. : *Le spectre étale réel d'un anneau est spatial*, Comptes rendus de l'Académie des Sciences, t. 290, série A-91, 1980.

[4] COSTE M. et ROY M.-F. : *La topologie du spectre réel* , Contemporary Mathematics, 1981.

[5] COSTE M. et COSTE-ROY M.-F. : *Topologie for real algebraic geometry*, A. Koch éditeur : *topos theoretic methods in geometry*, Various publications séries n° 30, Aarhus Universitet, 1979.

[6] COSTE M. : *Ensemble semi-algébriques et fonctions de Nash*, Prépublications de l'Université Paris-Nord, 1981.

[7] DELFS H. : *Kohomologie affine semi-algebraisches Raüme*, Thèse, 1980, Université de Regensburg.

[7'] LAFON J.-P. : *Algèbre locale* (à paraître).

[8] NAGATA : *Local rings*, Robert E. Krieyer publishing company, 1975.

[9] RAYNAUD : *Anneaux locaux henséliens*. Lecture notes in mathematics, Springer-Verlag, Vol. 169, 1970.

[10] TOGNOLI A. : *Algebraic geometry and Nash functions*. Istituto nazionale di alta matematica, Institutiones mathematicae, volume III, Academic Press, 1978.

[11] ZARISKI et SAMUEL P. : *Commutative algebra*.

B. P. 11573
Niamey
NIGER

Real holomorphy rings in real algebraic geometry

Heinz-Werner Schülting

Universität, Abt. Mathematik, Postfach 500500, 4600 Dortmund, BRD.

0. Introduction

The aim of this paper is to illustrate the role of the real holomorphy ring in real algebraic geometry from several points of view. In the following K is always the function field of a non-singular variety V over a real closed field R. The holomorphy ring is described as the set of those functions in K which are bounded on the real points of the domain of definition. A birational description of the partition of V_r (i.e. the set of rational points of V) into semialgebraic components is given in terms of the holomorphy ring. Further it is shown that every element of the holomorphy ring can be considered as a regular function on the real points of a suitable complete model of K. Finally the problem is discussed whether disjoint open and closed semialgebraic subsets of V_r can be separated by elements of the holomorphy ring.

1. Preliminaries

The (absolute) real holomorphy ring H_K of a formally real field K is defined as the intersection of all real valuation rings of K. Here a valuation ring is called real if its residue field is real. (For more details see (S)). H_K is a Prüfer domain and this implies that every homomorphism $f: H_K \to \mathbb{R}$ extends to a unique $\mathbb{R}$-valued place $\tilde{f}: K \to \mathbb{R} \cup \{\infty\}$. The elements $a \in H_K$ induce functions $\hat{a}: \text{Hom}(H_K, \mathbb{R}) \to (-n_a, n_a)$, where n_a is a suitable natural number. Thus one obtains an embedding

$$\mathrm{Hom}\,(H_K, \mathbb{R}) \to \bigtimes_{a \in H_K} (-n_a, n_a)\ , \quad f \to (\ldots, f(a), \ldots)\ ,$$

such that Hom $(H_K, \mathbb{R})$ becomes a compact Hausdorff space with respect to the induced topology. Denote by M_K the set of real places of K. The bijection Hom $(H_K, \mathbb{R}) \to M_K$, $f \to \tilde{f}$, carries the above topology over to M_K.

Another way to define this topology on M_K is as follows. Let X_K be the space of orderings of K. Every $P \in X_K$ induces a real place $\psi_P : K \to \mathbb{R} \cup \{\infty\}$, defined as

$$\psi_P(a) = \begin{cases} \infty\ , \text{if no } q \in \mathbb{Q} \text{ exists with } q \pm a \in P \\ \inf\ \{q \in \mathbb{Q} \mid q - a \in P\} \quad \text{otherwise} \end{cases}$$

The map $X_K \to M_K$, $P \to \psi_P$, is surjective and the corresponding quotient topology coincides with the topology defined above.

The ring H_K and the space M_K seem to be appropriate objects in the study of function fields over $\mathbb{R}$ but in the more general situation of a function field K over an arbitrary real closed field R one has to consider the relative holomorphy ring $H(K/R) = H_K \cdot R$, the compositum of H_K and R. At the first glance the right substitute for M_K seems to be the set of homomorphisms $\mathrm{Hom}_R(H(K/R), R)$ but generally this space provided with the canonical topology is totally disconnected. The following third description of M_K leads to a better generalization.

Given a place $\psi \in M_K$, define $\alpha_\psi := \psi^{-1}(\mathbb{R}_-) \cap H_K$ with $\mathbb{R}_- = \{ r \in \mathbb{R} \mid r \leq 0 \}$. Then α_ψ forms a prime ordering of H_K, i.e. an element of the real spectrum of H_K in the sense of Coste and Coste-Roy (CC); and the map $\psi \to \alpha_\psi$ is a homeomorphism from M_K

onto the set of closed points of $spec_r(H_K)$ (= the real spectrum of H_K), denoted by $(spec_r(H_K))_d$. Thus in the case of an arbitrary real closed constant field we replace M_K by $spec_r(H(K/R))$.

2. Let V be a non-singular projective variety over a real closed field R with formally real function field K and V_r be the set of rational points of V.
Since H(K/R) is integrally closed and since every valuation ring containing H_K is real (S), H(K/R) is the intersection of the real valuation rings of K containing R. Let Q_K be the sums of squares in K.

Lemma: $H(K/R) = \{f \in K \mid \text{ex. } r \in R, r \pm f \in Q_K\}$

Proof: For every ordering $P \in X_K$ the ring $B_P := \{f \in K \mid \text{ex. } r \in R, r \pm f \in P\}$ is a real valuation ring. Therefore a function $f \in H(K/R)$ is contained in the intersection of the rings B_P, $P \in X_K$. In other words for every $P \in X_K$ a real number r_P (necessarily positive) exists such that $r_P \pm f \in P$. The compactness of X_K then makes it possible to find a simultaneous bound r of f. On the other hand, for $r \in R$ and $f \in K$ with $r \pm f \in Q_K$ one obtains $1 \pm \frac{f}{r} \in Q_K$. This implies $\frac{f}{r} \in H_K$, hence $f \in H_K \cdot R$.
Applying the theorem of Artin ((A), Satz 11) we obtain

Theorem: Given $f \in K$ let U be a non-empty open subvariety of V such that f is defined on U. Then f is bounded on U_r by a number $s \in R$ iff $f \in H(K/R)$.

Proof: If f is bounded on U_r by $s \in R$ then s+f and s-f are positive definite on U_r and hence they are sums of squares. Conversely if f is not bounded, then for every $s \in R$ a point $P \in V_r$ exists with $|f(P)| > s$. This point P is the center of an ordering S, for which $s - f \notin S$ or $s + f \notin S$ (see for example (CC), Ch.8). Consequently $f \notin H(K/R)$.

3. Now let W be an affine subvariety of V which contains all the real (not necessarily closed) points of V (see (CT), p. 118) and let R[W] be the coordinate ring of W. Denote by O_V the ring of those functions which are regular in every real point of V. We shal give a birational description of the partition of V_r into semialgebraic components in terms of the holomorphy ring, which can be easily derived from a theorem of Coste and Coste-Roy.

Every point $P \in V_r$ induces a prime ordering $\alpha_P := \{ f \in R[W] \mid f(P) \leq 0 \} \in \mathrm{spec}_r(R[W])$. Coste and Coste-Roy proved that the map $P \to \alpha_P$ induces a bijection between the semialgebraic components of V_r and the connected components of $\mathrm{spec}_r(R[W])$ ((CC), Ch. 5). It remains to be shown that the components of $\mathrm{spec}_r(R[W])$ and those of $\mathrm{spec}_r(H(K/R))$ coincide in a canonical way. The holomorphy ring H(K/R) contains O_V. In fact, if B is a real valuation ring over R, the center P of B in V (i.e. the unique point $P \in V$ such that B dominates the local ring o_P) is real. This implies that B contains every function regular in P. One obtains the continuous restriction maps

$$X_K \simeq \mathrm{spec}_r(K) \xrightarrow{h_1} \mathrm{spec}_r(H(K/R)) \xrightarrow{h_2} \mathrm{spec}_r(R[W]) .$$

Let C be a connected component of $\mathrm{spec}_r(R[W])$ and $\tilde{C} = h_2^{-1}(C)$. The image of $h_2 \circ h_1$ is dense in $\mathrm{spec}_r(R[W])$ ((CC), 9.1). Hence $h_2^{-1}(C)$ is not empty.

Assume $\tilde{C}$ is not connected. There are closed non-empty subsets C_1 and C_2 of $\operatorname{spec}_r(H(K/R))$ with $\tilde{C} = C_1 \cup C_2$ and $C_1 \cap C_2 = \emptyset$. For $i \in \{1,2\}$ choose an element α of C_i. Let $k(\alpha)$ be the quotient field of $H(K/R)/_{\alpha \cap -\alpha}$ and $\pi : H(K/R) \to k(\alpha)$ the canonical homomorphism. Then π extends to a place $\mu : K \to k(\alpha) \cup \{\infty\}$. Let P be an ordering of K compatible with μ and $\bar{\alpha}$, where $\bar{\alpha}$ denotes the ordering of $k(\alpha)$ induced by $\pi(-\alpha)$. Since α contains $-P \cap H(K/R)$ it is adjacent to $-P \cap H(K/R)$, hence $-P \in h_1^{-1}(C_i)$. Thus the sets $h_1^{-1}(C_i)$ are not empty and $(h_2 \circ h_1)^{-1}(C)$ is the disjoint union of these sets. But Coste and Coste-Roy proved that this is impossible ((CC), 9.2).

Theorem: There is a natural bijection between the semialgebraic components of V_r and the connected components of $\operatorname{spec}_r(H(K/R))$.

4. The following theorem leads to a description of the holomorphy ring of a real function field K/R as the inductive limit of rings O_W, W running through all complete non-singular models of K.

Theorem: For every $f \in H(K/R)$ a non-singular complete R-variety W with function field K exists such that f is regular on the set of real points of W.

Proof: Let V be any non-singular complete R-variety which is a model of K. The function f can be considered as a rational function $\hat{f} : V \to \mathbb{A}_R^1 \to \mathbb{P}_R^1$. Applying Hironakas theorems on the resolution of singularities and of points of indeterminacy ((H), main theorems I, II) one obtains a complete non-singular R-variety W a birational R-morphism $\psi : W \to V$ and an R-morphism g represented by $\hat{f} \circ \psi$. Let $P \in W$ be a real point and let B be a real valuation ring of K with center P (see (<u>Ab</u>)). Assume $f \notin o_P$. This means $g(P) \notin \mathbb{A}^1_R$ and thus $\frac{1}{f}$ is contained in the maximal ideal of o_P. Hence $f \notin B$, a contradiction to $f \in H(K/R)$.

<u>Corollary</u>: H(K/R) is the intersection of the real discrete valuation rings of K over R.

<u>Proof</u>: Assume $f \in K$ is not contained in H(K/R). Let B be a real valuation ring of K with $f \notin B$ and consider the function $g = \frac{1}{1+f^2}$. Then g is contained in $H(K/R) \cap m$, where m denotes the maximal ideal of B. According to the theorem a complete non-singular model W of K exists, such that g is regular in the center P of B and consequently $g \in m_P$ (= maximal ideal of o_P). Since P is regular, a real discrete valuation ring D can be constructed ((<u>Ab</u>), Lemma 15) which dominates o_P. The maximal ideal of D contains g; hence $f \notin D$.

This corollary was also proved by E. Becker avoiding theorems about the existence of certain regular points.

The following corollary was suggested by L. Bröcker. Again let V be a complete non-singular variety with formally real function field K. Let B(V) be the set of pairs (W,ψ) with W being a complete non-singular R-variety and $\psi : W \to V$ a birational R-morphism. This set can be ordered in the following way: $(W,\psi) \geq (\tilde{W},\tilde{\psi})$ iff an R-morphism $\phi : W \to \tilde{W}$ exists represented by $\tilde{\psi}^{-1} \circ \psi$. In this way B(V) becomes a directed set (see (<u>Sh</u>), p. 45) and the rings O_W, $(W,\psi) \in B(V)$, form a direct system.

Corollary: i) $H(K/R) = \varinjlim O_W$

ii) $\mathrm{spec}_r(H(K/R)) = \varprojlim \mathrm{spec}_r(O_W)$

Proof: The first part follows directly from the Theorem and the second part is a conclusion from i) and (CC), prop. 2.4.

5. In a forthcoming paper in Math. Annalen, Mahe shows that disjoir open and closed subsets of the real spectrum of a ring A can be separated by a form $z \in W(A)$. On the other hand, if V is a complete $\mathbb{R}$-variety and $A = H(\mathbb{R}(V), \mathbb{R})$ or $A = O_r(V)$ it follows from the Stone-Weierstrass Approximation Theorem, that disjoint open and closed subsets of $\mathrm{spec}_r A$ can be separated by an element $f \in A$. We shall now discuss, whether this statement remains true when $\mathbb{R}$ is replaced by a real closed field R. More precisely we shall study the weaker form of the statement:

(*) Let R be a real closed field, K a real function field over R and $A, B \subset \mathrm{spec}_r(H(K/R))$ disjoint open and closed sets. Then one can find a function $f \in H(K/R)$ such that $-f \notin \alpha \; ; \; \alpha \in A$

$$f \notin \alpha \; ; \; \alpha \in B$$

Theorem: The statement (*) is true if and only if R is archimedean ordered.

Proof: The if-part can be derived from the Approximation Theorem, but the following proof gives a better insight into the role of the archimedean order (i.e. the role of $\mathbb{R}$).
Since every real valuation ring contains R the absolute holomorphy ring H_K and the relative holomorphy ring $H(K/R)$ coincide. Let $\tilde{A}$ be the pré-image of A under the continuous restriction map

$r: X_K \to \mathrm{spec}_r(H(K/R))$. The characteristic function $\chi_{\tilde{A}}: X_K \to \{\pm 1\}$ is continuous and the representation theorem of Becker and Bröcker ((BB), Th. 5.3) shows that $\chi_{\tilde{A}}$ can be represented by a quadratic form (i.e. $\chi_{\tilde{A}}$ is the total signature of a quadratic form). In fact, let $T \in K$ be a fan with finite Index $[K^{\cdot}:T^{\cdot}]$ and $X_T = \{P \in X_K \mid T \subset P\}$. Remember that a fan is a preordering T which satisfies the following condition: If P_1, P_2, P_3 are different orderings containing T then an ordering $P_4 \supset T$ different from P_1, P_2, P_3 exists with $P_1 \cap P_2 \cap P_3 \subset P_4$. According to ((Be), p. 37) at most two $\mathbb{R}$-valued places ψ with $\psi(T) \subset \mathbb{R}_+ \cup \{\infty\}$ exist, and if there are exactly two of such places ϕ, ψ, then the sets $\{P \in X_T \mid \psi(P) \subset \mathbb{R}_+ \cup \{\infty\}\}$ and $\{P \in X_T \mid \phi(P) \subset \mathbb{R}_+ \cup \{\infty\}\}$ have the same cardinality. Since the closure of $r(P)$, P an ordering, contains the place ψ_P, the function $\chi_{\tilde{A}}$ is constant or $\sum_{P \in X_T} \chi_{\tilde{A}}(P) = 0$.

In any case, $\chi_{\tilde{A}}$ satisfies the criterion of Becker and Bröcker; hence this map is represented by a form $z \in W(K)$. But even $g := \det(z)$ separates $\tilde{A}$ and $X_K \setminus \tilde{A}$, and $\mathrm{sign}_S(g) = \mathrm{sign}_P(g)$ whenever $\psi_P = \psi_S$. A theorem of Krull and Baer (see (K), Th. 2.5) now implies that g has an even value in the value group of every place $\phi \in M_K$. It follows $g \in H_K^* \cdot Q_K$ ((S), Cor. 3.3), which proves the first part.

We now assume that R is non-archimedean. We shall see that in this case the image of a fan T under the restriction map r may intersect more than two connected components of $\mathrm{spec}_r(H(K/R))$. Choose an infinitely small element $a \in R_+^{\cdot}$. Let $\lambda: R \to \mathbb{R} \cup \{\infty\}$ be the unique $\mathbb{R}$-valued place of R and $\tilde{\lambda}: R(X,Y) \to \mathbb{R}(X,Y) \cup \{\infty\}$ the canonical extension. Let Φ_Y be the unique place of $\mathbb{R}(X,Y)$ trivial on $\mathbb{R}(X)$ with $\Phi_Y(Y) = 0$ and let Φ_X be the unique real place of $\mathbb{R}(X)$ with $\Phi_X(X) = 0$. Consider the place $\phi := \Phi_X \circ \Phi_Y \circ \tilde{\lambda}$. The value group H of ϕ is $H = G \times \mathbb{Z} \times \mathbb{Z}$, where G is divisible, hence $H/_{2H} \simeq \mathbb{Z}/_{2\mathbb{Z}} \times \mathbb{Z}/_{2\mathbb{Z}}$.
It follows from the above-mentioned theorem of Krull and Baer

that exactly four orderings $P_1,\dots,P_4$ exist which induce the place ϕ, i.e. $\psi_{P_i} = \phi$ and it is well known that the intersection of these orderings is a fan T ((B), ex. 2.6).

Define $K := R(X,Y)(\sqrt{X^2-a}, \sqrt{Y^2-a}, \sqrt{1-X^2}, \sqrt{1-Y^2})$ and let T' be the preordering of K generated by T and the set

$$\{ \sqrt{\frac{X^2-a}{X^2}}, \sqrt{\frac{Y^2-a}{Y^2}}, \sqrt{1-X^2}, \sqrt{1-Y^2} \} .$$

Since $\phi(1-X^2) = \phi(1-Y^2) = \phi(\frac{X^2-a}{X^2}) = \phi(\frac{Y^2-a}{Y^2}) = 1$ each ordering P_i extends to 16 orderings of K and for every $i \in \{1,2,3,4\}$ exactly one extension $\tilde{P}_i$ of P_i contains T'. Further, there are exactly 16 extensions of ϕ, which correspond to the signs of $\sqrt{\frac{X^2-a}{X^2}}, \sqrt{\frac{Y^2-a}{Y^2}}, \sqrt{1-X^2}, \sqrt{1-Y^2}$ and that extension which maps all these square roots to 1 is induced by each of the orderings $\tilde{P}_i$. Thus, $T' = \tilde{P}_1 \cap \dots \cap \tilde{P}_4$ is a fan.

One easily proves that X and Y are units in H(K/R). Therefore the four sets $D(\pm X, \pm Y)$ don't intersect and hence they are open and closed. (For $a_1,\dots,a_n \in H(K/R)$ $D(a_1,\dots,a_n)$ denotes the open set $\{ \alpha \in \mathrm{spec}_r H(K/R) \mid a_i \notin \alpha \text{ for } 1 \leq i \leq n \}$). Each of these sets contains one of the prime orderings $\tilde{P}_i \cap H(K/R)$. Assume now that $\tilde{P}_1 \in D_1 = D(X,Y)$ and $\tilde{P}_2, \tilde{P}_3, \tilde{P}_4 \in D_2 = D(X,-Y) \cup D(-X,Y) \cup D(-X,-Y)$ and that D_1 and D_2 can be separated by an element $f \in H(K/R)$. This would imply f or -f is contained in $\tilde{P}_2 \cap \tilde{P}_3 \cap \tilde{P}_4 \setminus \tilde{P}_1$, a contradiction.

References

(A) E. Artin, Über die Zerlegung definiter Funktionen in Quadrate, Abh. Math. Sem. Univ. Hamburg 5 (1927), 100-115.

(Ab) S. Abyankar, On the valuations centered in a local domain, Amer. J. Math. 78 (1956), 321-348.

(Be) E. Becker, Hereditarily-pythagorean fields and orderings of higher level, IMPA Lecture Notes, Rio de Janeiro, 1978

(B) L. Bröcker, Characterization of fans and hereditarily pythagorean fields, Math. Z. 151 (1976), 149-163

(BB) E. Becker and L. Bröcker, On the description of the reduced Wittring, J. of Algebra 52 (1978), 328-346.

(CT) J.L. Colliot-Thélène, Formes quadratiques multiplicatives et variétés algébriques, Bull. Soc. Math. France, 106 (1978), 113-151

(CC) M. Coste and M.-F. Coste-Roy, La topologie du spectre reel, manuscript.

(H) H. Hironaka, Resolution of singularities of an algebraic variety over a field of characteristic zero, Ann. of Math., 79 (1964), 109-326.

(K) M. Knebusch, On the extension of real places, Comment. Math. Helv. 48 (1973), 354-369.

(S) H.W. Schülting, On real places of a field and their holomorphy ring, to appear in Comm. Algebra.

(Sh) I.R. Shafarevich, Lectures on minimal models and birational transformations of two dimensional schemes, Tata Institute of Fundamental Research, Bombay, 1966.

A Bound on the Order of $H_{n-1}^{(a)}(X,\mathbb{Z}/2)$ On a Real Algebraic Variety

R. SILHOL (Regensburg)

Introduction:

We propose here to give a bound on the order of the subgroup $H_{n-1}^{(a)}(X,\mathbb{Z}/2)$ of $H_{n-1}(X,\mathbb{Z}/2)$ generated by algebraic (n-1)-cycles, when X is a real algebraic variety. This bound will enable us to give a partial answer to questions raised in [1],[2],[11], in particular give explicit examples of n-dimensional tori ($n \geq 2$) with order of $H_{n-1}^{(a)}(X,\mathbb{Z}/2) \leq 2$ (improving in this way the result of [1]).

Notations:

Let X be a scheme over $\mathbb{R}$. We will assume throughout that X is geometrically integral, projective and smooth of dimension n.

We write $X = X(\mathbb{R})$, $\bar{X} = X\times_{\mathbb{R}}\mathbb{C}$ and $\bar{X} = \bar{X}(\mathbb{C})$. We will always assume $X \neq \emptyset$ and $\bar{X}$ endowed with its natural structure of analytic variety.

Finally we will note $G = \{1,\sigma\}$ the Galois group $G(\mathbb{C}|\mathbb{R})$ and consider its natural action on both $\bar{X}$ and $\bar{X}$ (we will call this Galois action).

We start with the following exact sequence of sheaves on $\bar{X}$:

$$(1) \qquad 0 \longrightarrow \mathbb{Z} \xrightarrow{2i} \mathcal{O} \xrightarrow{\exp\pi} \mathcal{O}^* \longrightarrow 0$$

where the first map is multiplication by 2i and $\mathcal{O}$ and $\mathcal{O}^*$ are the sheaves of holomorphic functions and invertibles holomorphic functions on $\bar{X}$.

Again, as in [10], this is not an exact sequence of sheaves of G-modules. To obtain this we must twist the action of G on $\mathbb{Z}$, that is apply a 1-Tate twist to the sheaf $\mathbb{Z}$.

In our case this can be simply obtained by composing the action of σ with the automorphism $x \longmapsto -x$ in the stalks.

We will denote $\mathbb{Z}(1)$ this twisted G-structure. Similarly we will denote $H^i(\bar{X},\mathbb{Z})(1)$ the G-module obtained from $H^i(\bar{X},\mathbb{Z})$ by changing the G-action to: $a \mapsto -\sigma(a)$.

On the Čech cocycle level one can explicitly describe the Galois action on $H^i(\bar{X},F)$ (where F is any sheaf of G-modules) by:

$$(U_{p_0 \ldots p_i}, f_{p_0 \ldots p_i}) \longmapsto (\sigma U_{p_0 \ldots p_i}, f^{\sigma}_{p_0 \ldots p_i})$$

From this we deduce in particular that:

$$H^i(\bar{X},\mathbb{Z}(1)) \cong H^i(\bar{X},\mathbb{Z})(1) .$$

Using this isomorphism we get the exact sequence of G-modules:

$$(2) \qquad 0 \to H^1(\bar{X},\mathbb{Z})(1) \to H^1(\bar{X},\mathcal{O}) \to \mathrm{Pic}_0(\bar{X}) \to 0$$

Note: This does not give $\mathrm{Pic}_0(\bar{X})$ the same G-module structure (or real structure) as in [10]. The real structure considered in [10] is, in the notations given here, in fact $\mathrm{Pic}_0(\bar{X})(1)$.

Applying Galois cohomology to (2) we get as in [10], an exact sequence:

$$(3) \qquad 0 \to H^1(\bar{X},\mathbb{Z})(1)^G \to H^1(\bar{X},\mathcal{O})^G \to \mathrm{Pic}_0(\bar{X})^G \to H^1(G,H^1(\bar{X},\mathbb{Z})(1)) \to 0 \quad .$$

The surjectivity of the last map coming from the fact that $H^1(\bar{X},\mathcal{O}) \cong H^1(X,\mathcal{O}_X) \otimes_{\mathbb{R}} \mathbb{C}$ i.e. is an induced G-module.

By considering Galois action on the sheaf of invertible holomorphic functions (resp. invertible regular functions) on $\bar{X}$ (resp. on $\bar{X}$) we can, in a natural way, give $\mathrm{Pic}(\bar{X})$ and $\mathrm{Pic}(\bar{X})$ G-module structures. We will for the moment denote these structures $\mathrm{Pic}(\bar{X})'$ and $\mathrm{Pic}(\bar{X})'$.

The classical GAGA isomorphism:

$$\mathrm{Pic}(\bar{X}) \cong \mathrm{Pic}(\bar{X})$$

that sends $\mathrm{Pic}_0(\bar{X})$ onto $\mathrm{Pic}_0(\bar{X})$, is obviously compatible with these structures and hence induces an isomorphism

$$\mathrm{Pic}(\bar{X})' \cong \mathrm{Pic}(\bar{X})' \quad .$$

On the other hand the map $\exp\pi : H^1(\bar{X},\mathcal{O}) \to \mathrm{Pic}(\bar{X})$ is also compatible with the natural G-structure on $H^1(\bar{X},\mathcal{O})$ and the G-structure $\mathrm{Pic}(\bar{X})'$.

Hence it makes sense to speak of the G-structures $\mathrm{Pic}_0(\bar X)'$ and $\mathrm{Pic}_0(\bar X)'$, which are isomorphic by GAGA.

The above argument also shows that $\mathrm{Pic}_0(\bar X)'$ has the same G-module structure as the one induced by the sequence (2).

Lemma 1: With X projective, smooth, irreducible and $X(\mathbb{R}) \neq \emptyset$ we have, if the G-structure is the one induced by (2), a natural injection:

$$\mathrm{Pic}_0(\bar X)^G \longrightarrow \mathrm{Pic}(X) .$$

From the above it is sufficient to prove that $\mathrm{Pic}(\bar X)^G \cong \mathrm{Pic}(X)$. This isomorphism is well known (and valid under far more general hypotheses) and we include a sketch of proof only because of the difficulty to give an explicit reference.

For this we can use the proof given by Knebusch ([6] lemma 2.7) in the case where X is a curve and note that this proof generalizes to our situation because (under the hypothesis $X \neq \emptyset$ and X smooth and irreducible) -1 is not a norm in $\mathbb{C}(\bar X)\,|\,\mathbb{R}(X)$ (this last statement is a direct consequence of the real Nullstellensatz).

Let Z be the free abelian group of algebraic (n-1)-cycles on X (as defined for example in [3] p.493 - see also [2] or [9]), and let $\bar Z$ be the group of algebraic (2n-2)-cycles on $\bar X$, where $n = \dim_{\mathbb{R}} X = \dim_{\mathbb{C}} \bar X$.

We call $H^{(a)}_{n-1}(X,\mathbb{Z}/2)$ the image of Z in $H_{n-1}(X,\mathbb{Z}/2)$, and $H^1_{(a)}(X,\mathbb{Z}/2)$ its Poincaré dual.

To every $c \in Z$ we associate its complexification in $\bar X$, $\gamma(c) = \bar c \in \bar Z$ and consider the following diagram:

$$(4) \qquad \begin{array}{ccc} \bar Z & \xrightarrow{\ h\ } & H^2(\bar X,\mathbb{Z}) \\ \uparrow \gamma & \nearrow \rho' & \\ Z & \xrightarrow[\ r\ ]{} & H^1_{(a)}(X,\mathbb{Z}/2) \end{array}$$

where h and r associate to a cycle its fundamental class (in the sense of Griffiths and Harris [4] p.61) and ρ' is the composed map $h\circ\gamma$ (this construction is similar, although not exactly identical, to the one in [3] p.494).

We define:

$$(5) \qquad A_0 = r(\ker \rho')$$

A_o factors, of course, in $H^1_{(a)}(X,\mathbb{Z}/2)$ (this is a $\mathbb{Z}/2$-vector space) so we get a decomposition:

(6) $\quad H^1_{(a)}(X,\mathbb{Z}/2) \cong A_o \times A_n$.

We have a canonical injection $Z \to \mathrm{Div}(X) \cong \mathrm{Div}(\bar{X})^G$. Combining the map defined in [3] §5.2 and Poincaré duality we get a map

$$\rho : \mathrm{Div}(X) \to H^1(X,\mathbb{Z}/2) .$$

By definition of this map we get a commutative diagram (see [3] p.494):

$$\begin{array}{ccc} Z & \xrightarrow{\quad r \quad} & \\ \downarrow & & \searrow \\ \mathrm{Div}(X) & \xrightarrow[\rho]{} & H^1(X,\mathbb{Z}/2) \end{array}$$

Recalling the fact that if $c \in Z$ has for image in $\mathrm{Div}(X)$ a principal divisor then $r(c)$ is zero (see for example [3]§5.13, [2] or [9]) we see that the preceding diagram induces:

$$\begin{array}{ccc} Z & \xrightarrow{\quad r \quad} & \\ \downarrow & & \searrow \\ \mathrm{Pic}(X) & \longrightarrow & H^1(X,\mathbb{Z}/2) \end{array}$$

or by definition of $H^1_{(a)}(X,\mathbb{Z}/2)$:

(7)
$$\begin{array}{ccc} Z & \xrightarrow{\quad r \quad} & \\ \downarrow & & \searrow \\ \mathrm{Pic}(X) & \longrightarrow & H^1_{(a)}(X,\mathbb{Z}/2) \end{array}$$

Combining (3), (7) and the injection of lemma 1 we get the following diagram:

$$\begin{array}{ccccccc} H^1(\bar{X},\mathcal{O})^G & \to & \mathrm{Pic}_o(\bar{X})^G & \to & H^1(G,H^1(\bar{X},\mathbb{Z})(1)) & \to & 0 \\ & & \downarrow & & & & \\ & & \mathrm{Pic}(X) & \longrightarrow & H^1_{(a)}(X,\mathbb{Z}/2) & \longrightarrow & 0 \end{array}$$

Noting that $H^1(\bar{X},\mathcal{O})^G$ is a $\mathbb{R}$- vector space we get that the composed map:

$$H^1(\bar{X},\mathcal{O})^G \to \mathrm{Pic}_o(\bar{X})^G \to \mathrm{Pic}(X) \longrightarrow H^1_{(a)}(X,\mathbb{Z}/2)$$

is the zero map.

Hence the map $\mathrm{Pic}_o(\bar{X})^G \longrightarrow H^1_{(a)}(X,\mathbb{Z}/2)$ induces a map:

(8) $\quad \varphi : H^1(G,H^1(\bar{X},\mathbb{Z})(1)) \longrightarrow H^1_{(a)}(X,\mathbb{Z}/2)$.

<u>Lemma 2</u> : With A_o as in (5) and φ as in (8) we have: $\mathrm{Im}\,\varphi = A_o$.

Recalling (7), the fact that $\operatorname{Im}\varphi \subset A_o$ follows from the definitions of φ and A_o and the fact that the fundamental class of a divisor is equal to its Chern class (see Griffiths and Harris [4] p.141).

To show that $A_o \subset \operatorname{Im}\varphi$, take a class in $A_o \subset H^1_{(a)}(X,\mathbb{Z}/2)$. By definition of A_o, there exists a representative $c \in Z$ of this class such that the image of $\gamma(c)$ in $\operatorname{Pic}(\bar{X})$ lies in $\operatorname{Pic}_o(\bar{X})$. The image is also by construction in $\operatorname{Pic}(\bar{X})^G$ and hence in

$$\operatorname{Pic}_o(\bar{X}) \cap \operatorname{Pic}(\bar{X})^G = \operatorname{Pic}_o(\bar{X})^G .$$

This proves the lemma.

<u>Proposition:</u> Let X be a geometrically integral, projective and smooth scheme over $\mathbb{R}$, let $X = X(\mathbb{R})$ be non empty, then:

$$\operatorname{Order} H^1_{(a)}(X,\mathbb{Z}/2) \le 2^{q-\lambda} + r + r_2$$

where q is the irregularity of X (i.e. $q = \dim_{\mathbb{R}} H^1(X,\mathcal{O}_X)$), λ is the Comessatti characteristic of $\bar{X}$ (i.e. such that $H^1(G,H^1(\bar{X},\mathbb{Z})) \cong (\mathbb{Z}/2)^{q-\lambda}$ -see [10] §4), $r = \operatorname{rank}(NS(\bar{X})^G)$ and r_2 is the $\mathbb{Z}/2$ dimension of the 2-torsion subgroup $[NS(\bar{X})^G]_2$ of $NS(\bar{X})^G$ (i.e. the kernel of multiplication by 2)

We consider again the diagram:

$$\begin{array}{ccc}
\bar{Z} & \xrightarrow{\;h\;} & H^2(\bar{X},\mathbb{Z}) \\
\uparrow{\scriptstyle\gamma} & \nearrow{\scriptstyle\rho'} & \\
Z & \xrightarrow{\;r\;} & H^1_{(a)}(X,\mathbb{Z}/2)
\end{array}$$

From the fact that $h(\bar{Z}) \cong NS(\bar{X})$ we get that :

(9) $\quad \rho'(Z) \subset NS(\bar{X})^G$.

Because $NS(\bar{X})$ is finitely generated we can write $\rho'(Z) \cong \mathbb{Z}^{r'} \times T$ where $r' \le r$ and T is a finite group. By the definition of A_o the preceding diagram induces a surjective map $f\colon \rho'(Z) \to H^1_{(a)}(X,\mathbb{Z}/2)/A_o \cong A_n$.

$H^1(X,\mathbb{Z}/2)$ being a $\mathbb{Z}/2$ vector space f factors to a map:

$$\mathbb{Z}^{r'} \times T/2T \to A_n \quad .$$

Now we have by (9) $T_2 \subset [NS(\bar{X})^G]_2$ and, because T is finite,

$$\operatorname{Order}(T_2) = \operatorname{Order}(T/2T) \quad .$$

Hence

$$\text{Order } (A_n) \le 2^{r + r_2} .$$

On the other hand from lemma 2 we have:

$$\text{Order } (A_o) \le \text{Order } H^1(G, H^1(\bar{X}, \mathbb{Z})(1)) .$$

By prop.2 of [10] and the fact that :

$$\text{rank}(H^1(\bar{X}, \mathbb{Z})^G = \text{rank}(H^1(\bar{X}, \mathbb{Z})(1)^G)$$

(see proof of prop.3 of [10]) we have an isomorphism:

$$H^1(G, H^1(\bar{X}, \mathbb{Z})(1)) \cong H^1(G, H^1(\bar{X}, \mathbb{Z})) \cong (\mathbb{Z}/2)^{q-\lambda} .$$

The proposition then follows from (6) .

Examples:

We are now ready to give examples of smooth connected real algebraic varieties with non-totally-algebraic (n-1)-homology.

We restrict to abelian varieties where we have $r_2 = 0$ and can explicitly compute the rank of $NS(\bar{X})$ in terms of $\text{End}_o(\bar{X}) = \text{End}(\bar{X}) \otimes \mathbb{Q}$.

Let ρ be the Rosati involution on $\text{End}_o(\bar{X})$. Then $NS_o(\bar{X}) = NS(\bar{X}) \otimes \mathbb{Q}$ can be identified with:

$\{a \in \text{End}_o(\bar{X}) \ / \ \rho(a) = a\}$ (see Mumford [8] p.190 and 208).

Let $\bar{X}$ be an abelian variety such that X is connected. This is equivalent to $q = \lambda$ ([10] prop.3).

1. If $\text{End}_o(\bar{X})$ is trivial, that is $\text{End}_o(\bar{X}) \cong \mathbb{Q}$ we get $r = 1$ and hence:

$$\text{Order}(H^1_{(a)}(X, \mathbb{Z}/2)) \le 2 .$$

For $\dim \bar{X} = q \ge 2$ this implies $H^1_{(a)}(X, \mathbb{Z}/2) \neq H^1(X, \mathbb{Z}/2)$ (this will be more generally the case every time $NS(\bar{X})$ is too small).

To construct explicit examples of this type, let Ω be a real symmetric positive definite matrix such that the coefficients (of the upper triangular half) are transcendental numbers algebraically independent over $\mathbb{Q}$, and take the abelian variety defined by the Riemann matrix:

$$[I_n \ , \ 1/2 \ I_n + i\Omega]$$

where I_n is the identity matrix. This clearly defines a real abelian variety; that the real part is connected follows from [10] prop.3 and 4; that $End_o(\bar{X}) \cong \mathbb{Q}$ follows from the construction and can be easily verified by direct computation.

2. In a different class of ideas we can consider the case when the multiplication algebra of $\bar{X}$ is of type I (that is with $End_o(\bar{X})$ isomorphic to a totally real number field) and when the action induced by σ on $NS(\bar{X})$ is not trivial.

An example of this type is the abelian variety defined by the Riemann matrix:

$$\begin{bmatrix} 1 & 0 & 1/2 + (1+\pi^2)\sqrt{d}\, i & -\pi\sqrt{d}\, i \\ 0 & 1 & -\pi\sqrt{d}\, i & 1/2 + \sqrt{d}\, i \end{bmatrix}$$

where d is a positive square free integer, $\pi = 3.14...$ and $i^2 = -1$ (see [10] §11). In this case we always have

$$\text{Order}(H^1_{(a)}(X,\mathbb{Z}/2)) \leq 2^{q/2}$$

(which gives for the explicit example above Order $H^1_{(a)}(X,\mathbb{Z}/2) \leq 2$).

In conclusion, I should like to express my thanks to the referees for their many helpful comments, and point out that Risler has written a paper, in which he also gives examples of surfaces where the homology is not all algebraic (these Proceedings).

Bibliography:

[1] R. Benedetti & A. Tognoli: Remarks and counterexamples in the theory of real algebraic vector bundles; These Proceedings.

[2] J. Bochnak, W. Kucharz, M. Shiota : The Divisor class group of some rings of global real analytic, Nash or Rational regular functions; These Proceedings.

[3] A. Borel & A. Haefliger: La classe d'homologie fondamentale d'un espace analytique; Bull. Soc. Math. France 89 (1961), p.461-513.

[4] P. Griffiths & J. Harris: Principles of Algebraic Geometry; A. Wiley Interscience Series New York 1978.

[5] A. Grothendieck: Technique de Descente et théorèmes d'existence en géométrie Algébrique VI; Séminaire Bourbaki 1961/62 Exposé no.236.

[6] M. Knebusch: On algebraic curves over Real closed fields I; Math. Z. 150 (1976) p.49-70.

[7] J. Milne: Étale Cohomology; Princeton University Press, Princeton 1980.

[8] D. Mumford: Abelian Varieties; Oxford University Press 1974.

[9] R. Silhol: Diviseurs sur les variétés algébriques réelles; to appear in Boll. U.M.I.

[10] R. Silhol: Real Algebraic varieties and the theory of Comessatti; to appear.

[11] A. Tognoli: Algebraic approximation of manifolds and spaces; Séminaire Bourbaki 1979/80 Exp.548

R. SILHOL
Fakultät für Mathematik
Universität Regensburg
Universitätsstraße 31
D-8400 Regensburg

LE PROBLEME D'ALGEBRISATION DES POLYEDRES EST LOCAL

par

ALBERTO TOGNOLI

<u>INTRODUCTION</u>. Les dernières années on a beaucoup étudié le problème de la caractérisation des polyèdres homéomorphes à une variété algébrique réelle (affine).

On démontre dans [2] le résultat suivant : un polyèdre T est homéomorphe à une variété algébrique si, et seulement si le compactifié d'Alexandrov de T a cette propriété. On réduit donc le problème au cas où T est compact.

Soit T un polyèdre compact homéomorphe à une variété algébrique ; en ce cas il est clair qu'il existe un sous-polyèdre S de dimension inférieure à celle de T tel que $T-S$ ait une structure de variété différentiable. De plus S a un voisinage U_S homéomorphe à une variété algébrique.

Le but de ce travail est de démontrer que si T remplit les deux conditions que nous avons données, T est alors homéomorphe à une variété algébrique.

1. UN THEOREME D'APPROXIMATION

a) Définitions et notations

Dans la suite, le mot : variété algébrique sous-entend variété algébrique affine, réelle, réduite (V, O_V) .

Le faisceau O_V est souvent négligé dans les notations parce que l'on ne considère que le cas réduit.

Les morphismes de variétés algébriques sont appelés applications algébriques ou régulières ; de même les fonctions rationnelles régulières sont appelées algébriques.

Une variété algébrique V de $\mathbb{R}^n$ est dite régulière en x si dans un voisinage U de x on a :
$U \cap V = \{\gamma \in U \mid P_1(\gamma) = \dots = P_q(\gamma) = 0\}$ où $q = n - \dim V$ et où $P_1, \dots, P_q$ sont des polynômes tels que $(dP_1)_x, \dots, (dP_q)_x$ soient linéairement indépendants.

On dit que V est régulière si elle est régulière en chaque point.

Pour $n, q \in \mathbb{N}$ on notera $G_{n,q}$ la variété de Grassmann des sous-variétés linéaires de dimension q dans $\mathbb{R}^n$.

Soit $\gamma_{n,q} = \{(G, \ell) \in G_{n,q} \times \mathbb{R}^n \mid G \ni \ell\}$; la projection canonique naturelle $\gamma_{n,q} \to G_{n,q}$ est appelée le fibré tautologique.

Les variétés $\gamma_{n,q}$, $G_{n,q}$ sont considérées munies de la structure usuelle de variété projective (et donc affine).

Soit $d' : G_{n,q} \times G_{n,q} \to \mathbb{R}$ une métrique qui induise la topologie usuelle et soit d la métrique euclidienne sur $\mathbb{R}^n$.

DEFINITION 1. Soit X une sous-variété différentiable $(= C^\infty)$ fermée de $\mathbb{R}^n$ et $X \supset V$. Etant donnée $\varepsilon > 0$ on dit que la sous-variété X' de $\mathbb{R}^n$ est une ε-approximation de X relativement à V s'il existe un difféomorphisme $h : X \to X'$ tel que :

i) $d(x, h(x)) < \varepsilon$, $x \in X$

ii) $d'(T_{X_x}, T_{X'_{h(x)}}) < \varepsilon$, $x \in X$ où T_{X_x} et $T_{X'_{h(x)}}$ sont les variétés linéaires tangentes à X, X' en x, $h(x)$

iii) $X \supset V$ et $h_{|V} = id$

Si la condition iii) n'est pas satisfaite on dit que $(X',h(V))$ est une ε-approximation de (X,V).

On dit que $(X',h(V))$ est une ε-approximation algébrique si X' et $h(V)$ sont des variétés algébriques et si X' est régulière.

Finalement, si (X,V) admet, pour chaque $\varepsilon > 0$, une ε-approximation algébrique, on dit que (X,V) est algébriquement approximable.

Si f est une fonction C^∞ sur $\mathbb{R}^n$, autrement dit si $f \in C^\infty(\mathbb{R}^n)$, on pose :

$$\|f\|_K^m = \sum_{|\alpha|\leq m} \frac{1}{\alpha!} \sup_{x\in K} |D^\alpha f(x)|$$

b) <u>Un théorème d'approximation à la Weierstrass</u>

Soit V une variété algébrique fermée de $\mathbb{R}^n$. On sait que, si pour chaque $x \in V$ la complexification analytique du germe V_x coïncide avec la complexification algébrique, alors l'idéal I_V des polynômes nuls sur V engendre l'idéal des germes des fonctions C^∞ qui sont zéro sur V_x, $x \in V$. Dans cette hypothèse V est appelée quasi-régulière.

On a le résultat suivant ([3]) :

<u>THEOREME 1</u>. Soient V une sous-variété algébrique quasi-régulière de $\mathbb{R}^n$ et $h \in C^\infty(\mathbb{R}^n)$ telle que $h_{|V}$ soit algébrique. Soit $K \subset \mathbb{R}^n$ un compact, $q \in \mathbb{N}$, $\varepsilon > 0$.

Alors il existe une fonction algébrique $g : \mathbb{R}^n \to \mathbb{R}$ telle que :

1) $\|f-g\|_K^q < \varepsilon$

2) $f_{|V} = g_{|V}$

On sait que l'approximation dans la norme $\| \ \|^q$, avec $q > 0$, n'est pas possible si V n'est pas quasi-régulière.

Malheureusement la condition de quasi-régularité n'est pas très naturelle. On se propose d'établir un théorème valable pour toutes les sous-variétés algébriques compactes V.

On commence d'abord par donner une définition et un résultat qu'on utilisera après.

Soit $f \in C^{\infty}(\mathbb{R}^n)$ telle que $f_{|V}$ soit algébrique.

DEFINITION 2. On dit que f admet une approximation algébrique locale, relativement à V, si pour chaque $x \in \mathbb{R}^n$ on peut trouver un voisinage U_x tel que pour chaque $\varepsilon > 0$, $q \in \mathbb{N}$ il existe $g_x : \mathbb{R}^n \to \mathbb{R}$ fonction algébrique qui satisfait les conditions :

$$g_{x|V} = f_{|V}, \quad \|f-g_x\|^q_{U_x} < \varepsilon .$$

Le résultat suivant est démontré dans [5].

LEMME 1. Soient X une variété algébrique et V une sous-variété algébrique compacte. Soit $\varphi : V \to V'$ une application algébrique, $V' = \varphi(V)$. Il existe alors une variété algébrique X' et une application algébrique $\psi : X \to X'$ telles que :

i) la variété algébrique X' contient $\varphi(V)$ et $\psi_{|V} = \varphi_{|V}$

ii) $\psi_{|X-V}$ est un isomorphisme algébrique sur $X'-\varphi(V)$.

On a le :

THEOREME 2. Soit V une sous-variété algébrique compacte de $\mathbb{R}^n$ et $f \in C^{\infty}(\mathbb{R}^n)$ telle que $f_{|V}$ soit algébrique. On a alors :

I) si K', K'', $K'' \cap V = \emptyset$ sont deux compacts de $\mathbb{R}^n$, pour chaque $\varepsilon > 0$, $q \in \mathbb{N}$ il existe une fonction régulière $g : \mathbb{R}^n \to \mathbb{R}$ telle que :

1) $\|f-g\|^o_{K'} < \varepsilon$, $\|f-g\|^q_{K''} < \varepsilon$, $g_{|V} = f_{|V}$

II) si f admet une approximation algébrique locale, relativement à V, on a aussi

2) $\|f-g\|^q_{K'} < \varepsilon$

Preuve. On sait (voir [3]) qu'il existe une fonction algébrique $F : \mathbb{R}^n \to \mathbb{R}$ telle que $F_{|V} = f_{|V}$.
Si f admet une approximation algébrique locale, relativement à V, par rapport aux fonctions $g_x : U_x \to \mathbb{R}$ alors $f-F$ a la même propriété par rapport aux $g_x - F$.
Donc il est suffisant de démontrer le théorème pour $f-F$. Dans la suite on supposera $f_{|V} = 0$.
Le lemme 1, dans le cas où $\varphi : V \to x_o$ est constante, affirme qu'il existe une variété algébrique X' et une application algébrique $\psi : \mathbb{R}^n \to X'$ telles que $\psi(V) = x_o$ et $\psi_{|\mathbb{R}^n - V}$ soit un isomorphisme algébrique sur $X' - \{x_o\}$.
On a $f_{|V} = 0$, donc l'application $f' = f \circ \psi^{-1} : X' \to \mathbb{R}$ est bien définie, continue et de classe C^∞ sur $X' - \{x_o\}$.

L'hypothèse dit que $K'' \cap V = \emptyset$, d'où $\psi(K'') \subset X' - \{x_o\}$.
Et donc, à cause du théorème de Weierstrass, on peut approcher f' avec un polynôme g', $g'(x_o) = 0$ tel que, si $g = g' \circ \psi$ la relation 1) soit satisfaite.

On suppose maintenant que f admet une approximation algébrique locale par rapport à V.

Soit $\mathfrak{U} = \{U_i\}$ un recouvrement ouvert de $V \cup K' \cup K''$ tel que sur chaque U_i, $U_i \subset \mathbb{R}^n$ on ait une approximation de $f_{|U_i}$ au sens de la définition 2. On peut supposer que $\mathfrak{U}$ est fini, $V \cup K' \cup K''$ étant compact. Soit $\{\rho_i\}$ une partition C^∞ de l'unité associée au recouvrement $\mathfrak{U}$.

On trouve alors des $\delta_i > 0$ de façon que, si les $h_i : U_i \to \mathbb{R}$ sont des fonctions C^∞ telles que $\|h_i - f\|^q_{U_i} < \delta_i$, on ait $\|\Sigma \rho_i h_i - f\|^q_{\cup U_i} < \varepsilon/2$. On déduit de l'hypothèse qu'on peut choisir les h_i algébriques telles que $h_{i|V} = 0$.

A cause du théorème de Weierstrass on peut approcher les fonctions ρ_i avec des polynômes $\hat{\rho}_i$ de façon que $g = \Sigma \hat{\rho}_i h_i$ satisfasse les conditions 1) et 2) du théorème, qui est ainsi démontré.

REMARQUE 1. Si, dans le théorème précédent, on suppose que V et $f_{|V}$ sont définies sur le corps K, $K \subset \mathbb{R}$, on peut supposer que l'approximation g est aussi définie sur K.

c) Le résultat principal

Soit (X,V) un couple formé par une variété X, compacte, différentiable, de $\mathbb{R}^n$ et par une variété algébrique V, $V \subset X$. On sait ([4]) que, en général, (X,V) n'a pas d'approximation algébrique relative à V.

On veut démontrer le

THEOREME 3. S'il existe une variété algébrique $X' \subset \mathbb{R}^n$ telle que X' coïncide, au voisinage de V, avec X, si X' est régulière pour tout $x \in V$ et si $n > 2 \dim X$, alors X admet une approximation algébrique $\hat{X}$ dans $\mathbb{R}^n$ relativement à V.

Preuve. On sait (voir [1]) que, pour approcher (X,V) il est suffisant d'approcher un couple d'équations $\gamma : U_X \to G_{n,n-q}$, $\gamma' : U_X \to \mathbb{R}^n$ où U_X est un voisinage de X dans $\mathbb{R}^n$. Il est facile de voir que l'existence de la variété algébrique X' implique que les fonctions qu'on doit approcher pour construire les équations algébriques $\hat{\gamma}$, $\hat{\gamma}'$ de $\hat{X}$ admettent des approximations algébriques locales relativement à V. On peut donc utiliser le théorème 2 au lieu du théorème 1 et répéter mot pour mot la démonstration du théorème 3 de [1].

REMARQUE 2. Une version plus forte du théorème 3 tout à fait semblable à celle donnée dans [1] est possible avec la même démonstration.

Soit T un polyèdre compact de dimension α et supposons qu'il existe un sous-polyèdre S tel que $\dim T > \dim S$ et que $T-S$ admette une structure de variété différentiable δ. On exprimera tout cela par la notation (T,S,δ) et on dira que le polyèdre T a son lieu singulier contenu dans S.

Etant donné (T,S,δ), on dira que T est algébrique au voisinage de S s'il existe un voisinage U_S de S, une variété algébrique U_S' et un homéomorphisme $f : U_S \to U_S'$ tels que :

i) $U_S' - f(S)$ est régulier

ii) $f : U_S - S \to U_S' - f(S)$ est un difféomorphisme.

On a alors :

<u>THEOREME 4</u>. Etant donné (T,S,δ), supposons que T soit compact. Si T est algébrique, au voisinage de S, alors il existe une variété algébrique T'' homéomorphe à T.

<u>Preuve</u>. Soit $f : U_S \to U'_S$ l'homéomorphisme entre le voisinage U_S de S et la variété algébrique U'_S.

On sait qu'il existe une désingularisation $\pi : \hat{U}_S \to U'_S$ et que $\pi_{|\pi^{-1}(U'_S - f(S))}$ est un isomorphisme algébrique.

Dans la suite on supposera que $S' = f(S)$ est l'ensemble des points singuliers de U'_S. On peut alors construire une variété différentielle compacte $\hat{T}$ avec les deux cartes $\hat{U}_S$ et $T-S$.

On remarque que $\hat{T}$ a une structure différentiable parce que $f : U_S - S \to U'_S - S'$ est un difféomorphisme. En plus on peut supposer que la variété $\hat{T}$ est plongée dans $\mathbb{R}^N$, et comme $\hat{T}$ contient $\hat{U}_S$ contenant lui-même $\pi^{-1}(S') \overset{\text{def.}}{=\!=} \hat{S}$, on a donc vérifié, pour le couple $(\hat{T},\hat{S})$, les hypothèses du théorème 3 ; on en déduit qu'il existe une approximation algébrique $(T',\hat{S})$ relative à $\hat{S}$).

On peut maintenant appliquer le résultat du lemme 1 à la variété T' et à l'application algébrique $\pi : \hat{S} \to S'$; on trouve ainsi une application algébrique $\psi : T' \to T''$ telle que $T'' \supset S'$, $\psi_{|\hat{S}} = \pi_{|\hat{S}}$ et $\psi_{|T'-\hat{S}}$ soit un isomorphisme.

Il est facile de voir que T'' est homéomorphe à T ; le théorème est donc démontré.

<u>REMARQUE 3</u>. La variété algébrique T'', homéomorphe à T, construite dans le théorème 4 contient la variété algébrique S' des singularités de U'_S.

REFERENCES

[1] A. Tognoli - "Algebraic approximation of manifolds and spaces" Sém. Bourbaki, n° 548 (1979-1980).

[2] S. Akbulut, H. King - "The topology of real algebraic sets with isolated singularities", Ann. Of Math.,(1981).

[3] A. Tognoli - "Algebraic geometry and Nash functions", Inst. Math. Vol. III, Acad. Press London and New-York.

[4] R. Benedetti, A. Tognoli - "On real algebraic vector bundles", Bull. Sc. Math., II, série 104 (1980).

[5] R. Benedetti, M. Dedo - "The topology of two dimensional algebraic varieties", à paraître dans Ann. Mat. Pura Appl.

[6] H. Whitney - "Differentiable manifolds", Ann. of Math., 37 (1936).